MEDICAL
INTELLIGENCE
UNIT

Interleukin-10

Francesco M. Marincola, M.D.
Immunogenetic Section
Department of Transfusion Medicine
Clinical Center, National Institutes of Health
Bethesda, Maryland, U.S.A.

CRC Press
Taylor & Francis Group
Boca Raton London New York

CRC Press is an imprint of the
Taylor & Francis Group, an **informa** business

Interleukin-10

Medical Intelligence Unit

CRC Press
Taylor & Francis Group
6000 Broken Sound Parkway NW, Suite 300
Boca Raton, FL 33487-2742

First issued in hardback 2019

ISBN-13: 978-1-58706-285-8 (hbk)

Visit the Taylor & Francis Web site at
http://www.taylorandfrancis.com

and the CRC Press Web site at
http://www.crcpress.com

Library of Congress Cataloging-in-Publication Data

Interleukin-10 / [edited by] Francesco M. Marincola.
p. ; cm. -- (Medical intelligence unit)
Includes bibliographical references and index.
ISBN 1-58706-285-2
1. Interleukin-10. I. Marincola, Francesco M. II. Series: Medical intelligence unit (Unnumbered : 2003)
[DNLM: 1. Interleukin-10--immunology. 2. Autoimmune Diseases --physiopathology. 3. Autoimmunity--physiology. QW 568 I603 2005]
QR185.8.I56I585 2005
616.97'8--dc22

2005028605

CONTENTS

IV. IL-10 in Infection and Inflammation

V. Role of IL-10 in Allergy

VI. Role of IL-10 in Autoimmune Pathology

VII. Role of IL-10 in Transplantation

VIII. Role of IL-10 in Inflammatory Disease

IX. Role of IL-10 in Cancer

EDITOR

Francesco M. Marincola
Immunogenetic Section
Department of Transfusion Medicine
Clinical Center, National Institutes of Health
Bethesda, Maryland, U.S.A.
Email: fmarincola@mail.cc.nih.gov
Chapter 9

CONTRIBUTORS

Julio Aliberti
Laboratory of Parasitic Diseases
National Institute of Allergy and Infectious Diseases
National Institutes of Health
Bethesda, Maryland, U.S.A.
Chapter 2

Lionel Almeras
Laboratoire d'Immunologie EA2686
Faculté de Medecine
Lille Cedex, France
Email: immunologie@univ-lille2.fr
Chapter 3A

Khusru Asadullah
Corporate Research Business Area Dermatology
Schering AG
Berlin, Germany
Email: khusru.asadullah@schering.de
Chapter 8B

Peter J. Barnes
Department of Thoracic Medicine
National Heart and Lung Institute
Imperial College
London, U.K.
Email: p.j.barnes@imperial.ac.uk
Chapter 5

Calogero Caruso
Gruppo di Studio sull'Immunosenescenza
Dipartimento di Biopatologia e Metodologie Biomediche
Università di Palermo, Italy
Email: marcoc@unipa.it
Chapter 3D

W. David Creery
Division of Molecular Virology and Immunology
Research Institute
Children's Hospital of Eastern Ontario
Ottawa, Ontario, Canada
Chapter 4

Jeffrey C. Edberg
Division of Clinical Immunology and Rheumatology
University of Alabama at Birmingham
Birmingham, Alabama, U.S.A.
Chapter 6

Carl G. Feng
Laboratory of Parasitic Diseases
National Institute of Allergy and Infectious Diseases
National Institutes of Health
Bethesda, Maryland, U.S.A.
Chapter 2

Shin-ichiro Fujii
Laboratory for Cellular Immunotherapy in Research Center for Allergy and Immunology
RIKEN Yokohama Institute
Tsurumi-ku
Yokohama, Japan
Email: fujiis@rcai.riken.jp
Chapter 1B

Maria Gerbase-DeLima
Immunogenetics Division
Department of Pediatrics
Universidade Federal de São Paulo (UNIFESP-EPM)
São Paulo, Brazil
Chapter 3C

Andrew W. Gibson
Division of Clinical Immunology and Rheumatology
University of Alabama at Birmingham
Birmingham, Alabama, U.S.A.
Chapter 6

Karl F. Hoffmann
Department of Pathology
University of Cambridge
Cambridge, U.K.
Chapter 2

W. Martin Howell
Molecular Pathology Laboratory
Southampton General Hospital
Southampton, U.K.
Email: m.howell@soton.ac.uk
Chapter 3B

Dragana Jankovic
Laboratory of Parasitic Diseases
National Institute of Allergy and Infectious Diseases
National Institutes of Health
Bethesda, Maryland, U.S.A.
Chapter 2

Robert P. Kimberly
Division of Clinical Immunology and Rheumatology
University of Alabama at Birmingham
Birmingham, Alabama, U.S.A.
Email: rpk@uab.edu
Chapter 6

Marko Kryworuchko
Division of Molecular Virology and Immunology
Research Institute
Children's Hospital of Eastern Ontario
Ottawa, Ontario, Canada
Chapter 4

Marika C. Kullberg
Laboratory of Parasitic Diseases
National Institute of Allergy and Infectious Diseases
National Institutes of Health
Bethesda, Maryland, U.S.A.
Chapter 2

Ashok Kumar
Division of Molecular Virology and Immunology
Research Institute
Children's Hospital of Eastern Ontario
Ottawa, Ontario, Canada
Email: akumar@uottawa.ca
Chapter 4

James Lindsay
Department of Gastroenterology
St. Mark's Hospital
The North West London Hospitals NHS Trust
Harrow, Middlesex, U.K.
Email:
james.lindsay@bartsandthelondon.nhs.uk
j.lindsay@doctors.org.uk
Chapter 8C

Domenico Lio
Gruppo di Studio
sull'Immunosenescenza
Dipartimento di Biopatologia
e Metodologie Biomediche
Università di Palermo, Italy
Chapter 3D

Michael T. Lotze
Molecular Medicine Institute
University of Pittsburgh School
of Medicine
Center for Biotechnology
and Bioengineering
Pittsburgh, Pennsylvania, U.S.A.
Email: lotzemt@upmc.edu
Chapter 1B

Simone Mocellin
Clinica Chirurgica II
Dipartimento di Scienze Oncologiche
e Chirurgiche
Università di Padova
Padova, Italy
Email: mocellins@hotmail.com
Chapters 1C, 8D

Andrey Morgun
Immunogenetics Division
Department of Pediatrics
Universidade Federal de São Paulo
(UNIFESP-EPM)
São Paulo, Brazil
Email: anamorgun@hotmail.com
Chapter 3C

Monica C. Panelli
Immunogenetics Section
Department of Transfusion Medicine,
Clinical Center
National Institutes of Health
Bethesda, Maryland, U.S.A.
Chapter 9

Pierluigi Pilati
Clinica Chirurgica II
Dipartimento di Scienze Oncologiche
e Chirurgiche
Università di Padova
Padova, Italy
Chapter 1C

Lionel Prin
Laboratoire d'Immunologie EA2686
Faculté de Medecine
Lille Cedex, France
Chapter 3A

Thomas Ritter
Institute of Medical Immunology
Charité-University Medicine Berlin
Campus Mitte
Berlin, Germany
Email: thomas.ritter@charite.de
Chapter 7

Carlo Riccardo Rossi
Clinica Chirurgica II
Dipartimento di Scienze Oncologiche
e Chirurgiche
Università di Padova
Padova, Italy
Chapter 1C

Martin Rudwaleit
Medical Department I, Rheumatology
Charité Medical School
Benjamin Franklin Campus
Berlin, Germany
Chapter 8A

Alan Sher
Laboratory of Parasitic Diseases
National Institute of Allergy
and Infectious Diseases
National Institutes of Health
Bethesda, Maryland, U.S.A.
Chapter 2

Natalia Shulzhenko
Immunogenetics Division
Department of Pediatrics
Universidade Federal de São Paulo
(UNIFESP-EPM)
São Paulo, Brazil
Chapter 3C

Joachim Sieper
Medical Department I, Rheumatology
Charité Medical School
Benjamin Franklin Campus
Berlin, Germany
Email: hjsieper@zedat.fu-berlin.de
Chapter 8A

Wolfram Sterry
Department of Dermatology
Schering AG
Berlin, Germany
Chapter 8B

Hans-Dieter Volk
Institute of Medical Immunology
Charité-University Medicine Berlin
Campus Mitte
Berlin, Germany
Chapters 7, 8B

Jianming Wu
Division of Clinical Immunology
and Rheumatology
University of Alabama at Birmingham
Birmingham, Alabama, U.S.A.
Chapter 6

Thomas A. Wynn
Laboratory of Parasitic Diseases
National Institute of Allergy
and Infectious Diseases
National Institutes of Health
Bethesda, Maryland, U.S.A.
Chapter 2

Alexander Zdanov
Macromolecular Crystallography
Laboratory
Center for Cancer Research
National Cancer Institute
Frederick, Maryland, U.S.A.
Chapter 1A

PREFACE

Interleukin-10 (IL-10) is regarded as an immune suppressant cytokine. This reputation is due to the experimental observation that IL-10 decreases the function of antigen presenting cells and T helper 1 type immune responses. Surprisingly, however, IL-10 has potent anti-cancer effects since most experimental models demonstrate immune-mediated anti-tumor effects whether IL-10 is locally (tumor transfectant models) or systemically (transgenic models, administration of soluble cytokine) provided. In addition, recent work has suggested that polymorphisms of the promoter region of IL-10 may segregate propulsions in high and low producers. Surprisingly, high IL-10 producers may be more predisposed to acquire cancer and other diseases. Finally, human observations demonstrate that systemic administration of IL-10 to normal volunteers is associated with very little toxicity and pro-inflammatory properties mediated through activation of effector cells of the innate immune response. Thus, the role that interleukin-10 plays in vivo in physiological or pathological conditions remains controversial. Thus, the need to summarize in a book the state of the science of this important cytokine.

I was lucky because not only I could solicit the help of several experts from all over the world but I was also able to have them generate their work in a very short amount of time (three months). This allowed the preparation of a contemporary book that through the magic of electronic publishing will be released within six months of its conception.

In this book, I tried to cover not only the molecular biology of IL-10 but also asked to summarize in general terms its postulated role in controlling adoptive and innate immunity. These general chapters will have to confront the more detailed information emerging from the following chapters that refer to specific experimental and clinical models. A relatively new area that I tried to emphasize was the role that polymorphism of the IL-10 regulatory regions may play in the context of autoimmune pathology, cancer, transplantation and aging. I thought that this could be an interesting component of the book that needed to be expanded into sub-chapters since among all cytokines IL-10 appears to be one in which genetic polymorphism is strongly associated with divergent phenotypes.

Finally, the role in infection, allergy, transplantation and inflammatory disease is discussed in classic separate chapters as increasing information continues to emerge in such fields. Last but not least a chapter was commissioned on the role of IL-10 in cancer since this is probably one of the most controversial areas.

Taken into prospective, this book is rich in information about IL-10 from several experts in the field. As a result, redundancy is a not easily avoidable feature. On the other hand, as I was reading the various chapters I realized with interest that at times subtle differences were noticeable in the way various authors interpreted data regarding overlapping subjects while at other times consensus emerged. This allowed a spontaneous and unbiased (since none of the authors had a chance to read contributions from others before submission) comparison of the different authors opinions that might contribute a deeper dimensions to this book. Thus, I learned to appreciate this apparent redundancy, and I hope that the readers will enjoy it as well.

In summary, this book offers a state of science overview of the biological role that IL-10 may play in the context of various diseases provided by several leaders in the field whom I take the opportunity to thank for their prompt and comprehensive contributions.

Francesco M. Marincola
Immunogenetic Section
Department of Transfusion Medicine
Clinical Center National Institutes of Health
Bethesda, Maryland, U.S.A.

CHAPTER 1A

Structure and Function of IL-10 and the IL-10 Receptor

Alexander Zdanov

Abstract

Interleukin-10 (IL-10) is a cytokine the primary function of which appears to be to limit and control inflammatory responses. It signals through interactions with two receptor chains: high affinity IL-10R1 and low affinity IL-10R2. Initially IL-10 interacts with IL-10R1, forming an intermediate complex with a binding site for the second receptor chain. Subsequent binding of the IL-10R2 completes the final complex. Human IL-10 is an intercalated dimer of two subunits, each consisting of 160 amino acid residues. Almost 85% of the residues of each subunit are involved in the formation of six α-helices, designated A-F. Helices E and F of each subunit mutually exchange with the helices E' and F' of another subunit, forming two compact six α-helix bundle domains of IL-10. The structure of the intermediate complex of IL-10 with an extracellular domain of IL-10R1 (sIL-10R1) consists of one molecule of IL-10 and two copies of the receptor bound to each domain of the IL-10 dimer. The molecule is positioned on the two-fold symmetry axis so that both domains of IL-10 and the sIL-10R1 moieties are exact copies of each other. The structures of both the free and receptor-bound IL-10 molecules are essentially the same. The structure of the sIL-10R1 molecule consists of two-domains, each having fibronection type III-like topology. Most of the interactions in the ligand/receptor interface are of polar nature, yet with two hydrophobic patches around the side chains of Tyr43 and Phe143 of the receptor. The position and structure of the binding site for the second receptor chain, IL-10R2, is still unclear.

Introduction

Interleukin-10 (IL-10) is a pleiotropic cytokine produced by activated T cells, B cells, monocytes/macrophages, mast cells and keranocytes (reviewed in ref. 1). IL-10 was first recognized as a cytokine synthesis inhibitory factor[1-3] due to its ability to suppress macrophages, inhibiting production of a number of cytokines, and an ability to serve as accessory cells for stimulation of T cell and natural killer cell function.[4-7] It also exhibits some stimulatory activity for proliferation and differentiation of B cells, mast cells, and T cells.[8] Biologically functional human IL-10 is a 36 kDa dimer[1,9,10] consisting of two 160 amino acid residue long polypeptide chains. In order to initiate signal transduction, IL-10 requires two receptor chains, IL-10R1[11] and IL-10R2.[12] Both chains contain extracellular, transmembrane and intracellular domains, and belong to class II or interferon receptor family.[13,14] That family is characterized by the presence of two particular disulfide bridges and the absence of the so called "WSXWS motif" in the C-terminal part of the extracellular domain.

A number of viral and cellular gene homologs of IL-10 have been discovered recently.[15-17] Viral homologs were found in the Epstein-Barr virus[18-20] (EBV), equine herpesvirus type 2,[21] Orf parapoxvirus,[22,23] human and simian cytomegaloviruses (CMV),[24,25] and Yaba-like disease

Interleukin-10, edited by Francesco M. Marincola. ©2006 Eurekah.com.

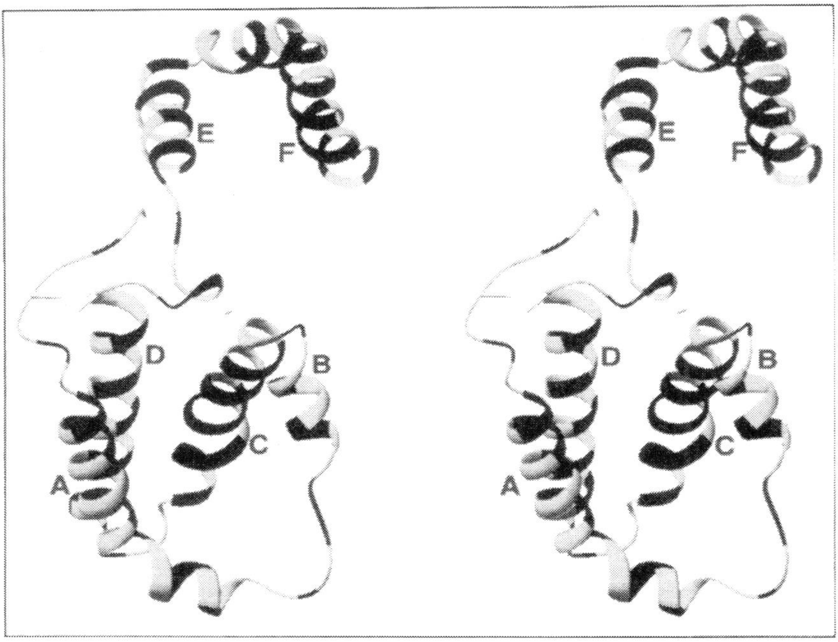

Figure 1A. Stereo ribbon diagram of the monomer of IL-10, hydrophobic residues are shown in red, all others in green, disulfide bridges are in yellow, helices are labeled A-F.

virus.[26] Although the amino acid identity of these proteins with human IL-10 varies between 23% to 85%, they all use IL-10 receptor system, and, as shown by crystal structures of EBV and CMV IL-10s,[27,28] they are very likely to be a dimers having three dimensional structures similar to that of IL-10. Cellular homologs include IL-19,[29] IL-20,[30] IL-22,[31,32] IL-24,[33] IL-26;[34] all have much lower amino acid sequence similarity with the human IL-10. Most of them are monomers having structures somewhat similar to the structure of a domain of IL-10;[35,36] the only exception is IL-26, which is likely a dimer.

This chapter describes three-dimensional structure of human IL-10, the interactions of this cytokine with the extracellular domain of IL-10R1 receptor, and possible stoichiometry of the signaling ternary complex.

Structure of Human IL-10

IL-10 is an intercalated dimer of two subunits,[37-39] each consisting of six amphipathic helices A-F (Fig. 1A). The polypeptide chains of each subunit contribute to both parts of the dimer. Helices A-D of one subunit form a distinctive six-helix domain with helices E' and F' through 180° rotation around two-fold axis (Fig. 1A). In addition, helices A, C, D, F' and A', C', D' and F of each domain form left-handed four-helix bundles which are signature elements of all helical cytokines.[40] The structure of the IL-10 subunit is stabilized by two intramolecular disulfide bridges (Fig. 1), Cys12-Cys108 and Cys62-Cys114. The disulfides hold together helices A, C and D, forming a frame with a long depression in the middle. The internal surface of the frame is very hydrophobic; therefore, when the amphipathic helices E' and F' cover the depression (Fig. 1B), almost all (86%) of the hydrophobic residues of IL-10 are involved in the formation of the intradomain hydrophobic cores. The domains are kept together by two flexible polypeptide links, separated by 15 Å from each other, allowing some degree of freedom to change the elbow angle between the domains. Comparison of crystal structures of human

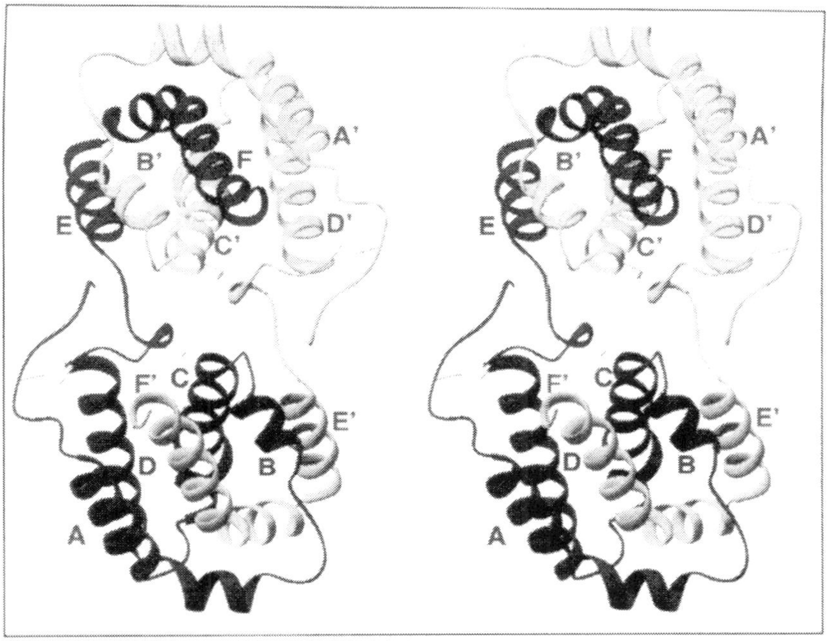

Figure 1B. Stereo diagram of the dimer of IL-10, subunits are shown in violet and green, helices of the symmetry related subunit are labeled as A'-F'.

IL-10 crystallized in different crystal forms,[37,39] Epstein-Barr virus protein BCRF1,[27] which is a very close homolog of human IL-10 (85% identity), and CMV IL-10,[28] indicated that the elbow angle may change easily, even due to different crystal packing.

Is Monomeric IL-10 Possible?

Is it possible that helices E' and F' belonging to one domain could swap with helices E and F of another domain so that two monomeric IL-10 molecules would form? The exchange of structural elements between molecular aggregates such as that observed in dimeric IL-10 is known as domain swapping. A recent review[41] on domain swapping identified about 40 proteins of known crystal structure, which satisfy certain conditions to be able to do the swapping, and IL-10 was placed on the list as one of the candidates. In principle, it should be possible to swap the helices E'-F' with E-F quite easily (Fig. 1). However, in practice the disulfide bridge Cys62-Cys114 restricts possible conformations of the loop DE, making it insufficiently long in order to allow the swapping to complete, unless the disulfide is reduced or serious distortions of the structure occur.[37,39] To overcome the problem, Josephson et al[42] extended the loop DE between Asn116 and Lys117 by insertion of additional six amino acids (GlyGlyGlySerGlyGly) and showed that the expressed protein folded as a monomer. Crystal structure of monomeric IL-10[43] confirmed that this form of the protein is very similar to a single domain of the IL-10 dimer. Further analysis showed that monomeric IL-10 was able to form a 1:1 complex with the IL-10 receptor 1 (IL-10R1), although with 60-fold reduced affinity. The protein still retained its biological activity, which was 10-fold lower as compared with the dimeric wild-type IL-10. There was also report[44] on the stability studies of the IL-10 dimer which showed that dimer/monomer transition induced either by 1.6 M guanidine hydrochloride or by pH 2.5 does take place and that the monomer created under such extreme conditions still retained 80% and 89% of its helical structure.

Ligand/Receptor Interactions

It is commonly accepted that since IL-10R1 is the high affinity receptor of IL-10, it should bind first,[11] forming a binding site for the second receptor. Subsequent interaction with the second receptor, IL-10R2,[12] completes the signaling ternary complex. Unfortunately, IL-10R2 is a low affinity receptor[12,45] and it is difficult, if not impossible, to obtain a stable ternary complex capable of crystallization. However, an intermediate binary complex of IL-10 and IL-10R1 is quite stable. It was also shown that in solution IL-10 and the soluble IL-10 receptor 1 (sIL-10R1) make a complex consisting of two IL-10 dimers and four sIL-10R1;[46,47] therefore the question of the exact stoichiometry of the signaling ternary complex was also open.

Crystal Structure of a Complex of Human IL-10 and Its Soluble Receptor

Crystals of the complex of IL-10 with a nonglycosylated mutant of the extracellular domain of IL-10R1 were obtained[48] and the crystal structure of IL-10/sIL-10R1 was solved[49] at 2.9 Å resolution. An asymmetric part of the crystal unit cell is composed of one domain of IL-10 and one molecule of sIL-10R1. 180° rotation around the crystallographic two-fold symmetry axis forms IL-10 dimer bound to two sIL-10R1 molecules (Fig. 2). The receptor molecules interacting with the same IL-10 dimer do not interact with each other, distance between their C-termini at the points of their likely entrance into cell membrane is 110 Å.

Structure of IL-10 Bound to Its Receptor

The structure of IL-10 bound to its receptor is practically the same as the structure of free IL-10; even the small change in the interdomain angle falls in the range of the values found previously for the human and viral IL-10s. The r.m.s. deviation for the Cα atoms that reside within the helices is only 0.6 Å, whereas loops AB and DE, as well as N and C-termini are more flexible and their conformations are more different. Human IL-10 was previously crystallized in different crystal forms and at different temperatures. It is interesting to note that the structure of free IL-10 in trigonal crystals and determined at the temperature 100 K (pdb code 2ilk[38]) is the closest to the receptor-bound IL-10.[49] The likely reason is the unique packing of the molecules in the crystal unit cell, where symmetry-related molecules of IL-10 may serve as surrogates of the receptor molecules.

Structure of the Soluble Receptor

sIL-10R1 structure[49] consists of N- and C-terminal domains (D1 and D2), each with fibronectin type III like topology.[50] Each domain is made of two β sheets (Fig. 2), formed by seven antiparallel β strands A,B,E and G,F,C,C' (Fig. 3), packed one against another in the form of a sandwich. The domains are comprised of residues 1-98 and 105-205, which are linked together by a short linkage 99-104, consisting of one helical turn and three residue β strand L, which is hydrogen bonded to both D1 and D2. Out of fourteen loops connecting the β strands, five loops located in the vicinity of the D1/D2 junction were found to be the most important[49] since they are involved in the ligand-receptor interactions. These are loops L2, L3 and L4 linking strands C-C', F-G, and G-L of the N-terminal domain D1 and L5 and L6, linking strands B-C and F-G of the C-terminal domain D2 respectively (Fig. 2, Fig. 3).

Ligand/Receptor Interface

Table I summarizes the interactions between IL-10 and sIL-10R1 in the range of 3.7Å. It is obvious that most of the contacts have polar nature. The ligand/receptor interface is formed by the residues coming from helix A, interhelical loop AB, and helix F of the IL-10, and from loops L2-L6 of the sIL-10R1. The interface can be divided into two interacting sites, Ia and Ib.[49]

The site Ia includes C-terminal part of helix A, loop AB and the middle part of helix F on the IL-10 side, and loops L2-L4 of the receptor, while site Ib includes the N-terminal and

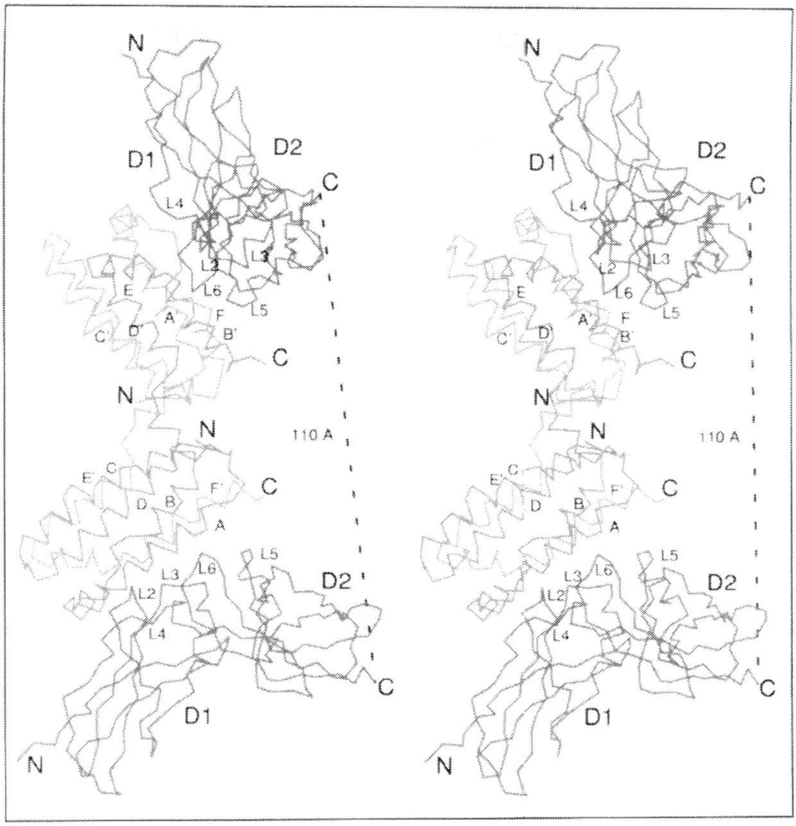

Figure 2. Stereo diagram of 1:2 IL-10/sIL-10R1 complex. IL-10 subunits are shown in violet and green, receptors are red, important receptor loops are labeled as L2-L6, N- and C-terminal domains of the receptor molecules as D1 and D2.

middle part of helix A, C-terminal part of helix F of the IL-10, and loops L5-L6 of the receptor (Table 1). Site Ia is certainly the primary binding site accounting for about 67% of the total buried surface of the interface.[49] It is centered around receptor residues Tyr43, Arg76 and Arg96, which make the majority of the interactions with IL-10. Burying of Tyr43 shields the largest surface area compared to any residues forming this interface.[49] Its side chain forms hydrogen bonds with the side chains of Lys138, Glu142 and main chain carbonyl oxygen of Asn45 through its hydroxyl group, while its aromatic ring penetrates a hydrophobic cavity made by side chains of Leu46 and Ile145 of IL-10, and the aliphatic parts of Arg76 and Arg96 of the receptor (Fig. 3). The guanidino groups of Arg76 and Arg96 make extensive hydrogen bonds with the ligand. The side chain of Arg76 adopts two alternative conformations; in the first one it interacts with Asp44 and Gln42, whereas in the second one it interacts with the main chain carbonyl oxygen of Gln38, and, through a bridging water Wat103, with Gln42. NH1 atom of Arg96 makes hydrogen bonds with Gln38 and the carbonyl oxygen of Ser141, while its NH2 interacts with the carboxyl group of Asp144, and through a bridging water Wat71, with Gln38 and Lys34. There is also a bridging water Wat88 in the site Ia that mediates interactions of IL-10 Lys34 with the carbonyl oxygen of Arg96, main chain nitrogen, and the hydroxyl group of Ser98 of the receptor. It is interesting to note that out of three water molecules involved in forming the ligand/receptor interface site Ia, Wat88 and Wat103 certainly

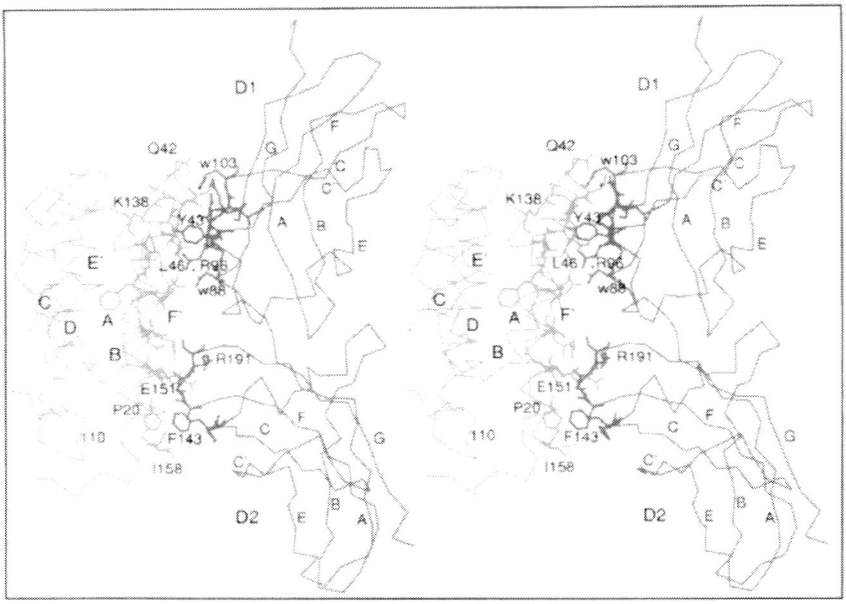

Figure 3. Interactions of one domain (green) of the IL-10 with the sIL-10R1 (red). Only residues included in Table 1 are shown, one letter amino acid residue code is used, not all residues are labeled. Water molecules 71, 88 and 103 found in the ligand/receptor interface are blue, only waters w88 and w103 are labeled. IL-10 helices are labeled as A-D and E'-F', β -strands of the receptor N- and C-terminal domains D1 and D2 are labeled as A-C and C'-G.

belong to the receptor molecule, while Wat71 is likely to belong to free IL-10. The latter water has a counterpart in Wat260 of IL-10 (pdb entry 2ILK) - the distance between the positions of Wat71 and Wat260 is only 1.7Å when free and receptor-bound IL-10 molecules are superimposed.

Site Ib may be considered to be secondary and it is centered around Arg27 and Glu151 of IL-10. The guanidinium group of Arg27 makes hydrogen bonds with the side chains of Asn148 and Glu151 of IL-10 and with the carbonyl oxygen of Ser190 of the receptor. In addition, the carboxyl group of Glu151 interacts with the hydroxyl group of Ser190 and with the side chain of Arg191 on the receptor side. The side chain of Arg24 of IL-10 also makes a hydrogen bond through its NE atom with the main chain carbonyl of Arg191. Hydrophobic contacts occur between the aromatic ring of the receptor Phe143 sandwiched between Pro20 and Ile158 (Fig. 3), the letter amino acid residue is not shown in Table 1 since the distance between its side chain and Phe143 is around 4 Å what is acceptable for hydrophobic interactions.

Therefore, even though most of the interactions in the ligand/receptor interface have polar nature, Tyr43 and Phe143 of the receptor are two hydrophobic residues, located at the top and bottom of the ligand/receptor interface (Fig. 3) and playing the role of hydrophobic locks that are likely to keep molecules together after they first recognized each other by long-range ionic interactions.

It is interesting to note that EBV IL-10, having very high identity (85%) toward human IL-10,[18-20] has about 1000-fold lower affinity toward IL-10R1.[3] Crystal structure of the IL-10/sIL-10R1 complex appears to explain this phenomenon. The hydrophobic pocket made by IL-10 residues Pro20 and Ile158 for the receptor side chain of Phe143 does not exist in EBV IL-10, in which residues 17-20 (human IL-10 numbering scheme) are deleted, and Ile158 is substituted by Ala. Because of the deletions at the N-terminus, the conformation of the main

Table 1. Amino acid residues of IL-10 and sIL-10R1 (pdb entry 1J7V[49]) in the range 3.7 D from each other. Secondary structure elements are given in accordance with ref. 37 for IL-10 and ref. 49 for sIL-10R1

IL-10		**sIL10R1**		
Amino Acid Residue	**Location**	**Amino Acid Residue**	**Location**	**Type of Interaction**
	Site Ia			
Lys34	Helix A	Arg96	Loop L4	Hydrogen bond through Wat71
		Arg96	Loop L4	Hydrogen bond through Wat88
		Ser98	Loop L4	Hydrogen bond through Wat88
Thr35	Helix A	Thr95	Loop L4	Close contact
Gln38	Helix A	Asn94	Loop L4	Close contact
		Thr95	Loop L4	Close contact
		Arg96	Loop L4	Hydrogen bond
		Arg96	Loop L4	Hydrogen bond through Wat71
Met39	Helix A	Asn94	Loop L4	Close contact
Gln42	Loop AB	Arg76	Loop L3	Close contact
		Arg76	Loop L3	Hydrogen bond through Wat103
Asp44	Loop AB	Gly44	Loop L2	Hydrogen bond
		Ile45	Loop L2	Hydrogen bond
		Arg76	Loop L2	Hydrogen bond
Asn45	Loop AB	Tyr43	Loop L2	Hydrogen bond
		Glu46	Loop L2	Close contact
Leu46	Loop AB	Tyr43	Loop L2	Hydrophobic
		Asn73	Loop L3	Close contact
Lys138	Helix F	Tyr43	Loop L2	Hydrogen bond
Ser141	Helix F	Arg96	Loop L4	Hydrogen bond
Glu142	Helix F	Tyr43	Loop L2	Hydrogen bond
Asp144	Helix F	Arg96	Loop L4	Hydrogen bond
Ile145	Helix F	Asn73	Loop L3	Close contact
		Arg96	Loop L4	Close contact
	Site Ib			
Pro20	Helix A	Phe143	Loop L5	Hydrophobic
		Ala189	Loop L6	Close contact
		Ser190	Loop L6	Close contact
Arg24	Helix A	Ala189	Loop L6	Close contact
		Ser190	Loop L6	Close contact
		Arg191	Loop L6	Hydrogen bond
Arg27	Helix A	Ser190	Loop L6	Hydrogen bond
		Arg191	Loop L6	Close contact
Asp28	Helix A	Ser192	Loop L6	Close contact
Glu151	Helix F	Ser190	Loop L6	Hydrogen bond
		Arg191	Loop L6	Hydrogen bond
Arg159	Helix F	His142	Loop L5	Close contact

chain in this area is quite different and in order for Phe143 of the receptor to get into even weak hydrophobic contacts with IL-10, some local conformational changes on the ligand side are necessary. In addition, the conformation of the loop AB involved in the formation of the site Ia is also different from what was found in the structure of EBV IL-10.[27] In other words, this would also require conformational change to take place, at the cost of additional energy.

Possible IL-10R2 Binding Sites

As already mentioned above, IL-10/sIL-10R1 complex is formed in solution by two dimers of IL-10 and four molecules of the sIL-10R1,[46] that is 2:4 stoichiometry. It is not clear what is the stoichiometry of the signaling complex in vivo, on the cell membrane, although it could also be 2:4. An asymmetric unit of the crystals of the human IL-10/sIL-10R1 complex contains one monomer of IL-10 and one molecule of sIL-10R1. A 1:2 complex (one IL-10 dimer bound to two molecules of sIL-10R1) is formed by application of the crystallographic symmetry. Further continuation of this analysis by application of the crystallographic translational symmetry along the axis *b* of the unit cell results in the 2:4 complex. The latter is stabilized in the crystals[49] by interactions of the D1 domain of one molecule of the 1:2 complex with IL-10 and D2 domain of the symmetry-mate molecule of the 1:2 complex. Josephson et al[49] suggested that this could be the complex which was seen in solution[46] and it gave them a clue of how a signaling ternary complex, that is the complex of IL-10 with both IL-10R1 and IL-10R2, might be organized. Since the key amino acids involved in the IL-10/sIL-10R1 interactions in sites Ia and Ib are mostly conserved between IL-10R1 and IL-10R2, they proposed an idea that at the first step both IL-10R1 and IL-20R2 may bind IL-10 by using the same site I, so that there will be a mixture of high affinity IL-10/IL-10R1 1:2 complexes and low affinity IL-10/IL-10R2 1:2 complexes. During the next step, these complexes interact with each other to produce a ternary (2:4) complex similar to the one seen in the crystals, and that one could be the signaling complex.[51] It is clear that when a receptor is bound to the membrane, then even weak interactions could become meaningful and from this point of view the events mentioned above could be happening. However, while in solution the IL-10/sIL-10R1 complex was found to be 2:4 with high degree of probability, the situation in the crystal is quite different. Crystal packing of IL-10/sIL-10R1 is such that ligand molecules make an "infinite" layers parallel to plane *ab* separated by parallel layers of the sIL-10R1 molecules. If we assume that a 1:2 complex lying on the two-fold symmetry axis is a minimal unit then millions of such units would interact with each other creating not just a 2:4 assembly but continuous "infinite" layers of 1:2 molecules interacting exactly in the same way as in 2:4 molecules. In other words, if a 1:2 molecules favored in solution to form 2:4 complexes found in the crystal, they inevitably must form higher aggregates in the solution also. Besides, a similar complex[28] of CMV IL-10 with the same nonglycosylated mutant of sIL-10R1 has absolutely different crystal packing and even though its 1:2 complex is very similar to human IL-10/sIL-10R1 both in terms of the structure of the receptor and its interaction with the ligand,[28] there is no 2:4 CMV IL-10/sIL-10R1 complex found in solution. In other words, it appears that question of the location of the IL-10R2 binding site and what is the structure of the ternary IL-10/IL-10R1/IL-10R2 complex still remain to be answered.

Conclusions

IL-10 is an intercalated dimer made up of two compact domains, each consisting of a six α-helix bundle. Each domain has extensive internal hydrophobic core that is additionally stabilized by two disulfide bridges. The overall structure of IL-10 is topologically similar to that of interferon-γ. Virally-encoded homologs of IL-10 mimic this structure even in the case when the amino acid sequence similarity is not high. Even though a naturally occurring monomer of IL-10 is unlikely to exist, an artificially designed IL-10 subunit with six extra amino acid residues inserted in the linker between helices D and E did fold as a monomer, and even exhibited some biological activity. The structure of the intermediate complex of IL-10 with sIL-10R1 consists of one IL-10 dimer and two receptor molecules, one bound to each domain. Most of the interactions on the ligand/receptor interface have polar nature enforced by two areas of hydrophobic interactions around the aromatic rings of Tyr43 and Phe143 of the receptor, located at the top and the bottom of the interface. The questions on the exact location of the binding site for the second receptor chain and on the structure of the ternary complex still need to be answered.

Acknowledgment

Dr. Alexander Wlodawer is thanked for critical reading the manuscript and for helpful remarks.

References

1. Fiorentino DF, Bond MW, Mosmann TR. Two types of mouse T helper cell: IV. Th2 clones secrete a factor that inhibits cytokine production by Th1 clones. J Exp Med 1989; 170:2081-2095.
2. Moore KW, O'Garra A, de Waal Malefyt R et al. Interleukin-10. Annu Rev Immunol 1993; 11:165-190.
3. Moore KW, de Waal MR, Coffman RL et al. Interleukin-10 and the interleukin-10 receptor. Annu Rev Immunol 2001; 19:683-765.
4. Hsu DH, Moore KW, Spits H. Differential effects of IL-4 and IL-10 on IL-2-induced IFN-gamma synthesis and lymphokine-activated killer activity. Int Immunol 1992; 4(5):563-569.
5. de Waal MR, Haanen J, Spits H et al. Interleukin 10 (IL-10) and viral IL-10 strongly reduce antigen-specific human T cell proliferation by diminishing the antigen-presenting capacity of monocytes via downregulation of class II major histocompatibility complex expression. J Exp Med 1991; 174(4):915-924.
6. Fiorentino DF, Zlotnik A, Vieira P et al. IL-10 acts on the antigen-presenting cell to inhibit cytokine production by Th1 cells. J Immunol 1991; 146(10):3444-3451.
7. Ding L, Shevach EM. IL-10 inhibits mitogen-induced T cell proliferation by selectively inhibiting macrophage costimulatory function. J Immunol 1992; 148(10):3133-3139.
8. Ho ASY, Moore KW. Interleukin-10 and its receptor. Therapeutic Immunology 1994; 1:173-185.
9. Tan JC, Indelicato SR, Narula SK et al. Characterization of interleukin-10 receptors on human and mouse cells. J Biol Chem 1993; 268:21053-21059.
10. Windsor WT, Syto R, Tsarbopoulos A et al. Disulfide bond assignments and secondary structure analysis of human and murine interleukin 10. Biochemistry 1993; 32:8807-8815.
11. Liu Y, Wei SHY, Ho ASY et al. Expression cloning and characterization of a human IL-10 receptor. J Immunol 1994; 152:1821-1829.
12. Kotenko SV, Krause CD, Izotova LS et al. Identification and functional characterization of a second chain of the interleukin-10 receptor complex. EMBO J 1997; 16(19):5894-5903.
13. Bazan JF. Structural design and molecular evolutuion of a cytokine receptor superfamily. Proc Natl Acad Sci USA 1990; 87:6934-6938.
14. Ho ASY, Liu Y, Khan TA et al. A receptor for interleukin 10 is related to interferon receptors. Proc Natl Acad Sci USA 1993; 90:11267-11271.
15. Kotenko SV. The family of IL10 related cytokines and their receptors: Related, but to what extent? Cytokine Growth Factor Rev 2002; 13:223-240.
16. Fickenscher H, Hor S, Kupers H et al. The interleukin-10 family of cytokines. Trends Immunol 2002; 23(2):89-96.
17. Dumoutier L, Renauld JC. Viral and cellular interleukin-10 (IL-10)-related cytokines: From structures to functions. Eur Cytokine Netw 2002; 13(1):5-15.
18. Moore KW, Vieira P, Fiorentino DF et al. Homology of cytokine synthesis inhibitory factor (IL-10) to the Epstein Barr Virus gene BCRFI. Science 1990; 248:1230-1234.
19. Hsu DH, de Waal Malefyt R, Fiorentino DF et al. Expression of interleukin-10 activity by Epstein-Barr virus protein BCRF1. Science 1990; 250(4982):830-832.
20. Vieira P, Malefyt RW, Dang M-N et al. Isolation and expression of human cytokine synthesis inhibitory factor cDNA clones: Homology to Epstein-Barr virus open reading frame BCRFI. Proc Natl Acad Sci USA 1991; 88:1172-1176.
21. Rode HJ, Janssen W, Rosen-Wolff A et al. The genome of equine herpesvirus type 2 harbors an interleukin 10 (IL10)-like gene. Virus Genes 1993; 7(1):111-116.
22. Fleming SB, McCaughan CA, Andrews AE et al. A homolog of interleukin-10 is encoded by the poxvirus orf virus. J Virol 1997; 71(6):4857-4861.
23. Fleming SB, Haig DM, Nettleton P et al. Sequence and functional analysis of a homolog of interleukin-10 encoded by the parapoxvirus orf virus. Virus Genes 2000; 21(1-2):85-95.

24. Kotenko SV, Saccani S, Izotova LS et al. Human cytomegalovirus harbors its own unique IL-10 homolog (cmvIL-10). Proc Natl Acad Sci USA 2000; 97(4):1695-1700.
25. Lockridge KM, Zhou SS, Kravitz RH et al. Primate cytomegaloviruses encode and express an IL-10-like protein. Virol 2000; 268(2):272-280.
26. Lee HJ, Essani K, Smith GL. The genome sequence of Yaba-like disease virus, a yatapoxvirus. Virol 2001; 281(2):170-192.
27. Zdanov A, Schalk-Hihi C, Menon S et al. Crystal structure of Epstein-Barr virus protein BCRF1, a homolog of cellular interleukin-10. J Mol Biol 1997; 268(2):460-467.
28. Jones BC, Logsdon NJ, Josephson K et al. Crystal structure of human cytomegalovirus IL-10 bound to soluble human IL-10R1. Proc Natl Acad Sci USA 2002; 99(14):9404-9409.
29. Gallagher G, Dickensheets H, Eskdale J et al. Cloning, expression and initial characterization of interleukin-19 (IL- 19), a novel homologue of human interleukin-10 (IL-10). Genes Immun 2000; 1(7):442-450.
30. Blumberg H, Conklin D, Xu WF et al. Interleukin 20: Discovery, receptor identification, and role in epidermal function. Cell 2001; 104(1):9-19.
31. Dumoutier L, Van Roost E, Colau D et al. Human interleukin-10-related T cell-derived inducible factor: Molecular cloning and functional characterization as an hepatocyte-stimulating factor. Proc Natl Acad Sci USA 2000; 97(18):10144-10149.
32. Xie MH, Aggarwal S, Ho WH et al. Interleukin (IL)-22, a novel human cytokine that signals through the interferon receptor-related proteins CRF2-4 and IL-22R. J Biol Chem 2000; 275(40):31335-31339.
33. Jiang H, Su ZZ, Lin JJ et al. The melanoma differentiation associated gene mda-7 suppresses cancer cell growth. Proc Natl Acad Sci USA 1996; 93(17):9160-9165.
34. Knappe A, Hor S, Wittmann S et al. Induction of a novel cellular homolog of interleukin-10, AK155, by transformation of T lymphocytes with herpesvirus saimiri. J Virol 2000; 74(8):3881-3887.
35. Chang C, Magracheva E, Kozlov S et al. Crystal structure of interleukin-19 defines a new subfamily of helical cytokines. J Biol Chem 2003; 278(5):3308-3313.
36. Nagem RAP, Colau D, Dumoutier L et al. Crystal structure of recombinant human interleukin-22. Structure 2002; 10:1051-1062.
37. Zdanov A, Schalk-Hihi C, Gustchina A et al. Crystal structure of interleukin-10 reveals the functional dimer with an unexpected topological similarity to interferon γ Structure 1995; 3(6):591-601.
38. Zdanov A, Schalk-Hihi C, Wlodawer A. Crystal structure of human interleukin-10 at 1.6 Å resolution and a model of a complex with its soluble receptor. Protein Sci 1996; 5(10):1955-1962.
39. Walter MR, Nagabhushan TL. Crystal structure of interleukin 10 reveals an interferon gamma-like fold. Biochemistry 1995; 34(38):12118-12125.
40. Presnell SR, Cohen FE. Topological distribution of four-α-helix bundles. Proc Natl Acad Sci USA 1989; 86:6592-6596.
41. Liu Y, Eisenberg D. 3D domain swapping: As domains continue to swap. Protein Sci 2002; 11(6):1285-1299.
42. Josephson K, DiGiacomo R, Indelicato SR et al. Design and analysis of an engineered human interleukin-10 monomer. J Biol Chem 2000; 275(18):13552-13557.
43. Josephson K, Jones BC, Walter LJ et al. Noncompetitive antibody neutralization of IL-10 revealed by protein engineering and x-ray crystallography. Structure (Camb) 2002; 10(7):981-987.
44. Syto R, Murgolo NJ, Braswell EH et al. Structural and biological stability of the human interleukin 10 homodimer. Biochemistry 1998; 37(48):16943-16951.
45. Spencer SD, Di Marco F, Hooley J et al. The orphan receptor CRF2-4 is an essential subunit of the interleukin 10 receptor. J Exp Med 1998; 187(4):571-578.
46. Tan JC, Braun S, Rong H et al. Characterization of recombinant extracellular domain of human interleukin-10 receptor. J Biol Chem 1995; 270(21):12906-12911.
47. Hoover DM, Schalk-Hihi C, Chou CC et al. Purification of receptor complexes of interleukin-10 stoichiometry and the importance of deglycosylation in their crystallization. Eur J Biochem 1999; 262(1):134-141.
48. Josephson K, McPherson DT, Walter MR. Purification, crystallization and preliminary X-ray diffraction of a complex between IL-10 and soluble IL-10R1. Acta Crystallogr D Biol Crystallogr 2001; 57(Pt 12):1908-1911.
49. Josephson K, Logsdon NJ, Walter MR. Crystal structure of the IL-10/IL-10R1 complex reveals a shared receptor binding site. Immunity 2001; 15(1):35-46.
50. Bork P, Holm L, Sander C. The Immunoglobulin Fold. Structural classification, sequence patterns and common core. J Mol Biol 1994; 242:309-320.
51. Walter MR. Strucure of interleukin-10/interleukin-10R1 complex: A paradigm for class 2 cytokine activation. Immunol Res 2002; 26(1-3):303-308.

CHAPTER 1B

IL-10 Effects on Innate Immunity

Shin-ichiro Fujii and Michael T. Lotze

Abstract

IL-10 is a dichotomous functional cytokine. It has characteristics of being both an immunosuppressive and immunostimulatory cytokine. In the setting of many infectious diseases including some viral infections, IL-10 administration prevents infected mice from progressing. IL-10 activates the cytolytic function of NK cells as well as that mediated by activated/memory CD8$^+$T cells. In a second stage following infection, IL-10 released from host cells often protects the host from secondary hyperinflammation, capable of damaging tissues, whereas endogenous IL-10 often imposes a suppressive function on CD4$^+$ T cells and antigen presenting cells. The IL-10 effects, modulating hyperinflammation can also induce regulatory T cells. In contrast, IL-10 production generated by or stimulated directly by some bacteria or protozoa infected DC may induce regulatory T cells, thereby helping microbes to aggressively invade host tissues. Indeed, IL-10$^{-/-}$ mice are protected from some infectious diseases. In the setting of tumor immunobiology, IL-10 expression appears to promote rejection of most experimental tumors. Thus, IL-10 release regulates innate immune mechanisms important for controlling microbes and playing a role in the chronic inflammatory response associated with tumorigenesis or secondary host defenses limiting progression of infection and host tissue damage.

The release of IL-10 during the acute or chronic inflammatory response regulates innate immune mechanisms following infection and specific immune responses to intracellular bacteria, fungi or protozoa. Systemic inflammatory responses occur after local infections including focal inflammation associated with cellulitis or pneumonia, systemic infections/sepsis, or non-infectious events including severe trauma, or burns, associated with pro-inflammatory cytokine release. Following infection, tissue damage induced directly by bacteria as well as the destruction mediated by the potent antimicrobial responses of the host-itself may cause significant damage to tissues, resulting in necrosis and loss of function. Such tissue damage leads to early cellular infiltration, progressing to granuloma formation during chronic-infection with resultant fibrosis and replacement of normal tissue.

During systemic inflammation, pro-inflammatory cytokines including TNF-α, IL-1, IL-18, IFN-γ and IL-12 play a central role in regulating the initiation and propagation of these events.[1,2] Release of such cytokines in patients infected with *T. gondii, P. chaubudi*, or certain strains of *T.cruzi* may cause substantial morbidity. The production of such pro-inflammatory cytokines by host cells stimulated by so called pattern recognition molecules capable of "sensing" bacterial or viral products is crucial to initiation of the innate and adaptive immune responses to infection.[1,2] These cytokines enhance the bactericidal capacity of phagocytes, recruiting additional innate cell populations to sites of infection and promote dendritic cell (DC) maturation which are capable of directing the ensuing specific immune response to the invading microbes.

Some bacterial pathogens have evolved mechanisms for modulating cytokine production by host cells, which modifies the host's subsequent immune response. Therefore, it is important to understand the suppressive or regulatory factors involved with inflammation following the inflammation associated with infection, sometimes paradoxically resulting in a state of

Interleukin-10, edited by Francesco M. Marincola. ©2006 Eurekah.com.

immunosuppression or immunoparalysis. Following instances of severe inflammation, IL-10 may play a critical role, helping to modulate the pathological consequences of local and systemic inflammatory responses. Infection models using IL-10$^{-/-}$ mice models suggest that the resultant immunopathology is apparently dependent upon Th1 or Tc1 type- T cell induction. IL-10 also inhibits Th2-mediated granulomas and fibrosis as secondary inflammatory effects in mouse models of schistomiasis as well as the immunopathology associated with the Th1 response.[3-5] To understand the interaction between the innate and adaptive immune responses against most pathogens, evaluation of the critical regulatory role played by IL-10 is essential.

The release of IL-10 may inhibit some antimicrobial responses during the initial inflammatory stage, but rescue the host from hyperinflammation or tissue damage elicited by host defense mechanisms following infection. In this chapter, we will discuss the role of IL-10 in regulating the response to infectious diseases and host defense by its ability to modulate innate immune mechanisms.

IL-10 and Innate Immunity in Infectious Diseases

The effects of IL-10 on innate immunity have been examined in several well-characterized infectious models in mice, including bacteria, fungus, and protozoan pathogens (*L. monocytogenes, C. albicans, L. major, and T. gondii*). The response to infection is demonstrated as a sequence of distinct intercellular interactions leading to induction of an array of anti-microbiocidal effector functions. Either enhanced or impaired immune responses are induced in various experimental situations, as a consequence of depletion or production/administration of IL-10 in vivo. The role of IL-10 in infectious disease models has been generally assessed by delivery of recombinant IL-10 (rIL-10), blocking IL-10 effects with treatment by neutralizing anti-IL-10 antibodies, or by using IL-10$^{-/-}$ mice created by targeted deletion. The observed responses are dependent on the levels of IL-10 receptors expression and signaling mediated by fully occupied receptors.

In *Listeria monocytogenes* models, elevation of IL-10 severely compromises resistance to infection in SCID mice,[6,7] while the enhanced innate response to Listeria in IL-10$^{-/-}$ or anti-IL-10mAb-treated mice leads to rapid control of Listeria infection.[8] IL-10$^{-/-}$ mice can eliminate an intravenous inoculation with *Candida albicans* in experimental acute systemic candidiasis.[9] Such enhancement of early innate responses also has been demonstrated in *T. gondii*, and *T.cruzi* infection.[10,11] However, even normal IL-10 levels tend to limit the effectiveness of the immune response to most infectious pathogens. The suppressive effects of endogenous IL-10 can be uncovered by reducing IL-10 levels. In contrast, the effects of altering IL-10 levels manifested at different phases of an immune/inflammatory response are quite complicated. In cerebral listeriosis models for example, IL-10$^{-/-}$ mice succumbed to listeriosis quite rapidly with endogenous IL-10 required to prevent a hyperinflammatory intracerebral immune response and resultant meningoencephalitis.[12] In a neurotropic mouse hepatitis virus model, IL-10$^{-/-}$ mice had more severe pathology and morbidity. Administration of rIL-10 however reduced lesion formation and mortality in a viral myocarditis model, with no increases in virus titers. Thus, IL-10 can prevent progression of the secondary hyperacute inflammation during the host defense. The relative lack of influence of IL-10 on antiviral responses may reflect the greater involvement of CD8$^+$ T cells in most antiviral responses.[13] However, vaccinia virus cannot replicate in IL-10$^{-/-}$ mice, suggesting a greater contribution of CD4$^+$T cells to the IL-10 response (either IL-10 or the IL-10 receptor).[2] Thus, different biological responses to IL-10 may be dependent on effector cells.

A strong correlation has been often observed when comparing the level of IL-10 and progression of several human infectious diseases, including visceral leishmaniasis,[14-17] malaria, filariasis,[18-21] leprosy,[21] tuberculosis,[22,23] candidiasis,[24] and *M. avium* infection.[25] In some diseases, decreased levels of IL-10 are related to successful drug treatment. However, in the setting of disease progression, high pathogen burdens are not simply related to the increase of IL-10 levels. Infected patients are prone to be "anergic" or "unresponsive" to the pathogenic antigen.

In such cases, IL-10 may inhibit protective immunity. Thus, "elevated IL-10 in vivo" or "IL-10-mediated unresponsiveness to antigen in vitro" may be seen in many chronic diseases as a poor prognostic factor. Thus, both IL-10 and IL-10 inhibitors may offer therapeutic promise in the treatment of either infectious diseases or infection-related immunopathologies. Supporting the application of the clinical promise of IL-10 in treating viral immunopathology, rIL-10 reversed liver fibrosis without increasing viral titers in patients with chronic hepatitis C infection.[26]

As above, IL-10 production usually imposes some limits on the effectiveness of immune responses to pathogens, especially innate immunity and adaptive Th1 responses. IL-10 appears to play a central role in the establishment of the balance between immune protection and pathology. To clarify the mechanism by which this IL-10 balance is achieved in various immune responses to pathogens, we have focused on the relationship between IL-10 and various types of immune cells, which mediate the regulation of innate immunity as described below.

Stimulatory Effects of IL-10 on NK Cells

IL-10 is a potent stimulator of NK cells (Fig. 1). While administration of IL-10 can induce long-term antigen specific anergy in $CD4^+T$ cells, it can also induce NK-cell activation and facilitate target-cell destruction in a dose-dependent manner. When the proliferative activity and cytotoxic activity of NK cells induced by IL-10 was tested, NK-cell proliferation was unaffected whereas, NK-cell cytotoxic function was markedly enhanced in vivo.[2,27-30] Since IL-10 amplifies the IL-2-induced proliferation of CD56-bright NK cells, the combination of IL-2 and IL-10 enhances lysis of Daudi but not K562 target cells.[31,32] The IL-10-mdiated enhancement of cytotoxic activity induced by IL-2 is associated with a significantly enhanced expression of perforin, granzyme, GM-CSF, TNF-α, inducible nitric oxide synthase, and IFN-γ. IL-18, particularly in conjunction with other cytokines including IL-10, is also a stimulator of NK cell activity.[33] NK cells pretreated with IL-10, either alone or in combination with other cytokines including IL-2, IL-18, and IL-12 are more efficient at lysing YAC-1 cells and other murine tumor targets. Activated NK cells enhance the clearance of pathogens and facilitate antigen acquisition from necrotic/apoptotic cells to activated antigen-presenting cells (APCs) for cross-priming, providing a link between innate immunity and the adaptive immune response.

IL-10 also plays a key role in graft-versus-host disease (GVHD) following bone marrow transplantation (BMT) through its interaction with allogeneic NK cells. Because of the extensive effects of IL-10 on NK-cell function, it could be important to monitor IL-10 following transplantation. High IL-10 serum levels in patients following BMT with GVHD are predictive of poor survival.[30,34] High levels of IL-10 protein sometimes enhances rejection, although lower levels of IL-10 protected against GVHD-associated lethality. NKT cell derived IL-10 may be essential for the differentiation of antigen-specific T regulatory cells observed in systemic tolerance.[35] NKT cells activation by the CD1-ligand mimic, α-galactosylceramide (αGalCer) is dictated by its expression on various types of APCs. DCs presenting this molecule skews the subsequent immune response to one with a Th1 phenotype whereas nonDC as APC induce anergy.[36] In fact, repeated injection of αGalCer induced anergy and rescued NOD SCID mice from diabetes mellitus, possibly due to IL-4 and IL-10 production.[37]

Suppressive Effects on $CD4^+$ T Cells, but not $CD8^+$ T Cells

IL-10 Acts on $CD8^+T$ Cells to Stimulate Survival

APCs and lymphocytes are the primary targets of IL-10. IL-10 suppresses proinflammatory cytokine production and the antigen presenting capacity of monocytes/macrophages and dendritic cells.[2] IL-10 costimulates proliferation and differentiation of B cells,[38] which is important in inducing production of antibodies in the setting of intestinal parasites, neutralization of bacterial toxins, and in local mucosal defense. In addition, the direct immunological effects of

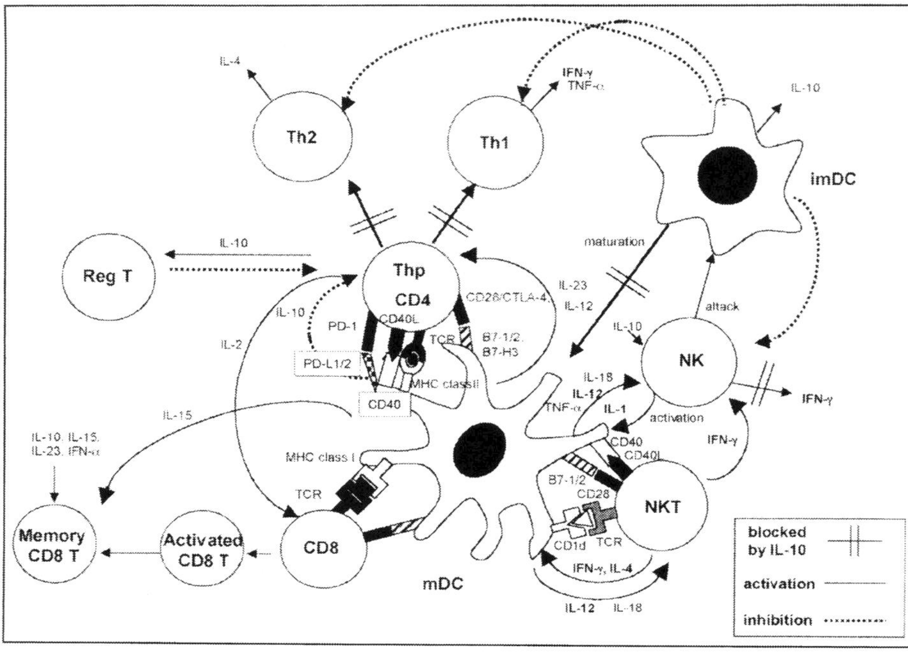

Figure 1. IL-10 is a pleiotrophic cytokine regulating immune effector cells and modulating inflammation to limit host tissue injury. IL-10 has two diametrically opposed and separate functional activities on effector cells. IL-10 stimulates and maintains the cytolytic and effector function of NK cells and CD8⁺T cells. In contrast, it limits naïve CD4⁺T cell activation and Th1 cell evolution, inducing abrogates the maturation of DCs, resulting in downregulation of costimulatory molecules and IL-12 production. These desparate effects of IL-10, orchestrate the balance of stimulatory and suppressive effects for many types of effector cells, dependent on the phase of the immune/inflammatory response.

IL-10 on these populations include the regulation of Th1/Th2 balance. IL-10 sometimes promotes the development of type 2 cytokines by inhibiting the IFN-γ production of CD4⁺ and CD8⁺ T lymphocytes particularly via suppression of IL-12 synthesis in accessory and costimulatory molecule-expressing APCs.[39-41] Thus, IL-10 can be a substantial suppressor of the cellular immune response.

Besides the dominant indirect impact via the APCs, IL-10 exerts suppressive effects on naïve T cells (Fig. 1).[2,39-41] In contrast, IL-10 directly enhances not only NK cell but also CD8⁺ cytotoxic T lymphocytes (CTL) cytotoxicity.[42-45] IL-10 has stimulatory effects on CD8⁺ T cells and induces their recruitment, cytotoxic activity, and proliferation. IL-10 does not exert potent direct inhibitory effects on CD8⁺T cells, but rather can activate CD8⁺T cells under many conditions.

IL-10 enhances the proliferative responses of murine IL-2- and IL-4-activated CD8⁺T cells and rescues T cells from apoptotic cell death.[46,47] As such, IL-10 enhances the effects of antitumor CD8⁺T cells in vivo, leading to the reduced growth of immunogenic tumors.[45,48-50] Administration of high doses of IL-10 injections mediates rejection of tumors with contrasting effects on CD4⁺ and CD8⁺ T cells that result in either immune dampening or immune potentiation of these individual cell types in situ, respectively.[45] Both interferon family members, IL-10 and IFN-γ, may have immunostimulatory or immunoregulatory effects leading to exacerbation of GVHD or down-regulation of this posttransplantation complications.[34,51] In some

GVHD models, IL-10 may cause exacerbation probably through effects mediated by both NK and activated CD8⁺ T cells. IL-10 promotes IL-2-independent growth of activated CD8⁺T cells in the absence of additional apparent costimulatory signals in autologous and allogeneic GVHD.[34,51]

IL-10 Induces Nonresponsiveness in CD4⁺ T Cells

As described above, IL-10 positively stimulates CD8⁺ T cells under some conditions. In contrast, IL-10 has suppressive effects on CD4⁺T cells. IL-10 directly inhibits proliferation and cytokine/chemokine-synthesis of CD4⁺T cells. This includes IL-2, IFN-γ, IL-4 and IL-5 production as well as CXCR4 production and chemotaxis in response to the CXCR4 ligand SDF1.[2,52-54] IL-10 inhibits both the Th1-type and the Th2-type responses of CD4⁺T cells (Fig. 1). Naïve CD4⁺T cells are the major targets of IL-10; whereas activated and memory T cells seem to be rather insensitive.[32] The presence of IL-10 during the activation of CD4⁺T cells results in the development of a regulatory phenotype. Such regulatory cells are characterized by weak proliferation, a failure to produce IL-2, and expression of a specific cytokine profile (IL-10⁺, IFN-γ⁺, IL-4⁻, IL-5⁻) after repeated stimulation.[32,55-57] IL-10 strongly inhibits cytokine production and proliferation of CD4⁺T cells or CD4⁺T cell clones via its downregulatory effects on APC function. Either CD4⁺ T cells cultured in the presence of IL-10, or IL-10-treated APCs become nonresponsive and fail to proliferate or produce inflammatory cytokines. The role of IL-10 in induction and maintenance of nonresponsiveness or anergy were studied in anti-tumor cell responses, parasitic infection, and HIV infection.[59-67] IL-10-mediated anergy may be associated with induction of regulatory T cells (Tr1 cells) that produce high levels of IL-10 and can suppress antigen-specific responses in vivo and in vitro.[68,69] This effect may not be dependent on soluble mediators but on cell-cell direct interactions. The influence of IL-10 directly on CD4⁺T cells or indirectly on APC is more important in vivo for the generation of Tr1 cells.

T cell anergy can be induced when naive CD4⁺ or CD8⁺ T cells are cultured in the presence of APCs pretreated with IL-10 that downregulate expression of costimulatory molecules.[2] In addition, purified T cells activated with anti-CD3 mAbs in the presence of IL-10 also become nonresponsive. Generally, anergy is defined as a long-lasting inability of antigen-specific T cells to proliferate upon rechallenge with antigen-loaded fully competent APCs in two functional assays; proliferation and IL-2 production. The anergic state of IL-10-treated T cells cannot be reversed by the addition of exogenous IL-2 (or IL-15) or stimulation by anti-CD3 mAbs and anti-CD28mAb activities,[2,56] indicating that IL-10-anergized cells are not equivalent to those induced by costimulatory blockade. IL-10-anergized T cells acquire regulatory activity and suppress the proliferation of T cells with the same peptide specificity via a cell-cell contact and apparently cytokine-independent mechanism, because IL-10-anergized cells fail to produce cytokines. When IL-10-anergized T cells are forced to proliferate by CD3 mAb crosslinking, they display a profile of cytokine productions, which are distinct from those of classical Th1 or Th2 cells. IL-10-mediated anergy can be associated with induction of a population of Tr1 cells that produce high levels of IL-10 and can suppress antigen-specific responses in vivo and in vitro.[30,70,71]

Biological Characteristics of Two Types of Regulatory T Cells; IL-10 Producing Tr1 Cells and CD4⁺CD25⁺ T Cells

Tr1 cells suppress and regulate immune responses in vitro and in vivo, via the production of IL-10 and TGF-β, in the absence of the T cell growth factors IL-2 and IL-4.[69,72] In this intermediate stage, the T cells have already acquired the capacity to suppress T cell responses by suppressive molecules expressed on the cell surface but not through the production of cytokines. Interestingly, human Tr1 cells produce small amounts of IFN-γ (whereas murine Tr1 cells do not). Proliferation of Tr1 cells is limited with either polyclonal or antigen-specific TCR activation following the addition of IL-2 or IL-15.[73]

In the resting state, Tr1 cell clones reveal the same levels of T cell activation markers as those of normal activated T cells. They constitutively express high levels of IL-2/IL-15Rβ and γ common chains, as well as a vast repertoire of chemokine receptors.[73] Interestingly, Tr1 cells (but not Th1 or Th2 cell clones) express the homing receptor to lymph nodes CCR7.[73] Tr1 cells can regulate the responses of naïve and memory T cells in vitro and in vivo as well as suppress both Th1 and Th2 cell-mediated pathology through the production of IL-10 and TGF-β (Fig. 1).[72] Mycobacterium-induced production of immunosuppressive cytokines may also contribute to the generation of regulatory T cells that down-regulate immune activation. For example, aerosol treatment of mice with killed *Mycobacterium vaccae* induces TGF-β-production.[74,75] *B. pertussis* exploits IL-10 to down-regulate the host immune response.[1] *Bordetella filamentous* hemagglutinin (FHA) induces IL-10 production by DC,[76] in turn promoting naïve T cells to develop into regulatory cells capable of suppressing IFN-γ production by antigen-specific T cells. Thus, bacterial exploitation of host cell capacity to produce immunosuppressive cytokines, particularly IL-10, provides an effective means for invading microbes to modulate host defense mechanisms and evade immune recognition. IL-10 production by bacteria or protozoa infected DC induced regulatory T cells, which have the potential to secrete IL-10.

$CD4^+CD25^+$ T cells are distinct from IL-10-producing Tr1 cells. They do not produce IL-10, but can produce TGF-β. $CD4^+CD25^+$ T cells are generated in the thymus, and are thought to arise via "altered negative selection" by self-peptides.[77,78] The suppressive activity of $CD4^+CD25^+$ T cells is related to their ability to inhibit IL-2 production and promote cell cycle arrest in both $CD4^+$ and $CD8^+$ T cells. This requires direct cell-cell contact, and may involve signals through CTLA-4 and/or the glucocorticoid-induced TNF-α receptor. A role for IL-10 in the suppression mediated by these cells is controversial.

Interaction between IL-10 and Antigen Presenting Cells

APC can directly produce many cytokines and chemokines during the innate immune response as well as promote effective antigen presentation in acquired adaptive immunity. IL-10 inhibits the expression of cytokines, soluble mediators and cell surface molecules by monocytes, macrophages, and myeloid DC (Fig. 1).[39,40,79-81] IL-10 is able to prevent monocyte differentiation into myeloid DC, which are the most important APCs, especially for primary immune responses. IL-10 inhibits the inflammatory or specific cellular immune response, and these effects are regulated with the induction of tolerance in adaptive immunity. IL-10 modulates the resultant cytokine cascade which develops in vivo. It inhibits the production of proinflammatory mediators by monocytes and macrophages including endotoxin- and IFN-γ-induced release of IL-1α, IL-1β, IL-6, IL-8, IL-10 itself, IL-12, IL-18, M-CSF, G-CSF, GM-CSF, TNF, LIF and PAF.[2,82] Moreover, IL-10 inhibits the production of IL-12 by monocytes or myeloid DC, which is an essential mediator for the development of specific cellular immune defense. In the innate immune response of the parasitic protozoa models,[83] macrophages and DC are impaired in the production of IL-12 and migration, but not equally. For example, IL-12 production is actively suppressed in macrophages infected with *L. major* or *T. gondii*, whereas DC produce IL-12p40 in response to the same parasites, both in vitro and in vivo.[83-85] However, endogenous agonists such as CD40L or IFN-γ are necessary as costimuli to induce high levels of IL-12 production. Either decrease of IL-12 or increase of IL-10 are observed in infected DC, and they inhibit the resultant T cell immune response.

Effects of IL-10 on Monocytes, Macrophages

IL-10 upregulates expression of the fMLP receptor, the PAF receptor, and the CCR1, CCR2, and CCR5 receptors on monocyes, making these cells more responsive to chemotactic factors and thereby more susceptible to HIV infection (e.g., IL-12 inhibits CCR5 expression by monocytes).[2,86-89] IL-10 enables monocytes conversion to a macrophage phenotype not only by directly suppressing function, but also by enhancing the production of anti-inflammatory molecules including interleukin-1 receptor antagonist (IL-1RA or receptors) including the soluble p55 and p75 TNFR and the IL-1RI and IL-1RII receptors.[90-95] Pretreatment of monocytes with chemokines including MCP1-4 suppressed production of IL-12p70 as a result of the induction of endogenous IL-10 production.[96] IL-10 inhibited production of prostaglandin E_2 (PGE_2), through downregulation of cyclooxygenase 2 (COX-2) expression.[97-99] Despite down regulation of T cell responses, IL-10 up-regulated the phagocytic activity of monocytes, macrophages, and immature DC via up-regulation of expression of IgG-Fc receptors (CD16, CD32, and CD64) as well as scavenger receptors (CD163 and CD14).[100]

Dying cells use both recruitment and "eat-me" signals for phagocyte attraction and recognition. Annexin I is a phosphatidylserine [PS] binding soluble protein. PS apprears to be a key molecule "exposed" on the surface of apoptotic cells that facilitates their phagocytic recognition.[101] During enhancement of the phagocytic activity of monocytes/macrophages for opsonized particles, bacteria, or fungi, Annexin I reduces the ability of cells to kill ingested organisms by decreasing generation of superoxide anion (O_2^-) and nitric oxide (NO). This mechanism involves inhibition of endogenous cytokine (TNF-α, IFN-γ) synthesis. Ligation of CD23 or CD64 induces expression of IL-10 by monocytes. IL-10 also downregulates expression of TLR4, the signal transducing receptor for LPS, and enhances expression of CD14, CD16, CD64, and CD163, a scavenger receptor that is downregulated by LPS, IFN-γ and TNF-α.[100,102] Collectively, these observations indicate that IL-10 induces differentiation of a macrophage-like cell, but limits ongoing immune responses and inflammation, and contributes to clearances of the infection via enhanced phagocytosis. Both IL-10 and TGF-β are produced by murine macrophages after Leishmania infection in vitro, promoting replication within macrophages and are important factors for determining in vivo susceptibility to infection.[103,104] After infection, IL-10 producing APC become competent for protozoa or virus replication.

Effects of IL-10 on Different Types of DC

Mouse and human DC are defined by their ability to activate and prime naïve resting T cells and initiate immune responses. As IL-10 inhibited production of IL-12 and expression of costimulatory molecules by various types of DC,[105-108] its expression correlated with inhibition of primary alloantigen-specific T cell responses. IL-10 treatment of DC can induce or contribute to a state of anergy in allo-antigen- or peptide-antigen-activated $CD4^+$ and $CD8^+$ T cells.[2,39,40]

Immature DC cultured human monocytes with GM-CSF and IL-4 for 6 days can be matured by LPS, CD40 ligand, or TNF-α to secrete IL-12 and induce differentiation of naïve T cells to Th1 cells. The peptide-specific T cells induced by immature DCs, but not mature DCs were indeed able to suppress the effector function of IFN-γ secreting cells.[109] By adding IL-10 in culture with GM-CSF and IL-4, maturation of DC is inhibited (Fig. 1). Rather, this caused differentiation of the immature DC into macrophage-like cells that expressed reduced levels of costimulatory molecules and MHC class II, but increased expression of MCSF and MCSFR, did not produce IL-12, and exhibited enhanced phagocytosis.[2] In contrast, IL-10 did not affect mature monocyte-derived DC; these cells may have lost IL-10R1 expression.[110,111]

IL-10 also affects $CD4^+CD11c^-IL\text{-}3R^+$ plasmacytoid DC. They can produce large amounts of IFN-α after exposure to virus, and furthermore they differentiate in vitro with IL-3 and CD40L into plasmacytoid DCs that can support differentiation of Th2 cells. IL-10 induced apoptosis of both freshly isolated and cultured plasmacytoid DC.[112] Myeloid DCs produce large amounts of IL-12 and induce strong Th1 and CTL response, while plasmacytoid DCs do not produce large amounts of IL-12, and induce Th2 responses, or the generation of IL-10-producing $CD8^+$T suppressor cells. Unlike IL-3- and CD40L-induced plasmacytoid DCs, which promote Th2 responses, viral-induced plasmacytoid DCs promote naïve T cells to produce both IFN-γ and IL-10.[113]

Particular DC populations producing IL-10 have been identified in Payer's patches in intestine, liver, lung and monocyte-derived DC.[2,114] In these organs, IL-10 plays a role to lead to tolerance in several ways: IL-10 can directly suppress T cells; IL-10 may be required to induce regulatory T cells; or IL-10 can act on DCs to decrease their function or make them tolerogenic. They are associated with development of either Th2 responses or hyporesponsiveness. In general, the effects of IL-10 on DC are consistent with inhibition of Th1 inflammatory responses and can be activated by inhibitory effects on "inflammation-inducing DC" or by induction of anti-inflammatory T cell populations by IL-10-producing DC. In the innate immunity of infectious disease, IL-10 is secreted by DC through microbial Toll-like receptor activators, such as LPS, gram positive bacterial peptide glycan (PGN), and yeast zymosan.[115] Especially, two TLR2 agonists; PGN and yeast zymosan produced IL-10 in CD40-dependent manner. In contrast, IL-12 synthesis was seen in murine splenic DCs after injection of a soluble extract of *T. gondii* tachyzoites (STAg). The response was short-term and could not be released by a second injection of STAg for a period of 1 week after initial priming.[83] This paralysis of the DC IL-12 response induced by STAg does not require IL-10 but instead appears to depend on the induction of lipoxin A4 (LXA4), a product of arachadonic metabolism. Infected DC produced IL-10 and reduced IL-12 leading to induction of IL-10 producing T cells. This balance help to reduce the innate immunity.

In several tumor models, IL-10 expressed within the tumors could drive rejection and elimination of the tumor, whereas the EBV expressed homologue, vIL-10 promoted tumor growth.[58,116] In some models, IL-10 can promote the generation of nominally tolerogenic regulatory T cells.[117] In this setting, tumor-infiltrating DCs (TIDCs) are largely immature, and are able to present tumor antigen, but they are refractory to stimulation with a combination of LPS, IFN-γ, and anti-CD40 antibody. DC paralysis in this setting can be reversed by CpG plus anti-IL-10 R treatment. This combination, but not CpG alone, had a potent therapeutic antitumor effect and induce immune memory.[118] In most tumor settings, IL-10 production has interestingly been associated with improved outcome in murine models and in predicting melanoma responsiveness to IL-10 therapy.[119]

References

1. Hornef MW, Wick MJ, Rhen M et al. Bacterial strategies for overcoming host innate and adaptive immune responses. Nat Immunol 2002; 3:1033-1040.
2. Moore KW, de Waal Malefyt R, Coffman RL et al. Interleukin-10 and the interleukin-10 receptor. Annu Rev Immunol 2001; 19:683-765.
3. Flores-Villanueva PO, Zheng XX, Strom TB et al. Recombinant IL-10 and IL-10/Fc treatment down-regulate egg antigen-specific delayed hypersensitivity reactions and egg granuloma formation in schistosomiasis. J Immunol 1996; 156:3315-3320.
4. Hoffmann KF, Cheever AW, Wynn TA. IL-10 and the dangers of immune polarization: Excessive type 1 and type 2 cytokine responses induce distinct forms of lethal immunopathology in murine schistosomiasis. J Immunol 2000; 164:6406-6416.
5. Wynn TA, Morawetz R, Scharton-Kersten T et al. Analysis of granuloma formation in double cytokine-deficient mice reveals a central role for IL-10 in polarizing both T helper cell 1- and T helper cell 2-type cytokine responses in vivo. J Immunol 1997; 159:5014-5023.
6. Kelly JP, Bancroft GJ. Administration of interleukin-10 abolishes innate resistance to Listeria monocytogenes. Eur J Immunol 1996; 26:356-364.
7. Tripp CS, Beckerman KP, Unanue ER. Immune complexes inhibit antimicrobial responses through interleukin-10 production. Effects in severe combined immunodeficient mice during Listeria infection. J Clin Invest 1995; 95:1628-1634.
8. Dai WJ, Kohler G, Brombacher F. Both innate and acquired immunity to Listeria monocytogenes infection are increased in IL-10-deficient mice. J Immunol 1997; 158:2259-2267.
9. Vazquez-Torres A, Jones-Carson J, Wagner RD et al. Early resistance of interleukin-10 knockout mice to acute systemic candidiasis. Infect Immun 1999; 67:670-674.
10. Neyer LE, Grunig G, Fort M et al. Role of interleukin-10 in regulation of T-cell-dependent and T-cell-independent mechanisms of resistance to Toxoplasma gondii. Infect Immun 1997; 65:1675-1682.
11. Grazzinelli RT, Wysocka M, Hieny S et al. In the absence of endogenous IL-10, mice acutely infected with Toxoplasma gondii succumb to a lethal immune response dependent on $CD4^+$ T cells and accompanied by overproduction of IL-12, IFN-γ and TNF-α. J Immunol 1996; 157:798-805.

12. Deckert M, Soltek S, Geginat G et al. Endogenous interleukin-10 is required for prevention of a hyperinflammatory intracerebral immune response in Listeria monocytogenes meningoencephalitis. Infect Immunity 2001; 69:4561-4571.
13. van Den Broek M, Bachmann MF, Kohler G et al. IL-4 and IL-10 antagonize IL-12-mediated protection against acute vaccinia virus infection with a limited role of IFN-γ and nitric oxide synthetase 2. J Immunol 2000; 164:371-378.
14. Gasim S, Elhassan AM, Khalil EA et al. High levels of plasma-IL-10 and expression of IL-10 by keratinocytes during visceral leishmaniasis predict subsequent development of postkala-azar dermal leishmaniasis. Clin Exp Immunol 1998; 111:64-69.
15. Ghalib HW, Piuvezam MR, Skeiky YA et al. Interleukin 10 production correlates with pathology in human Leishmania donovani infections. J Clin Invest 1993; 92:324-329.
16. Holaday BJ, Pompeu MM, Jeronimo S et al. Potential role for interleukin-10 in the immunosuppression associated with kala azar. J Clin Invest 1993; 92:2626-2632.
17. Karp CL, el-Safi SH, Wynn TA et al. In vivo cytokine profiles in patients with kala-azar. Marked elevation of both interleukin-10 and interferon-γ. J Clin Invest 1993; 91:1644-1648.
18. Mahanty S, Mollis SN, Ravichandran M et al. High levels of spontaneous and parasite antigen-driven interleukin-10 production are associated with antigen-specific hyporesponsiveness in human lymphatic filariasis. J Infect Dis 1996; 173:769-773.
19. Mahanty S, Ravichandran M, Raman U et al. Regulation of parasite antigen-driven immune responses by interleukin-10 and IL-12 in lymphatic filariasis. Infect Immun 1997; 65:1742-1747.
20. King CL, Mahanty S, Kumaraswami V et al. Cytokine control of parasite-specific anergy in human lymphatic filariasis. Preferential induction of a regulatory T helper type 2 lymphocyte subset. J Clin Invest 1993; 92:1667-1673.
21. Yamamura M, Uyemura K, Deans RJ et al. Defining protective responses to pathogens: Cytokineprofiles in leprosy lesions. Science 1991; 254:277-279.
22. Zhang M, Gong J, Iyer DV et al. T cell cytokine responses in persons with tuberculosis and human immunodeficiency virus infection. J Clin Invest 1994; 94:2435-2442.
23. Gerosa F, Nisii C, Righetti S et al. $CD4^+$ T cell clones producing both interferon-γ and interleukin-10 predominate in bronchoalveolar lavages of active pulmonary tuberculosis patients. Clin Immunol 1999; 92:224-234.
24. Roilides E, Sein T, Schaufele R et al. Increased serum concentrations of interleukin-10 in patients with hepatosplenic candidiasis. J Infect Dis 1998; 178:589-592.
25. Muller F, Aukrust P, Lien E et al. Enhanced interleukin-10 production in response to Mycobacterium avium products in mononuclear cells from patients with human immunodeficiency virus infection. J Infect Dis 1998; 177:586-594.
26. Nelson DR, Lauwers GY, Lau JY et al. Interleukin 10 treatment reduces fibrosis in patients with chronic hepatitis C: A pilot trial of interferon nonresponders. Gastroenterology 2000; 118:655-660.
27. Cavaillon JM. Pro-versus anti-inflammatory cytokines: Myth or reality. Cell Mol Biol 2001; 47:695-702.
28. Zheng LM, Ojcius DM, Garaud F et al. Interleukin-10 inhibits tumor metastasis through an NK cell-dependent mechanism. J Exp Med 1996; 184:579-584.
29. Kundu N, Beaty TL, Jackson MJ et al. Antimetastatic and antitumor activities of interleukin-10 in a murine model of breast cancer. J Natl Cancer Inst 1996; 88:536-541.
30. Mocellin S, Panelli MC, Wang E et al. The dual role of IL-10. Trends immunol 2003; 24:36-43.
31. Carson WE, Lindemann MJ, Baiocchi R et al. The functional characterization of interleukin-10 receptor expression on human natural killer cells. Blood 1995; 85:3577-3585.
32. Asadullah K, Sterry W, Volk HD. Interleukin-10 therapy-review of a new approach. Pharmacol Rev 2003; 55:241-269.
33. Cai G, Kastelein RA, Hunter CA. IL-10 enhances NK cell proliferation, cytotoxicity and production of IFN-γ when combined with IL-18. Eur J Immunol 1999; 29:2658-2665.
34. Blazar BR, Taylor PA, Panoskaltsis-Mortari A et al. Interleukin-10 dose-dependent regulation of $CD4^+$ and $CD8^+$ T cell-mediated graft-versus-host disease. Transplantation 1998; 66:1220-1229.
35. Sonoda KH, Faunce DE, Taniguchi M et al. NKT cell-derived IL-10 is essential for the differentiation of antigen-specific T regulatory cells in systemic tolerance. J Immunol 2001; 166:42-50.
36. Fujii SI, Shimizu K, Kronenberg M et al. Prolonged IFN-γ-producing NKT response induced with α-galactosylceramide-loaded DCs. Nat Immunol 2002; 3:867-874.
37. Hong S, Wilson MT, Serizawa I et al. The natural killer T-cell ligand α-galactosylceramide prevents autoimmune diabetes in nonobese diabetic mice. Nat Med 2001; 7:1052-1056.
38. Romagnani S. Biology o fhuman TH1 and TH2 cells. J Clin Immunol 1995; 15:121-129.

39. Steinbrink K, Jonuleit H, Muller G et al. Interleukin-10-treated human dendritic cells induce a melanoma-antigen-specific anergy in CD8⁺ T cells resulting in a failure to lyse tumor cells. Blood 1999; 93:1634-1642.
40. Steinbrink K, Granulich E, Kubsch S et al. CD4⁺ and CD8⁺ anergic T cells induced by interleukin-10-treated human dendritic cells display antigen-specific suppressor activity. Blood 2002; 99:2468-2476.
41. Groux H, Bigler M, de Vries JE et al. Interleukin-10 induces a long-term antigen-specific anergic state in human CD4⁺ T cells. J Exp Med 1996; 184:19-29.
42. Groux H, Bigler M, de Vries JE et al. Inhibitory and stimulatory effects of IL-10 on human CD8⁺ T cells. J Immunol 1998; 160:3188-3193.
43. Santin AD, Hermonat PL, Ravaggi A et al. Interleukin-10 increases Th1 cytokine production and cytotoxic potential in human papillomavirus-specific CD8⁺ cytotoxic T lymphocytes. J Virol 2000; 74:4729-4737.
44. Groux H, Cottrez F, Rouleau M et al. A transgenic model to analyze the immunoregulatory role of IL-10 secreted by antigen-presenting cells. J Immunol 1999; 162:1723-1729.
45. Fujii SI, Shimizu K, Shimizu T et al. Interleukin-10 promotes the maintenance of antitumor CD8⁺ T-cell effector function in situ. Blood 2001; 98:2143-2151.
46. Taga K, Cherney B, Tosato G. Interleukin-10 inhibits apoptotic cell death in human T cells starved of IL-2. Int Immunol 1993; 5:1599-1608.
47. Cohen SB, Crawley JB, Kahan MC et al. Interleukin-10 rescues T cells from apoptotic cell death: Association with an up-regulation of bcl-2. Immunology 1997; 92:1-5.
48. Berman RM, Suzuki T, Tahara H et al. Systemic administration of cellular IL-10 induces an effective, specific, and long-lived immune response against established tumors in mice. J Immunol 1996; 157:231-238.
49. Cervenak L, Morbidelli L, Donati D et al. Abolished angiogenicity and tumorigenicity of Burkitt lymphoma by interleukin-10. Blood 2000; 96:2568-2573.
50. Zheng LM, Ojcius DM, Garaud F et al. Interleukin-10 inhibits tumor metastasis through an NK cell-dependent mechanism. J Exp Med 1996; 184:579-584.
51. Qian S, Li W, Li Y et al. Systemic administration of cellular interleukin-10 can exacerbate cardiac allograft rejection in mice. Transplantation 1996; 62:1709-1714.
52. de Waal, MR Yssel H, de Vries JE. Direct effects of IL-10 on subsets of human CD4⁺ T cell clones and resting T cells. J Immunol 1993; 150:4754-4765.
53. Taga K, Mostowski H, Tosato G. Human interleukin-10 can directly inhibit T-cell growth. Blood 1993; 81:2964-2971.
54. Schandene L, Alonso-Vega C, Willems F et al. B7/CD28-dependent IL-5 production by human resting T cells is inhibited by IL-10. J Immunol 1994; 152:4368-4374.
55. Zeller JC, Panoskaltsis-Mortari A, Murphy WJ et al. Induction of CD4⁺ T cell alloantigen-specific hyporesponsiveness by IL-10 and TGF-beta. J Immunol 1999; 163:3684-3691.
56. Jonuleit H, Schmitt E, Schuler G et al. Induction of interleukin-10-producing, nonproliferating CD4⁺ T cells with regulatory properties by repetitive stimulation with allogeneic immature human dendritic cells. J Exp Med 2000; 192:1213-1222.
57. Levings MK, Sangregorio R, Roncarolo MG. Human CD25⁺ CD4⁺ T regulatory cells suppress naïve and memory T cell proliferation and can be expanded in vitro without loss of function. J Exp Med 2001; 193:1295-1302.
58. Suzuki T, Tahara H, Narula S et al. Viral interleukin-10 (vIL-10), the human herpes virus 4 cellular IL-10 homologue, induces local anergy to allogeneic and syngeneic tumors. J Exp Med 1995; 182:477-486.
59. Enk AH, Angeloni VL, Udey MC et al. Inhibition of Langerhans cell antigen-presenting function by IL-10. A role for IL-10 in induction of tolerance. J Immunol 1993; 151:2390-2398.
60. Enk AH, Saloga J, Becker D et al. Induction of hapten-specific tolerance by interleukin 10 in vivo. J Exp Med 1994; 179:1397-1402.
61. Flores-Villanueva PO, Chikunguwo SM, Harris TS et al. Role of IL-10 on antigen-presenting cell function for Schistosomal egg-specific monoclonal T helper cell responses in vitro and in vivo. J Immunol 1993; 151:3192-3198.
62. Flores Villanueva PO, Reiser H, Stadecker MJ. Regulation of T helper cell responses in experimental murine schistosomiasis by IL-10. Effect on expression of B7 and B7-2 costimulatory molecules by macrophages. J Immunol 1994; 153:5190-5199.
63. Becker JC, Czerny C, Brocker EB. Maintenance of clonal anergy by endogenously produced IL-10. Int Immunol 1994; 6:1605-1612.
64. King CL, Medhat A, Malhotra I et al. Cytokine control of parasite-specific anergy in human urinary schistosomiasis. IL-10 modulates lymphocyte reactivity. J Immunol 1996; 156:4715-4721.

65. Schols D, De Clercq E. Human immunodeficiency virus type 1 gp120 induces anergy in human peripheral blood lymphocytes by inducing interleukin-10 production. J Virol 1996; 70:4953-4960.
66. Sundstedt A, Holden I, Rosendahl A et al. Immunoregulatory role of IL-10 during superantigen-induced by poresponsiveness in vivo. J Immunol 1997; 158:180-186.
67. Shreedhar VK, Pride MW, Sun Y et al. Origin and characteristics of ultraviolet-B radiation-induced suppressor T lymphocytes. J Immunol 1998; 161:1327-1335.
68. Asseman C, Mauze S, Leach MW et al. An essential role for interleukin 10 in the function of regulatory T cells that inhibit intestinal inflammation. J Exp Med 1999; 190:995-1004.
69. Groux H, O'Garra A, Bingler M et al. A CD4+ T-cell subset inhibits antigen-specific T-cell responses and prevents colitis. Nature 1997; 389:737-742.
70. Roncarolo MG, Bacchetta R, Bordignon C et al. Type 1 T regulatory cells. Immunol Rev 2001; 182:68-79.
71. Levings MK, Sangregorio R, Galbiati F et al. IFN-α and IL-10 induce the differentiation of human type 1 T regulatory cells. J Immunol 2001; 166:5530-5539.
72. Levings MK, Bacchetta R, Schulz U et al. The role of IL-10 and TGF-β in the differentiation and effector function of T regulatory cells. Int Arch Allergy Immunol 2002; 129:263-276.
73. Bacchetta R, Sartirana C, Levings MK et al. Growth and expansion of human regulatory type 1 cells are independent from TCR activation but require exogenous cytokines. Eur J Immunol 2002; 32:2237-2245.
74. VanHeyningen TK, Collins HL, Russell DG. IL-6 produced by macrophages infected with Mycobacterium species suppresses T cell responses. J Immunol 1997; 158:330-337.
75. Zuany-Amorim C, Sawicka E, Manlius C et al. Suppression of airway eosinophilia by killed Mycobacterium vaccae-induced allergen-specific regulatory T-cells. Nat Med 2002; 8:625-629.
76. McGuirk P, McCann C, Mills KH. Pathogen-specific T regulatory 1 cells induced in the respiratory tract by a bacterial molecule that stimulates interleukin 10 production by dendritic cells: A novel strategy for evasion of protective T helper 1 responses by Bordetella pertussis. J Exp Med 2002; 195:221-231.
77. Sakaguchi S, Sakaguchi N, Shimizu J et al. Immunologic tolerance maintained by CD25+ CD4+ regulatory T cells: Their common role in controlling autoimmunity, tumor immunity, and transplantation tolerance. Immunol Rev 2001; 182:18-32.
78. Shevach EM. CD4+ CD25+ suppressor T cells: More questions than answers. Nat Rev Immunol 2002; 2:389-400.
79. Wilkes DS, Neimeier M, Mathur PN et al. Effect of human lung allograft alveolar macrophages on IgG production: Immunoregulatory role of interleukin-10, transforming growth factor-β, and interleukin-6. Am J Respir Cell Mol Biol 1995; 13:621-28.
80. Park DR, Skerrett SJ. IL-10 enhances the growth of Legionella pneumophila in human mononuclear phagocytes and reverses the protective effect of IFN-γ: Differential responses of blood monocytes and alveolar macrophages. J Immunol 1996; 157:2528-38.
81. Berkman N, John M, Roesems G et al. Inhibition of macrophage inflammatory protein-1 alpha expression by IL-10. Differential sensitivities in human blood monocytes and alveolar macrophages. J Immunol 1995; 155:4412-4418.
82. de Waal Malefyt R, Figdor CG, Huijbens R et al. Effects of IL-13 on phenotype, cytokine production, and cytotoxic function of human monocytes. Comparison with IL-4 and modulation by IFN-γ or IL-10. J Immunol 1993; 151:6370-6381.
83. Sacks D, Sher A. Evasion of innate immunity by parasitic protozoa. Nat Immunol 2002; 3:1041-1047.
84. von Stebut E, Belkaid Y, Jakob T et al. Uptake of Leishmania major amastigotes results in activation and interleukin 12 release from murine skin-derived dendritic cells: Implications for the intiation of anti-Leishmania immunity. J Exp Med 1998; 188:1547-1552.
85. Marovich MA, McDowell MA, Thomas EK et al. IL-12p70 production by Leishmania major-harboring human dendritic cells is a CD40/CD40 ligand-dependent process. J Immunol 2000; 164:5858-5865.
86. Andrew DP, Chang MS, McNinch J et al. STCP-1 (MDC) CC chemokine acts specifically on chronically activated Th2 lymphocytes and is produced by monocytes on stimulation with Th2 cytokines IL-4 and IL-13. J Immunol 1998; 161:5027-5038.
87. Sozzani S, Ghezzi S, Iannolo G et al. Interleukin 10 increases CCR5 expression and HIV infection in human monocytes. J Exp Med 1998; 187:439-444.
88. Thivierge M, Parent JL, Stankova J et al. Modulation of formyl peptide receptor expression by IL-10 in human monocytes and neutrophils. J Immunol 1999; 162:3590-3595.
89. Wang J, Guan E, Roderiquez G et al. Inhibition of CCR5 expression by IL-12 through induction of β-chemokines in human T lymphocytes. J Immunol 1999; 163:5763-5769.

90. Hart PH, Hunt EK, Bonder CS et al. Regulation of surface and soluble TNF receptor expression on human monocytes and synovial fluid macrophages by IL-4 and IL-10. J Immunol 1996; 157:3672-3680.
91. Joyce DA, Steer JH. IL-4, IL-10 and IFN-γ have distinct, but interacting,effects on differentiation-induced changes in TNF-α and TNF receptor release by cultured human monocytes. Cytokine 1996; 8:49-57.
92. Linderholm M, Ahlm C, Settergren B et al. Elevated plasma levels of tumor necrosis factor (TNF)-α, soluble TNF receptors, interleukin (IL)-6, and IL-10 in patients with hemorrhagic fever with renal syndrome. J Infect Dis 1996; 173:38-43.
93. Dickensheets HL, Freeman SL, Smith MF et al. Interleukin-10 upregulates tumor necrosis factor receptor type-II (p75) gene expression in endotoxin-stimulated human monocytes. Blood 1997; 90:4162-4171.
94. Jenkins JK, Malyak M, Arend WP. The effects of interleukin-10 on interleukin-1 receptor antagonist and interleukin-1β production in human monocytes and neutrophils. Lymphokine Cytokine Res 1994; 13:47-54.
95. Dickensheets HL, Donnelly RP. IFN-gamma and IL-10 inhibit induction of IL-1 receptor type I and type II gene expression by IL-4 and IL-13 in human monocytes. J Immunol 1997; 159:6226-6233.
96. Braun MC, Lahey E, Kelsall BL. Selective suppression of IL-12 production by chemoattractants. J Immunol 2000; 164:3009-3017.
97. Niiro H, Otsuka T, Kuga S et al. IL-10 inhibits prostaglandin E_2 production by lipopolysaccharide-stimulated monocytes. Int Immunol 1994; 6:661-664.
98. Niiro H, Otsuka T, Tanabe T et al. Inhibition by interleukin-10 of inducible cyclooxygenase expression in lipopolysaccharide-stimulated monocytes:its underlying mechanism in comparison with interleukin-4. Blood 1995; 85:3736-3745.
99. Mertz PM, DeWitt DL, Stetler-Stevenson WG et al. Interleukin 10 suppression of monocyte prostaglandin H synthase-2. Mechanism of inhibition of prostaglandin-dependent matrix metalloproteinase production. J Biol Chem 1994; 269:21322-21329.
100. Muzio M, Bosisio D, Polentarutti N et al. Differential expression and regulation of toll-like receptors (TLR) in human leukocytes: Selective expression of TLR3 in dendritic cells. J Immunol 2000; 164:5998-6004.
101. Ravichandran KS. "Recruitment signals" from apoptotic cells: Invitation to a quiet meal. Cell 2003; 113:817-820.
102. Buechler C, Ritter M, Orso E et al. Regulation of scavenger receptor CD163 expression in human monocytes and macrophages by pro- and antiinflammatory stimuli. J Leukoc Biol 2000; 67:97-103.
103. Kane MM, Mosser DM. The role of IL-10 in promoting disease progression in Leishmaniasis. J Immunol 2001; 166:1141-1147.
104. Barral A, Barral-Netto M, Yong EC et al. Transforming growth factor b as a virulence mechanism for Leishmania braziliensis. Proc Natl Acad Sci USA 1993; 90:3442-3446.
105. Macatonia SE, Doherty TM, Knight SC et al. Differential effect of interleukin 10 on dendritic cell-induced T cell proliferation and INF-γ production. J Immunol 1993; 150:3755-3765.
106. Rea D, van Kooten C, van Meijgaarden KE et al. Glucocorticoids transform CD40-trigering of dendritic cells into an alternative activation pathway resulting in antigen-presenting cells that secrete IL-10. Blood 2000; 95:3162-3167.
107. Penna G, Adorini L. Alpha, 25-dihydroxyvitamin D3 inhibits differentiation, maturation, activation, and survival of dendritic cells leading to impaired alloreactive T cell activation. J Immunol 2000; 164:2405-2411.
108. Matyszak MK, Citterio S, Rescigno M et al. Differential effects of corticosteroids during different stages of dendritic cell maturation. Eur J Immunol 2000; 30:1233-1242.
109. Dhodapkar MV, Steinman RM. Antigen-bearing immature dendritic cells induce peptide-specific $CD8^+$ regulatory T cells in vivo in human. Blood 2002; 100:174-177.
110. Allavena P, Piemonti L, Longoni D et al. IL-10 prevents the differentiation of monocytes to dendritic cells but promotes their maturation to macrophages. Eur J Immunol 1998; 28:359-369.
111. Fortsch D, Rollinghoff M, Stenger S. IL-10 converts human dendritic cells into macrophage-like cells with increased antibacterial activity against virulent Mycobacterium tuberculosis. J Immunol 2000; 165:978-987.
112. Rissoan MC, Soumelis V, Kadowaki N et al. Reciprocal control of T helper cell and dendritic cell differentiation. Science 1999; 283:1183-1186.
113. Liu YJ. Dendritic cell subsets and lineages, and their functions in innate and adaptive immunity. Cell 2001; 106:259-262.
114. Steinman RM, Nussenzweig MC. Avoiding horror autotoxicus: The importance of dendritic cells in peripheral T cell tolerance. Proc Natl Acad Sci USA 2002; 99:351-358.

115. Qi H, Denning TL, Soong L. Differential induction of interleukin-10 and interleukin-12 in dendritic cells by microbial toll-like receptor activators and skewing of T-cell cytokine profiles. Infec Immun 2003; 71:3337-3342.
116. Kaufman HL, Rao JB, Irvine KR et al. Inteleukin-10 enhances the therapeutic effectiveness of a recombinant poxvirus-based vaccine in an experimental murine tumor model. J Immunother 1999; 22:489-496.
117. Castro AG, Neighbors M, Hurst SD et al. Anti-interleukin-10 receptor monoclonal antibody is an adjuvant for T helper cell type 1 responses to soluble antigen only in the presence of lipopolysaccharide. J Exp Med 2000; 192:1529-1534.
118. Vicari AP, Chiodoni C, Vaure C et al. Reversal of tumor-induced dendritic cell paralysis by CpG immunostimulatory oligonucleotide and anti-interleukin-10 receptor antibody. J Exp Med 2002; 196:541-549.
119. Wang E, Miller LD, Ohnmacht GA et al. Prospective molecular profiling of melanoma metastases suggests classifiers of immune responsiveness. Cancer Res 2002; 62:3581-3586.

CHAPTER 1C

Interleukin-10 and Adaptive Immune Response

Simone Mocellin, Carlo Riccardo Rossi and Pierluigi Pilati

Introduction

IL-10 is a pleiotropic cytokine that is produced by a variety of hematopoietic cells (e.g., CD4+ T cells, NK cells, macrophages, dendritic cells)[1-4] and regulates the function of several adaptive immunity-related cells. Although generally considered an immunosuppressive molecule, IL-10 has been demonstrated to possess immunostimulating properties in several in vitro and in vivo models.

These very different outcomes are believed to depend upon experimental conditions (e.g., in vitro/in vivo), dominant effector system involved in a given immune response (humoral immunity, Th1 type immunity, mixed innate/adaptive immunity), timing of IL-10 production/administration, and IL-10 dose and/or location of expression. However, the cascade of molecular events underlying IL-10 ultimate effect on adaptive immune responses are still largely unexplored.

Here we review the current knowledge regarding IL-10 functions with respect to the cell types involved in adaptive immunity, emphasize new insights on IL-10 molecular/cellular targets, and summarize the available data on the relationship between IL-10 and some pathological conditions such as infectious diseases, autoimmunity, allergy, cancer and transplantation. Taken together, currently available data lead to the conclusion that IL-10, while suppressing Th1 and some Th2 type immune response, favours humoral immunity and CTL functions by acting both directly on relevant cells and indirectly through its influence on other immune cells such as T regulatory cells, dendritic cells, and NK cells.

Effects on B Cells and Immunoglobulin Production

Studies involving administration of IL-10 protein, IL-10 gene delivery, IL-10 transgenic animals or inhibition of IL-10 activity by neutralizing mAb or gene knock-out animals suggest that the in vivo role of IL-10 in B cell function is limited.[5-7] For instance, plasma immunoglobulin levels are normal in IL-10 -/- mice.[8] An important exception is represented by a model of antibody-dependent autoimmunity such as systemic lupus erythematosus (SLE), in which IL-10 neutralization does affect disease onset.[9]

IL-10 effects on survival, proliferation and differentiation of human B cells are much more evident.[7] IL-10 enhances survival of normal B cells, which correlates with increased expression of the anti-apoptotic protein bcl-2.[10,11] IL-10 is also a potent cofactor for B cell proliferation:[12,13] this phenomenon is further promoted by IL-2, which correlates with IL-10-induced overexpression of the high-affinity IL-2 receptor on B cells.[14]

Finally, B cell-derived and exogenous IL-10 promotes B cell differentiation and isotype switching.[15,16] Long-term culture of B cells stimulated by either anti-CD40 or follicular dendritic cells and IL-10 results in differentiation of B cells into plasma cells, and IL-10 acts

Interleukin-10, edited by Francesco M. Marincola. ©2006 Eurekah.com.

synergistically with CD27/CD70 signals to induce plasma cell differentiation from $CD27^+$ memory B cells.[17] Accordingly, IL-10 induces IgA and IgG production by B cells of patients suffering from IgA deficiency and X-linked hyper-IgM-syndrome, respectively.[18]

Direct Effects on T Cells

Because of its ability to reduce the production of IL-2, IFNγ and TNFα by murine Th1 cells, IL-10 was initially named cytokine-synthesis inhibitory factor (CSIF).[1] In humans, IL-10 inhibits the production of IL-2 and IFNγ by Th1 cells[3,19] as well as proliferation and chemotaxis of T lymphocytes.[20,21] Moreover, both $CD8^+$ and $CD4^+$ T cell allogeneic responses are blunted in vitro by IL-10.[22-24] As in vitro it shows potent anti-inflammatory/immunosuppressive properties and stimulates B cell function,[13] IL-10 has been often referred to as Th2 type cytokine[25] with the physiological role to terminate T cell-mediated immunity and start humoral immune response. However, as the knowledge on IL-10 function progresses, its biological activity appears to be more complex and a body of evidence has accumulated that IL-10 can have both immunosuppressive and immunostimulating effects on adaptive immunity cell mediators.

IL-10 Receptor Signal Transduction Pathway

The IL-10/IL-10R interaction engages tyrosine kinases of the Jak family (Fig. 1), which in turn leads to the tyrosine phosphorilation and activation of the latent transcription factors STAT1 and STAT3.[26,27] The molecular link between STAT activation and IL-10 immunomodulatory effects on T cells is not well characterized. Some data suggest that, in analogy with the macrophage model,[28] suppressor of cytokine signalling (SOCS)-1 and SOCS-3 might be two of the genes targeted by IL-10-activated STAT proteins.[29] These genes potently impair cytokine production by immune cells and are therefore potential regulators of immune responses.[30] As SOCS genes have also been reported to be part of a negative feed-back mechanism regulating the IL-10/STAT pathway,[31,32] their role as mediators of IL-10 activities is not yet defined.

Nuclear factor kappa beta (NFκB) is a well known transcription factor that regulates the expression of a large number of immune-related genes (e.g., cytokines).[33] Interestingly, IL-10 effects on NFκB are opposite depending on the target cell: NFκB inhibition is observed in macrophages and $CD4^+$ T cells,[34] whereas activation is shown in $CD8^+$ T cells,[35] which suggests a significant diversification in the reaction of different cell types to this cytokine (see below).

Recently, the inhibition of the CD28 costimulatory pathway has been proposed as a main mechanism for IL-10-mediated suppression of T cell activation[36] (Fig. 1). The IL-10R/CD28 association appears to be similar to the described association between CTLA- 4 (a well known negative regulator of T-cell function) and the TCR complex ζ chain in T cells.[37] Remarkably, this IL-10 inhibitory effect is selective for CD28 costimulatory signal-dependent T cell responses, as purified $CD45RO^+$ (memory/effector) T cell activity is not affected.[36] This may imply that naïve but not antigen-experienced T cells are the main target of IL-10 inhibition, with important consequences in the interpretation of IL-10 ultimate effect on adaptive immune responses.

Peripheral Anergy

The mechanisms responsible for peripheral tolerance are usually classified into those acting directly on the responding T cell (intrinsic tolerance) and those mediated by additional subsets of immune cells (extrinsic tolerance), including DC and regulatory T cells.[38]

Anergy, a main feature of tolerance, can be defined as a long-lasting inability of antigen-specific T cells to proliferate upon rechallenge with fully competent antigen presenting cells (APC). Classically, anergy is induced when a T cell meets its antigen in the absence of costimulation, and the resulting state of nonresponsiveness can be fully reversed by the addition of exogenous IL-2.[39] Alternatively, T cell anergy can be induced when $CD4^+$ or $CD8^+$ T cells are activated in

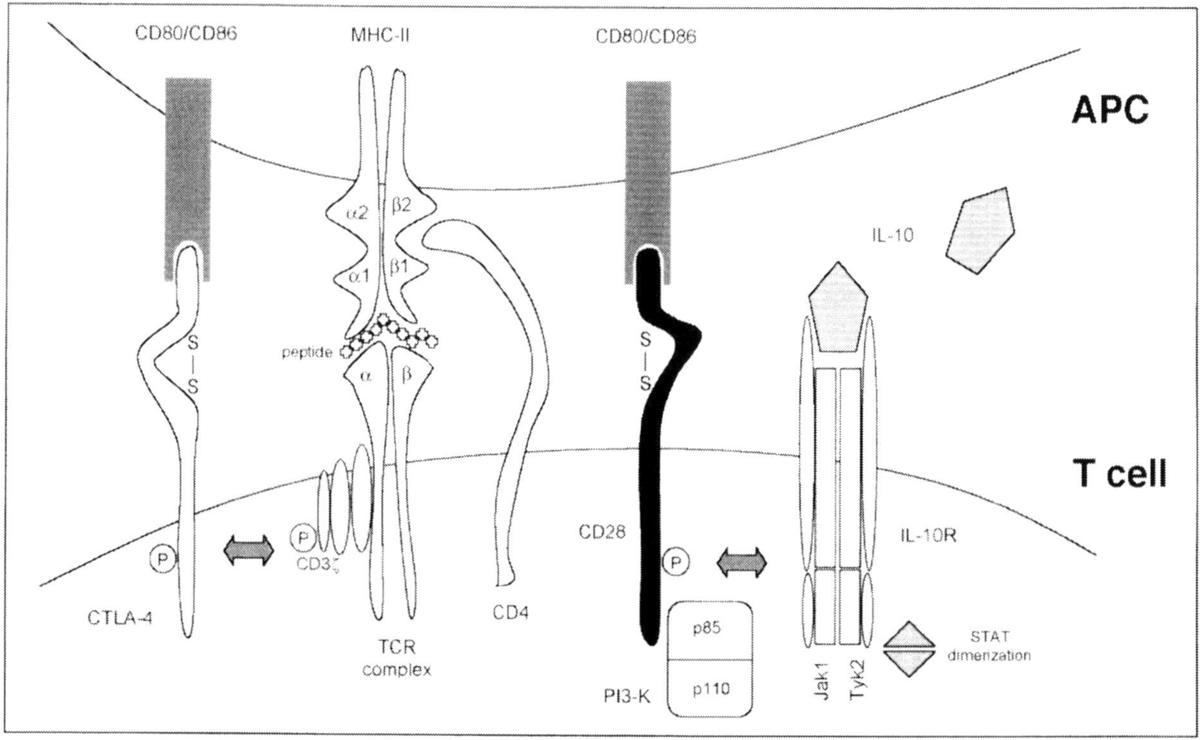

Figure 1. The molecular basis of T cell suppression by IL-10. Stimulation of CD28 by CD80/86 surface costimulatory molecules expressed by antigen presenting cells (APC) leads to tyrosine phosphorylation of CD28. Ligation of IL-10 receptor (IL-10R) at the time of CD28 stimulation inhibits tyrosine phosphorylation of CD28. As a consecutive event for signal transduction, phosphatidyl-inositol 3-kinase (PI3-K) binds to CD28 by its p85 subunit. IL-10 exerts its biological functions through the activation of Jak1 and Tyk2, the members of the receptor-associated Janus tyrosine kinases family and STAT1 and STAT3. The engagement of IL-10R by IL-10 inhibits CD28 both tyrosine phosphorylation and phosphatidylinositol 3-kinase (PI3-K) binding. Similarly, CTLA-4, a negative regulator of T cell function, associates with the TCR complex ζ chain in T cells and utilizes the tyrosine phosphatase SHP-2 to dephosphorylate CD3ζ chain.

the presence of IL-10 or with APC that have been previously treated with IL-10.[22,40] This effect of IL-10 is partly due to the ability of IL-10 to downregulate costimulatory molecules on APC (see below), but also involves direct effects on the T cells themselves. As a matter of fact, purified T cells activated with anti-CD3 mAb in the presence of IL-10 also become nonresponsive.[22] Interestingly, the anergic state of IL-10-treated T cells cannot be reversed by the addition of exogenous IL-2 (or IL-15), indicating that IL-10-anergized cells are not equivalent to those induced by costimulation blockade.[22] However, reversibility of IL-10 immunosuppressive properties by other cytokines remains debated. In an in vitro model of allergen-specific immunotherapy, $CD4^+$ T cell proliferation and cytokine production initially inhibited by IL-10 is fully recovered after the administration of IL-2,[36] which underlines the importance of tissue microenvironment cytokine balance in determining the ultimate outcome of an immune response.

The Other Side of the Coin

Despite the above mentioned findings, the immune suppressive activity of IL-10 is not universal and the ultimate effect of this cytokine cannot be evaluated outside the context of other relevant functions. Although the importance of CD28 signalling in $CD4^+$ T cells is well established both in vivo and in vitro,[41] the role of CD28 in the activation of $CD8^+$ T cells is less clear.[42,43] Therefore, the above mentioned inhibition of CD28 signal transduction pathway by IL-10 may affect more selectively the $CD4^+$ T cell subset rather than $CD8^+$ cytotoxic lymphocyte, which could be of relevance in the interpretation of IL-10 ultimate effect on some adaptive immunity responses. Moreover, recent data suggest that CD28 activity is required during the process of $CD8^+$ T cell tolerance,[44] further supports the idea that the tolerogenic effect of IL-10 may be limited to the $CD4^+$ T cell subset.

Immunostimulating properties of IL-10 have been shown in several in vitro and in vivo models. Leukocyte recruitment is enhanced by IL-10 via both chemotaxis[45-47] and induction of endothelial cell adhesion molecule expression.[48-50] Also in this case, a differential effect on T cell subsets has been reported, only $CD8^+$ cell migration being increased by IL-10.[21,45] Even original properties suggested for this cytokine, such as inhibition of APC function, T cell proliferation and Th1 type cytokine secretion have been revisited in the context of the immune response to human papilloma virus-related cervix carcinoma.[51] In this tumor model IL-10 synergistically enhances IL-2-induced in vitro expansion of tumor specific CTL and cytotoxic activity. Other investigators have reported similar results,[52-56] all of them underscoring the preferential activation of $CD8^+$ T cells.

These preclinical data indicate that, under certain conditions, IL-10 can have stimulatory effects on innate immunity cell mediators. Knowledge of potential proinflammatory effects of IL-10 in humans in vivo is scarce. In healthy volunteers, intravenous administration of human recombinant IL-10 given 1 h after the induction of endotoxaemia produced proinflammatory effects by enhancing release of IFNγ, IP-10 and other monokines induced by IFNγ.[57] In addition, IL-10 induced activation of cytotoxic T and NK cells, as reflected by increased levels of granzyme-B. In a separate study in which IL-10 was given to patients with acute myelogenous leukaemia, IL-10 induced elevated serum levels of TNFα and IL-1.[58] These effects are opposite to the in vitro observations,[34] emphasizing that in vitro experiments do not necessarily reflect the in vivo biological activity of IL-10.

Effects on Immune Cells Regulating T Cell Activities

Dendritic Cells

It is universally acknowledged that T cells activity is strictly dependent upon their interaction with dendritic cells (DC).[59] Therefore, it is not surprising that IL-10 immunosuppressive effects on T cells are believed to be largely mediated by the effects of this cytokine on DC.[60-63] IL-10 impairs the antigen presenting properties of DC by reducing their expression of HLA class II molecules,[64,65] intercellular adhesion molecules (e.g., ICAM-1)[66,67] and costimulatory molecules[68-71] (i.e., CD80/B7-1 and CD86/B7.2), which correlates with its ability to impair primary alloantigen-specific T cell responses.[23 24] Of note, as class I HLA expression by DC is not reported to be downregulated by IL-10,[66] cross priming of CTL might not be as much affected by this cytokine as it is that of $CD4^+$ T cells.

These observations have been extended to different experimental models and have shown that IL-10-conditioned DC can induce a state of anergy in alloantigen- or peptide-activated T cells.[22,40,72-76] Finally, like T cells, DC can be the target of but also produce IL-10.[77,78] In particular, it has been proposed that IL-10-producing DC may be involved in the generation of T regulatory cells with immunosuppressive function (see below).

Although the inhibition of antigen presentation is a well documented result of IL-10 activity on DC, antigen uptake by APC appears to be promoted by this cytokine.[79,80] This, combined with the IL-10 induced inhibition of DC migration,[81-83] might represent an important function of IL-10 in an early phase of DC maturation, when immature DC must accumulate in the relevant arena and be loaded with antigens from damaged tissues.

T Regulatory Cells

Regulatory $CD4^+$ T cell subsets have been described as major mediators of peripheral tolerance by inhibiting adaptive immunity and/or inflammatory pathologies[84,85] through the regulation of both Th1- and Th2 type immune responses. For several years, investigators have been interested in how regulatory $CD4^+$ T cells contribute to the induction of peripheral tolerance, either via expression of inhibitory cell surface molecules ($CD4^+/CD25^+$ T cells) or the production of immunoregulatory cytokines (Type 1 T regulatory (Tr1) cells).[86] The latter type of T cells are defined by their ability to produce high levels of IL-10 and TGF-β.[87] The production and action of these two cytokines are interrelated and likely involve a positive feed-back loop, in which IL-10 enhances expression of TGF-β and vice versa.[88] In vitro data demonstrate that exogenous/transgenic IL-10 stimulates the production of TGF-β.[89,90] Lymphocytes from IL-10 deficient mice produce less TGF-β than cells from wild-type animals.[90] In addition, the protective effect of IL-10 in a colitis model is due to its ability to create an environment which is permissive for TGF-β production, which ultimately mediates the suppressive effects.[91] In fact, IL-10 enhances production of TGF-β and also controls the ability of target cells to respond to TGF-β. This involves the IL-10-mediated restoration of the expression of TGF-βR2 on recently activated T cells that usually downregulate this receptor and become insensitive to the inhibitory effects of TGF-β.[92] On the other side, TGF-β can promote the production of IL-10. For example, treatment of colitis by administration of a plasmid encoding TGF-β provides a beneficial effect via stimulating the production of IL-10.[93] These two cytokines mediate Tr1 cell physiological function to suppress pathological immune responses (e.g., allergy, autoimmune disease), although T regulatory cells can also be involved in the pathogenesis of certain diseases, as they can suppress immune reaction to antigens from tumors and pathogens.

Finally, it has been proposed that IL-10 exerts a direct role in the differentiation of regulatory $CD4^+$ T cells.[87] In particular, the differentiation of Tr1 cells is likely controlled by tolerogenic dendritic cells that promote IL-10 production and may express tolerogenic costimulatory molecules.[94,95]

NK Cells

An integral link between innate and adaptive immunity has been largely demonstrated.[96-98] The early participation of NK cells in response to pathogen/cancer invasion might influence the subsequent development of an adaptive immune response, perhaps providing cues to indicate a "nonself"/"dangerous" encounter. Furthermore, an aberrant innate reaction to self tissue might promote an autoimmune disease, as demonstrated by the fact that NK cells are required for the development of experimental autoimmune myasthenia gravis in mice.[99] Therefore, adaptive immunity cannot longer be studied without taking into consideration the influence exerted by innate immunity mediators on antigen-specific T cell activity.

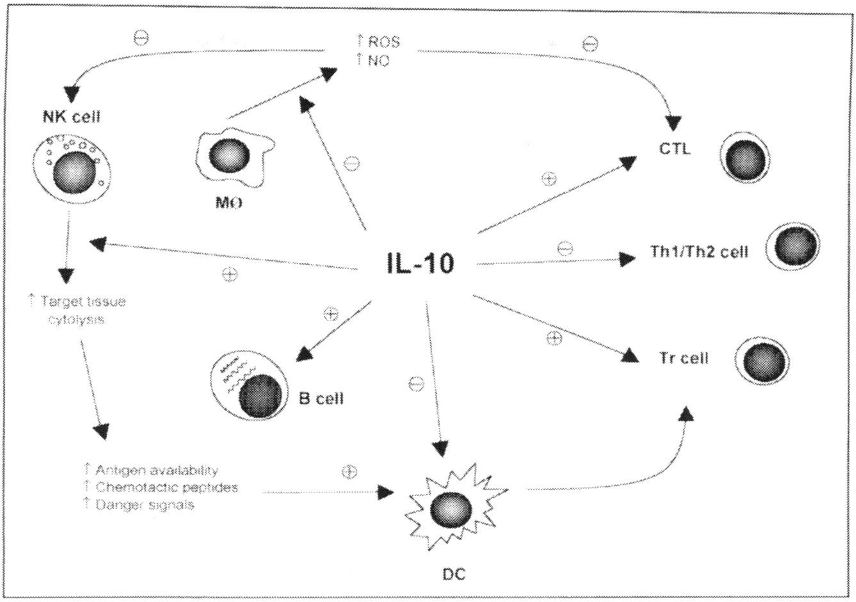

Figure 2. IL-10 effects on relationship with innate and adaptive immune response cell mediators. ROS: reactive oxygen species; NO: nitric oxide; NK cell: natural killer cell; MØ: macrophage; DC: dendritic cell; Tr cell: T regulatory cell; Th1/Th2 cell: T helper 1/T helper 2 cell; CTL: cytotoxic T lymphocytes.

Although IL-10 negatively affects IFNγ and TNFα production by NK cells in vitro [34] we and others have reported on the stimulating effects of this cytokine on NK cell cytotoxicity both in vitro and in preclinical models.[100,101]

These observations support the hypothesis that, at an early stage of the immune response, IL-10 might induce destruction of "abnormal" cells by stimulating the innate arm of the immune system, which in turn would lead to secondary beneficial effects on the incoming adaptive immune response[97] In fact, NK cell-mediated cytolysis of target cells would provide DC with adequate amounts of relevant antigens,[102] chemotaxis-related molecules (e.g., chemotactic peptides)[103] and danger signal molecules (e.g., heat shock proteins, double stranded DNA),[104] which ultimately initialize the process of DC maturation (Fig. 2). Of note, IL-10, by augmenting the expression of toll-like receptors on the monocyte lineage,[105] can enhance their sensibility to danger signal molecular mediators.[106] Upon addition of a secondary stimulus (e.g., IL-2, IL-12) in secondary lymphatic organs,[107] the balance is then shifted towards full maturation of DC, with consequent production of costimulatory molecules and cytokines with proliferative/activating effects on naïve T cells.[108]

IL-10 and Disease

Infectious Diseases

Several lines of preclinical evidence have demonstrated that innate and adaptive immune responses to a wide range of bacterial pathogens are blunted by IL-10 (reviewed in ref. 34). In animal models, administration of recombinant IL-10 or the use of IL-10 transgenic mice nearly always impairs the anti-pathogen response. On the other hand, reduction/elimination of IL-10 by neutralizing mAb or the use of IL10-knock-out mice enhances the clearance/control of the infection.

In humans, a strong correlation exists between IL-10 plasma/tissue levels and numerous chronic infectious diseases, such as malaria,[109] visceral leishmaniasis,[110] leprosy,[111] candidiasis[112] and AIDS.[113] Despite these observations, IL-10 might play an important physiological role in mitigating the inflammatory phenomena accompanying the immune response to pathogens. In fact, the potent anti-microbial mechanisms operated by the immune system can also cause significant collateral damage to the host. Destruction of even small areas of critical tissues (e.g., cardiac muscle, central/peripheral nervous system) can be more harmful than the infection itself. A successful immune response must strike a balance between protection from pathogen invasion and pathology of the tissues, and IL-10 might play a pivotal role in the establishment of this equilibrium.

Autoimmunity

Given its effects on T regulatory cells and DC, IL-10 could play a significant role in tolerance to self-antigens.[22,114] In line with the above reported IL-10 effect on Th1 cells and the notion that Th1 cytokines (e.g., TNFα, IL-2, IFNγ) play a pivotal role in the pathogenesis of rheumatoid arthritis (RA),[115,116] IL-10 administration to animals before and/or after induction of disease results protective.[117-119] Furthermore, one of the most important evidences supporting of IL-10 immunosuppressive activity comes from IL-10 knock-out mice experiments, in which animals develop a chronic inflammatory bowel disease that is considered the experimental equivalent of Chron's disease and ulcerative colitis occurring in humans.[8] Oral administration of *Lattococcus lactis* secreting IL-10 prevented enterocolitis in IL10 -/- mice.[120] Encouraging results have also been obtained in humans: in fact, both intravenous and subcutaneous administration of IL-10 in patients with Chron's disease reduced bowel inflammation, even in case of steroid-resistance.[121] These findings strengthen the hypothesis that, when the dominant pathogenetic mechanism is Th1 cell hyperactivity, IL-10 can be beneficial due to its ability to dampen the function of these T cell subset.

As regards other autoimmune diseases, reports on IL-10 role are more controversial, likely because their pathogenesis is sustained by the dysregulation of multiple immune mechanisms.

For instance, attempts to treat experimental allergic encephalitis (EAE) with recombinant IL-10 has yielded conflicting results: systemic treatment of animals with IL-10 partially inhibits disease progression in EAE induced by active immunization with central nervous system-derived antigens, but only if treatment is initiated at the time of initial immunization.[122,123] Injection of IL-10 does not suppress EAE after active immunization, and exacerbates the disease in an adoptive transfer model.[124] These findings can be interpreted bearing in mind that IL-10 can impair the effectiveness of vaccination due to its ability to interfere with antigen presentation to naïve T cells by APC, whereas, in the presence of antigen-experienced T cells (adoptively transferred into the target animal), IL-10 can enhance the immune response against self antigens by stimulating the activity of effector CTL.

In the nonobese diabetic mouse model, IDDM spontaneously develops, both $CD4^+$ and $CD8^+$ T cells being involved in the destruction of pancreatic islets β-cells.[125] Systemic administration of IL-10 late in mice life is followed by protection from IDDM [126,127] In contrast, IL-10 production by transgenic pancreatic β-cells accelerated the onset of diabetes in mice.[48,128] As antigen presentation mainly occurs in secondary lymphatic organs,[102] these experimental observations, together with those made by others in tumor models (see below), support the idea that, when IL-10 is expressed locally, its immunosuppressive effect on DC cells is overwhelmed by the stimulatory activity on CTL.

Allergy

Mucosa infiltration by eosinophils, basophils and mast cells, together with abnormal production of IgE are the key elements in the determinism of asthma and other allergic diseases. Th2 cells secreting IL-4, IL-5 and IL-13 induce and amplify the allergic response by enhancing recruitment and activation of the above mentioned inflammatory cells and by increasing the production of IgE.[129,130] IL-10 can inhibit cytokine production by Th2 cells and induce Th2 cell anergy[131,132] by acting directly on these cells and through other immune cells such as DC[133] and Tr1 cells.[134] In vivo, its administration had beneficial consequences in animals with allergic diseases.[6,135,136]

Cancer

As IL-10 inhibits antigen presentation by DC, potentially preventing T cells from mounting an effective immune response against cancer, several authors raised the teleological hypothesis that IL-10 is an immunosuppressive molecule secreted by tumors to escape from immune surveillance.[137,138] Some in vivo and in vivo tumor models support this hypothesis. $CD8^+$ T cells can be anergized towards melanoma associated antigens when stimulated with IL-10-conditioned DC.[40,74,75,139] Yet, a Lewis lung carcinoma cell line grows more rapidly in a transgenic mouse expressing IL-10 under control of an IL-2 promoter than in nontransgenic control mice,[140] supposedly by suppressing both T cell and APC function.[141] However, a large body of preclinical evidence is in contrast with the above mentioned reports. Transfection of mouse carcinomas[142,143] and melanoma[144] cell lines with IL-10 elicits loss of tumorigenicity and increases immunogenicity accompanied by a strong lymphocyte and antibody-dependent immune memory.[145] Yet, exogenous IL-10 administration can mediate regression of melanoma and breast cancer metastases in various preclinical models.[146-151] Although some authors using *scid* models have linked IL-10 antitumor effect to enhanced NK cell activity,[147,149] others have demonstrated that it depends on either $CD8^+$[151] or $CD4^+$[152] T cell activity.[151]

In humans, only a few reports are available on the role of IL-10 in tumor immunology. A negative correlation between circulating levels of IL-10 and prognosis was observed in patients with solid or haematological malignancies. Nonetheless, this association might reflect the bulk of IL-10-producing tumor cells in advanced stages. Using quantitative real time-PCR[153] and high-throughput cDNA microarray, we analyzed the gene profile of transit melanoma metastases in patients undergoing vaccination with TAA-derived peptides plus IL-2.[154] In pretreatment samples, IL-10 was overexpressed in responding versus progressing lesions.[155] A follow-up study on an independent patient population confirmed this finding by identifying a positive correlation between clinical regression and IL-10 protein levels in tumor cells from FNA obtained before therapy.[156] Thus, it could be speculated that the presence of high levels of IL-10 in situ might precondition the tumor microenvironment to the anticancer effects of systemic vaccination. As TIA-1, a gene encoding a cytotoxicity-related protein,[157] was also overexpressed in responding metastases and IL-10 increases TIA-1 expression by NK cells,[100] our results support the hypothesis that IL-10 might exert a permissive activity on vaccine-induced adaptive immunity by increasing NK cell antitumor function[158] (Fig. 2).

Transplantation

In view of its in vitro immunosuppressive functions, IL-10 has been extensively studied in transplantation models. However, in vivo results are not always univocal.

IL-10 treatment of bone marrow transplantation (BMT) recipient animals sustained opposite results depending upon cytokine dosage, lower doses being protective and higher amounts exacerbating rejection.[159] In organ transplantation models, IL-10 administration prior to grafting enhanced graft survival[160-163] and was correlated with impaired APC function.[163] In contrast, IL-10 administration at or after the time of grafting had no or little impact or even enhanced rejection.[164-166] Overall, the ultimate effect of IL-10 on BMT and organ allografts is the result of the complex combination of timing, kinetics and amounts of cytokine as well as the relative impact of different IL-10 activities in each particular experimental model. Considering the above reported data, it may be inferred that, before T cell priming by APC, IL-10 net effect is immunosuppression, while after antigen presentation the dominant IL-10 effect on adaptive immunity is stimulation. However, more experimental observations are needed to confirm this hypothesis.

Table 1. Effects of IL-10 on immune cells

	IL-10 Dominant Effect	Comments
Adaptive Immunity Cell Mediators		
B cells	Stimulatory	IL-10 worsens the outcome of diseases mediated by B cell hyperactivity (e.g., SLE)
CTL	Stimulatory	Together with that of NK cells, CTL stimulation may explain in vivo beneficial effects of IL-10 in tumor models
Th1 cells	Inhibitory	IL-10 administration is beneficial in disease models characterized by Th1 cell hyperactivity (e.g., RA, psoriasis)
Th2 cells	Inhibitory	IL-10 administration is beneficial in disease models characterized by Th2 cell hyperactivity (e.g., allergy)
Immune Regulatory Cells		
T regulatory cells	Stimulatory	IL-10 induces the maturation of these cells, which mediate $CD4^+$ T cell peripheral anergy
Dendritic cells	Inhibitory	Suppression of antigen presentation with consequent impairment of naïve T cell activation
NK cells	Stimulatory	An increased NK cell mediated cytolysis favors CTL activity

Conclusions

In view of its ability to differently affect the function of several immune cell subsets, IL-10 plays a pivotal role in the modulation of adaptive immunity. Currently, IL-10 cannot be categorized solely as a Th2 type cytokine, as its activities span inhibition of some T cell mediated responses (i.e., Th1 and some Th2 type immunity) as well as stimulation of B cells and CTL. The ability to promote opposite results on different adaptive immune responses hinges upon IL-10 direct effects on relevant cells as well its influence on immune cells involved in the modulation of adaptive immunity outcome (e.g., T regulatory cells, dendritic cells, NK cells). As a matter of fact, in vivo IL-10 function can only be appreciated within the frame of an integrated and dynamic view of the immune system network, including both innate and adaptive immunity.

Revisited under these premises, IL-10 role in several experimental findings apparently conflicting may result less enigmatic. For instance, as summarized in Table 1, IL-10 ability of suppressing Th1 and Th2 cells can account for the inhibitory effects on disease models (e.g., RA, psoriasis, allergy) dominated by the pathological hyperactivity of these types of immune response. On the other side, as it links and enhances some innate (NK cells) and adaptive (CTL) immunity cell mediators, IL-10 can be understandably associated with tumor regression.

Although some advances have been recently made in the knowledge of IL-10 molecular and cellular targets, as we have above outlined, the cascade of molecular events leading to IL-10-modulated immune phenomena is still largely unexplored, and further efforts are warranted to fully dissect its pleiotropic functions in vivo.

References

1. Fiorentino DF, Bond MW, Mosmann TR et al. Two types of mouse T helper cell. IV. Th2 clones secrete a factor that inhibits cytokine production by Th1 clones. J Exp Med 1989; 1706:2081-2095.
2. de Waal, Malefyt R et al. Interleukin 10(IL-10) inhibits cytokine synthesis by human monocytes: An autoregulatory role of IL-10 produced by monocytes. J Exp Med 1991; 174(5):1209-20.
3. Del Prete G et al. Human IL-10 is produced by both type 1 helper (Th1) and type 2 helper (Th2) T cell clones and inhibits their antigen-specific proliferation and cytokine production. J Immunol 1993; 150:353-356.
4. Mehrotra PT et al. Production of IL-10 by human natural killer cells stimulated with IL-2 and/or IL-12. J Immunol 1998; 160(6):2637-44.
5. Groux H et al. A transgenic model to analyze the immunoregulatory role of IL-10 secreted by antigen-presenting cells. J Immunol 1999; 162:1723-1729.
6. Stampfli M et al. Interleukin-10 gene transfer to the airway regulates allergic mucosal sensitization in mice. Am J Respir Cell Mol Biol 1999; 21:586-596.
7. Beebe A, Cua D, de Waal Malefyt R. The role of interleukin-10 in autoimmune disease: Systemic lupus erythematosus (SLE) and multiple sclerosis (MS). Cytokine Growth Factor Rev 2002; 13(4-5):403-412.
8. Kuhn R et al. Interleukin-10-deficient mice develop chronic enterocolitis. Cell 1993; 75(2):263-274.
9. Ishida H et al. Continuous administration of anti-IL-10 antibodies delays onset of autoimmunity in NZB/W F1 mice. J Exp Med 1994; 179:305-310.
10. Levy Y, JC Brouet, Interleukin-10 prevents spontaneous death of germinal center B cells by induction of the bcl-2 protein. J Clin Invest 1994; 93(1):424-8.
11. Itoh K, Hirohata S. The role of IL-10 in human B cell activation, proliferation, and differentiation. J Immunol 1995; 154(9):4341-4350.
12. Saeland S et al. Human B cell precursors proliferate and express CD23 after CD40 ligation. J Exp Med 1993; 178:113-120.
13. Rousset F et al. Interleukin 10 is a potent growth and differentiation factor for activated human B lymphocytes. Proc Natl Acad Sci USA 1992; 89(5):1890-3.
14. Fluckiger A et al. Interleukin 10 (IL-10) upregulates functional high affinity IL-2 receptor on normal and leukemic B lymphocytes. J Exp Med 1993; 178:1473-1481.
15. Burdin N et al. Endogenous IL-6 and IL-10 contribute to the differentiation of CD40-activated human B lymphocytes. J Immunol 1995; 154:2533-2544.
16. Malisan F et al. Interleukin-10 induces immunoglobulin G isotype switch recombination in human CD40-activated naive B lymphocytes. J Exp Med 1996; 183:937-947.
17. Agematsu K et al. Generation of plasma cells from peripheral blood memory B cells: Synergistic effect of interleukin-10 and CD27/CD70 interaction. Blood 1998; 91:173-180.
18. Briere F. et al. Interleukin 10 induces B lymphocytes from IgA-deficient patients to secrete IgA. J Clin Invest 1994; 94:97-104.1
19. de Waal Malefyt R, Yssel H, de Vries J. Direct effects of IL-10 on subsets of human CD4+ T cell clones and resting T cells. Specific inhibition of IL-2 production and proliferation. J Immunol 1993; 150(11):4754-4765.
20. Taga K, Mostowski H, Tosato G. Human interleukin-10 can directly inhibit T-cell growth. Blood 1993; 81(11):2964-71.
21. Jinquan T, et al. CXC chemokine receptor 4 expression and stromal cell-derived factor 1alpha-induced chemotaxis in CD4+ T lymphocytes are regulated by interleukin-4 and interleukin-10. Immunology 2000; 99:402-410.
22. Groux H et al. Interleukin-10 induces a long-term antigen-specific anergic state in human CD4+ T cells. J Exp Med 1996; 184(1):19-29.
23. Caux C et al. Interleukin 10 inhibits T cell alloreaction induced by human dendritic cells. Int Immunol 1994; 6(8):1177-1185.
24. Bejarano M et al. Interleukin 10 inhibits allogeneic proliferative and cytotoxic T cell responses generated in primary mixed lymphocyte cultures. Int Immunol 1992; 4(12):1389-1397.
25. Romagnani S. T-cell subsets (Th1 versurs Th2). Ann Allergy Asthma Immunol 2000; 85(1):9-18.
26. Finbloom D, Winestock K. IL-10 induces the tyrosine phosphorylation of tyk2 and jak1 and the differential assembly of STAT1 and STAT3 complexes in human T cells and monocytes. J Immunol 1995; 155(3):1079-1090.
27. Lai C et al. Receptors for interleukin(IL)-10 and IL-6-type cytokines use similar signaling mechanisms for inducing transcription through IL-6 response elements. J Biol Chem 1996; 271(24):13968-13975.

28. Ito S et al. Interleukin-10 inhibits expression of both interferon-alpha and interferon-gamma induced genes by suppressing tyrosine phosphorilation of STAT1. Blood 1999; 93(5):1456-1463.
29. Matsumoto A et al. A role of suppressor of cytokine signaling 3 (SOCS3/CIS3/SSI3) in CD28-mediated interleukin 2 production. J Exp Med 2003; 197(4):425-436.
30. Alexander W. Suppressors of cytokine signalling (SOCS) in the immune system. Nat Rev Immunol 2002; 2(6):410-416.
31. Suzuki A et al. CIS3/SOCS3/SSI3 plays a negative regulatory role in STAT3 activation and intestinal inflammation. J Exp Med 2001; 193(4):471-481.
32. Ding Y et al. Suppressor of cytokine signaling 1 inhibits IL-10-mediated immune responses. J Immunol 2003; 170(3):1383-1391.
33. Hanada T, Yoshimura A. Regulation of cytokine signaling and inflammation. Cytokine Frowth Factor Rev 2002; 13(4-5):413-421.
34. Moore KW et al. Interleukin-10 and the interleukin-10 receptor. Annu Rev Immunol 2001; 19:683-765.
35. Hurme M et al. Effect of interleukin-10 on NF-kB and AP-1 activities in interleukin-2 dependent CD8 T lymphoblasts. Immunol Lett 1994; 42(3):129-133.
36. Akdis CA, Blaser K. Mechanisms of interleukin-10-mediated immune suppression. Immunology 2001; 103(2):131-6.
37. Lee K, Chuang E, Griffin M. Molecular basis of T cell inactivation by CTLA-4. Science 1998; 282:2262-2266.
38. Walker L, Abbas A. The enemy within: Keeping self-reactive T cells at bay in the periphery. Nat Rev Immunol 2001; 2(1):11-19.
39. Jenkins M et al. Inhibition of antigen-specific proliferation of type 1 murine T cell clones after stimulation with immobilized anti-CD3 monoclonal antibody. J Immunol 1990; 144(1):16-22.
40. Steinbrink K et al. CD4(+) and CD8(+) anergic T cells induced by interleukin-10-treated human dendritic cells display antigen-specific suppressor activity. Blood 2002; 99:2468-2476.
41. Lenschow D, Walunas T, Bluestone J. CD28/B7 system of T cell costimulation. Annu Rev Immunol 1996; 14:233-258.
42. Goldstein J et al. Purified MHC class I and peptide complexes activate naive CD8+ T cells independently of the CD28/B7 and LFA-1/ICAM-1 costimulatory interactions. J Immunol 1998; 160(7):3180-3187.
43. Wang B et al. Naive CD8+ T cells do not require costimulation for proliferation and differentiation into cytotoxic effector cells. J Immunol 2000; 164(3):1216-1222.
44. Vacchio M, Hodes R. CD28 costimulation is required for in vivo induction of peripheral tolerance in CD8 T cells. J Exp Med 2003; 197(1):p 19-26.
45. Jinquan T et al. Human interleukin 10 is a chemoattractant for CD8+ T lymphocytes and an inhibitor of IL-8-induced CD4+ T lymphocyte migration. J Immunol 1993; 151(9):4545-4551.
46. Sozzani S et al. Interleukin 10 increases CCR5 expression and HIV infection in human monocytes. J Exp Med 1998; 187(3):439-444.
47. Hedrick J et al. Characterization of a novel CC chemokine, HCC-4, whose expression is increased by interleukin-10. Blood 1998; 91:4242-4247.
48. Wogensen L, Huang X, Sarvetnick N. Leukocyte extravasation into the pancreatic tissue in transgenic mice expressing interleukin 10 in the islets of Langerhans. J Exp Med 1993; 178(1):175-185.
49. Fiehn C, Paleolog EM, Feldmann M. Selective enhancement of endothelial cell VCAM-1 expression by interleukin-10 in the presence of activated leucocytes. Immunology 1997; 91(4):565-71.
50. Vora M, Romero LI, Karasek MA. Interleukin-10 induces E-selectin on small and large blood vessel endothelial cells. J Exp Med 1996; 184(3):821-9.
51. Santin AD et al. Interleukin-10 increases Th1 cytokine production and cytotoxic potential in human papillomavirus-specific CD8(+) cytotoxic T lymphocytes. J Virol 2000; 74(10):4729-37.
52. Groux H et al. Inhibitory and stimulatory effects of IL-10 on human CD8+ T cells. J Immunol 1998; 160(7):3188-93.
53. Schwarz MA et al. Stimulation of cytolytic activity by interleukin-10. J Immunother Emphasis Tumor Immunol 1994; 16(2):95-104.
54. Rowbottom A et al. IL-10 induced CD8 cell proliferation. Immunology 1999; 98:80-89.
55. Chen W, Zlotnik A. IL-10: A novel cytotoxic T cell differentiation factor. J Immunol 1991; 147(2):528-534.
56. Ebert EC, IL-10 enhances IL-2-induced proliferation and cytotoxicity by human intestinal lymphocytes. Clin Exp Immunol 2000; 119(3):426-32.
57. Lauw FN et al. Proinflammatory effects of IL-10 during human endotoxemia. J Immunol 2000; 165(5):2783-9.

58. Tao M et al. In vivo effects of IL-4, IL-10 and amifostine on cytokine production in patients with acute myelogenous leukemia. Leuk Lymphoma 2001; 41(1-2):161-168.
59. Lanzavecchia A, Sallusto F. Regulation of T cell immunity by dendritic cells. Cell 2001; 106(3):263-266.
60. De Waal-Malefyt R et al. Interleukin 10 (IL-10) and viral IL-10 strongly reduce antigen-specific human T cell proliferation by diminishing the antigen-presenting capcity of monocytes via downregulation of class II major histocompatibility complex expression. J Exp Med 1991; 174(4):915-924.
61. Fiorentino DF et al. IL-10 acts on the antigen-presenting cell to inhibit cytokine production by Thl cells. J Immunol 1991; 146(10):3444-51.
62. Beissert S, Hosoi J, Grabbe S et al. IL-10 inhibits tumor antigen presentation by epidermal antigen-presenting cells. J Immunol 1995; 154(3):1280-1286.
63. De Smedt T, Van Mechelen M, De Becker G et al. Effect of interleukin-10 on dendritic cell maturation and function. Eur J Immunol 1997; 27(5):1229-1235.
64. Peguet-Navarro J et al. Interleukin-10 inhibits the primary allogeneic T cell response to human epidermal Langerhans cells. Eur J Immunol 1994; 24(4):884-891.
65. Koch F et al. High level IL-12 production by murine dendritic cells: upregulation via MHC class II and CD40 molecules and downregulation by IL-4 and IL-10. J Exp Med 1996; 184(2):741-6.
66. Chang C, Furue M, Tamaki K. Selective regulation of ICAM-1 and major histocompatibility complex class I and II molecule expression on epidermal Langerhans cells by some of the cytokines released by keratinocytes and T cells. Eur J Immunol 1994; 24(11):2889-2895.
67. Chatelain R et al. IL-10 inhibits ICAM-1 expression on human Langerhans cells but not on keratinocytes, dermal endothelial cells or fibroblasts. Arch Dermatol Res 1998; 290(9):477-482.
68. Mitra R et al. Psoriatic skin-derived dendritic cell function is inhibited by exogenous IL-10. Differential modulation of B7-1 (CD80) and B7-2 (CD86) expression. J Immunol 1995; 154(6):2668-2677.
69. Chang CH, Furue M, Tamaki K. B7-1 expression of Langerhans cells is up-regulated by proinflammatory cytokines, and is down-regulated by interferon-gamma or by interleukin- 10. Eur J Immunol 1995; 25(2):394-8.
70. Ozawa H et al. Interferon-gamma and interleukin-10 inhibit antigen presentation by Langerhans cells for T helper type 1 cells by suppressing their CD80 (B7-1) expression. Eur J Immunol 1996; 26(3):648-652.
71. Buelens C et al. Interleukin-10 differentially regulates B7-1 (CD80) and B7-2 (CD86) expression on human peripheral blood dendritic cells. Eur J Immunol 1995; 25(9):2668-2672.
72. Steinbrink K et al. Induction of tolerance by IL-10-treated dendritic cells. J Immunol 1997; 159(10):4772-4780.
73. Takayama T, Tahara H, Thomson AW. Transduction of dendritic cell progenitors with a retroviral vector encoding viral interleukin-10 and enhanced green fluorescent protein allows purification of potentially tolerogenic antigen-presenting cells. Transplantation 1999; 68(12):1903-1909.
74. Enk AH et al. Dendritic cells as mediators of tumor-induced tolerance in metastatic melanoma. Int J Cancer 1997; 73(3):309-16.
75. Steinbrink K, Jonuleit H, Muller G et al. Interleukin-10-treated human dendritic cells induce a melanoma-antigen-specific anergy in CD8(+) T cells resulting in a failure to lyse tumor cells. Blood 1999; 93(5):1634-1642.
76. Zeller J et al. Induction of CD4+ T cell alloantigen-specific hyporesponsiveness by IL-10 and TGF-beta. J Immunol 1999; 163:3684-3691.
77. Corinti S et al. Regulatory activity of autocrine IL-10 on dendritic cell functions. J Immunol 2001; 166(7):4312-4318.
78. De Saint-Vis B et al. The cytokine profile expressed by human dendritic cells is dependent on cell subtype and mode of activation. J Immunol 1998; 160(4):1666-1676.
79. Allavena P, Piemonti L, Longoni D et al. IL-10 prevents the differentiation of monocytes to dendritic cells but promotes their maturation to macrophages. Eur J Immunol 1998; 28(1):359-369.
80. Morel A et al. Split activity of interleukin-10 on antigen capture and antigen presentation by human dendritic cells: Definition of a maturative step. Eur J Immunol 1997; 27(1):26-34.
81. D'Amico G et al. Uncoupling of inflammatory chemokine receptors by IL-10: Generation of functional decoys. Nat Immunol 2000; 1(5):387-391.
82. Demangel C, Bertolino P, Britton W. Autocrine IL-10 impairs dendritic cell (DC)-derived immune responses to mycobacterial infection by suppressing DC trafficking to draining lymph nodes and local IL-12 production. Eur J Immunol 2002; 32(4):994-1002.
83. Wang B, Amerio P, Sauder D. Role of cytokines in epidermal Langerhans cell migration. J Leukoc Biol 1999; 66(1):33-39.

84. Groux H, O'Garra A, Bigler M et al. CD4+ T-cell subset inhibits antigen-specific T-cell responses and prevents colitis. Nature 1997; 389(6652):737-742.
85. Schwartz R, T cell anergy. Annu Rev Immunol 2003; 21:305-334.
86. Walker L, A Abbas. The enemy within: Keeping self-reactive T cells at bay in the periphery. Nat Rev Immunol 2002; 2(1):11-19.
87. Roncarolo M et al. Type 1 T regulatory cells. Immunol Rev 2001; 182:68-79.
88. Levings M et al. The role of IL-10 and TGF-b in the differentiation and effector function of T regulatory cells. Int Arch Allergy Immunol 2002; 129:263-276.
89. Seder R et al. Factors involved in the differentiation of TGF-beta-producing cells from naive CD4+ T cells: IL-4 and IFN-gamma have opposing effects, while TGF-beta positively regulates its own production. J Immunol 1998; 160(12):5719-5728.
90. De Winter H et al. Regulation of mucosal immune responses by recombinant interleukin 10 produced by intestinal epithelial cells in mice. Gastroenterology 2002; 122:1829-1841.
91. Fuss I et al. The interrelated roles of TGF-beta and IL-10 in the regulation of experimental colitis. J Immunol 2002; 168(2):900-908.
92. Cottre F, Groux H. Regulation of TGF-beta response during T cell activation is modulated by IL-10. J Immunol 2001; 167(2):773-778.
93. Kitani A et al. Treatment of experimental (trinitrobenzene sulfonic acid,) colitis by intranasal administration of transforming growth factor (TGF)-beta1 plasmid: TGF-beta1-mediated suppression of T helper cell type 1 response occurs by interleukin(IL)-10 induction and IL-10 receptor beta2 chain downregulation. J Exp Med 2000; 192:41-52.
94. Akbari O, DeKruyff R, Umetsu D. Pulmonary dendritic cells producing IL-10 mediate tolerance induced by respiratory exposure to antigen. Nat Immunol 2001; 2(8):725-731.
95. Wakkach A et al. Characterization of dendritic cells that induce tolerance and T regulatory 1 cell differentiation in vivo. Immunity 2003; 18(5):605-617.
96. Fearon D Locksley R. The instructive role of innate immunity in the acquired immune response. Science 1996; 272:50-53.
97. Kelly JM et al, Induction of tumor-specific T cell memory by NK cell-mediated tumor rejection. Nat Immunol 2002; 3(1):83-90.
98. Gerosa F et al. Reciprocal activating interaction between natural killer cells and dendritic cells. J Exp Med 2002; 195(3):327-333.
99. Shi FD et al. Natural killer cells determine the outcome of B cell-mediated autoimmunity. Nat Immunol 2000; 1(3):245-51.
100. Mocellin S et al. The dual role of IL-10. Trends Immunol 2003; 24(1):36-43.
101. Shibata Y et al. Immunoregulatory roles of IL-10 in innate immunity: IL-10 inhibits macrophage production of IFN-gamma-inducing factors but enhances NK cell production of IFN-gamma. J Immunol 1998; 161(8):4283-8.
102. Ochsenbein A. Principles of tumor imunosurveillance and implications for immunotherapy. Cancer Gene Therapy 2002; 9:1043-1055.
103. Le Y, Murphy PM, Wang JM. Formyl-peptide receptors revisited. Trends Immunol 2002; 23(11):541-8.
104. Gallucci S, Matzinger P. Danger signals: SOS to the immune system. Curr Opin Immunol 2001; 13(1):114-9.
105. Flo T et al. Differential expression of toll-like receptor 2 in human cells. J Leukoc Biol 2001; 69(3):474-481.
106. Vabulas R et al. The endoplasmic reticulum-resident heat shock protein gp96 activates dendritic cells via the toll-like receptor 2/4 pathway. J Biol Chem 2002; 277(23):20847-20853.
107. Belardelli F, Ferrantini M. Cytokines as a link between innate and adaptive antitumor immunity. Trends Immunol 2002; 23:201-208.
108. Panelli MC et al. Gene-expression profiling of the response of peripheral blood mononuclear cells and melanoma metastases to systemic IL-2 administration. Genome Biol 2002; 3(7):RESEARCH0035.
109. Peyron F et al. High levels of circulating IL-10 in human malaria. Clin Exp Immunol 1994; 95:300-303.
110. Ghalib H et al. Interleukin 10 production correlates with pathology in human Leishmania donovani infections. J Clin Invest 1993; 92(1):324-329.
111. Yamamura M et al. Defining protective responses to pathogens: cytokine profiles in leprosy lesions. Science 1991; 254:277-279.
112. Roilides E et al. Increased serum concentratrions of interleukin-10 in patients with hepatosplenic candidiasis. J Infect Dis 1998; 178(2):589-592.

113. Clerici M et al. Immunoendocrinologic abnormalities in human immunodeficiency virus infection. Ann N Y Acad Sci 2000; 917:956-961.
114. Asseman C et al. An essential role for interleukin 10 in the function of regulatory T cells that inhibit intestinal inflammation. J Exp Med 1999; 190:995-1004.
115. Katsikis P et al. 179 5, 1994; 1517-1527.
116. Andreakos E et al. Cytokines and anti-cytokine biologicals in autoimmunity: Present and future. Cytokine Growth Factor Rev 2002; 13(4-5):299-313.
117. Kasama T et al. Interleukin-10 expression and chemokine regulation during the evolution of murine type II collagen-induced arthritis. J Clin Invest 1995; 95(6):2868-2876.
118. Walmsley M et al. Interleukin-10 inhibition of the progression of established collagen-induced arthritis. Arthritis Rheum 1996; 39:495-503.
119. Joosten L et al. Role of interleukin-4 and interleukin-10 in murine collagen-induced arthritis. Protective effect of interleukin-4 and interleukin-10 treatment in cartilage destruction. Arthritis Rheum 1997; 40:249-260.
120. Steidler L et al. Treatment of murine colitis by Lactococcus lactis secreting interleukin- 10. Science 2000; 289(5483):1352-5.
121. Fedorak N et al. Recombinant human interleukin 10 in the treatment of patients with mild to moderately active Crohn's disease. The Interleukin 10 Inflammatory Bowel Disease Cooperative Study Group. Gastroenterology 2000; 119(6):1473-1482.
122. Rott FB, Cash E. Interleukin-10 prevents experimental allergic encephalomyelitis in rats. Eur J Immunol 1994; 24:1434-1440.
123. Nagelkerken LB, Blauw M, Tielmans. IL-4 abrogates the inhibitory effect of IL-10 on the development of experimental allergic encephalomyelitis in SJL mice. Int Immunol 1997; 9:1243-1251.
124. Croxford J et al. Cytokine gene therapy in experimental allergic encephalomyelitis by injection of plasmid DNA-cationic liposome complex into the central nervous system. J Immunol 1998; 160:5181-5187.
125. Wong F, Janeway C. Insulin-dependent diabetes mellitus and its animal models. Curr Opin Immunol 1999; 11(6):643-647.
126. Pennline K, Roque-Gaffney E, Monahan M. Recombinant human IL-10 prevents the onset of diabetes in the nonobese diabetic (NOD) mouse. Clin Immunol Immunopathol 1994; 71:169-175.
127. Zheng X et al. A noncytolytic IL-10/Fc fusion protein prevents diabetes, blocks autoimmunity and promotes suppressor phenomena in NOD mice. J Immunol 1997; 158:4507-4513.
128. Wogensen L, Lee M, Sarvetnick N. Production of interleukin 10 by islet cells accelerates immune-mediated destruction of beta cells in nonobese diabetic mice. J Exp Med 1994; 179:1379-1384.
129. Umetsu D, DeKruyff R. Th1 and Th2 CD4+ T cells in human allergic diseases. J Allergy Clin Immunol 1997; 100:1-6.
130. Robinson D et al. Evidence for a predominant Th2-type bronchoalveolar lavage T-lymphocyte population in atopic asthma. New Engl J Med 1992; 326:298-304.
131. Grunig G et al. Interleukin-10 is a natural suppressor of cytokine production and inflammation in a murine model of allergic bronchopulmonary apergillosis. J Exp Med 1997; 185(6):1089-1099.
132. Adachi M et al. IL-10 induces a Th2 cell tolerance in allergic asthma. Int Arch Allergy Immunol 1999; 118(2-4):391-394.
133. Bellinghausen I et al. Inhibition of human allergic T-cell responses by IL-10-treated dendritic cells: Differences from hydrocortisone-treated dendritic cells. J Allergy Clin Immunol 2001; 108(2):242-249.
134. Herrick C, Bottomly K. To respond or not to respond: T cells in allergic asthma. Nat Rev Immunol 2003; 3(5):405-412.
135. Zuany-Amorim C et al. Interleukin-10 inhibits antigen-induced cellular recruitment into the airways of sensitized mice. J Clin Invest 1995; 95:2644-2651.
136. Hall G et al. Suppression of allergen reactive Th2 mediated responses and pulmunary eosinophilia by intranasal administration of an immunodominant peptide is linked to IL-10 production. Vaccine 2003; 21(5-6):549-561.
137. Marincola FM et al. Escape of human solid tumors from T-cell recognition: Molecular mechanisms and functional significance. Adv Immunol 2000; 74:181-273.
138. Pawelec G et al. Escape mechanisms in tumor immunity: A year 2000 update. Crit Rev Oncog 2000; 11(2):97-133.
139. Yang A, Lattime E. Tumor-induced interleukin 10 suppresses the ability of splenic dendritic cells to stimulate CD4 and CD8 T-cell responses. Cancer Res 2003; 63(9):2150-2157.
140. Hagenbaugh A et al. Altered immune reponses in interleukin 10 transgenic mice. J Exp Med 1997; 185(12):2101-2110.

141. Sharma S et al. T cell-derived IL-10 promotes lung cancer growth by suppressing both T cell and APC function. J Immunol 1999; 163(9):5020-5028.
142. Giovarelli M et al. Local release of IL-10 by transfected mouse mammary adenocarcinoma cells does not suppress but enhances antitumor reaction and elicits a strong cytotoxic lymphocyte and antibody-dependent immune memory. J Immunol 1995; 155(6):3112-23.
143. Adris S et al. IL-10 expression by CT26 colon carcinoma cells inhibits their malignant phenotype and induces a T cell-mediated tumor rejection in the context of a systemic Th2 response. Gene Ther 1999; 6(10):1705-12.
144. Gerard CM et al. Loss of tumorigenicity and increased immunogenicity induced by interleukin-10 gene transfer in B16 melanoma cells. Hum Gene Ther 1996; 7(1):23-31.
145. Barth Jr RJ, Coppola MA, Green WR. In vivo effects of locally secreted IL-10 on the murine antitumor immune response. Ann Surg Oncol 1996; 3(4):381-6.
146. Zheng LM et al. Interleukin-10 inhibits tumor metastasis through an NK cell-dependent mechanism. J Exp Med 1996; 184(2):579-84.
147. Kundu N et al. Antimetastatic and antitumor activities of interleukin 10 in a murine model of breast cancer. J Natl Cancer Inst 1996; 88(8):536-41.
148. Kaufman HL et al. Interleukin-10 enhances the therapeutic effectiveness of a recombinant poxvirus-based vaccine in an experimental murine tumor model. J Immunother 1999; 22(6):489-96.
149. Huang S, Ullrich SE, Bar-Eli M. Regulation of tumor growth and metastasis by interleukin-10: The melanoma experience. J Interferon Cytokine Res 1999; 19(7):697-703.
150. Berman RM et al. Systemic administration of cellular IL-10 induces an effective, specific, and long-lived immune response against established tumors in mice. J Immunol 1996; 157(1):231-8.
151. Fujii S et al. Interleukin-10 promotes the maintenance of antitumor CD8(+) T-cell effector function in situ. Blood 2001; 98(7):2143-2151.
152. Segal BD, Glass E, Shevach. Cutting edge: IL-10-producing CD4+ T cells mediate tumor rejection. J Immunol 2002; 168(1):1-4.
153. Mocellin S et al. Quantitative real time PCR: A powerful ally in cancer research. Tends in Molecular Medicine 2003; 9(5): 189-195.
154. Rosenberg SA et al. Immunologic and therapeutic evaluation of a synthetic peptide vaccine for the treatment of patients with metastatic melanoma. Nat Med 1998; 4(3):321-7.
155. Mocellin S et al. Kinetics of cytokine expression in melanoma metastases classifies immune responsiveness. Int J Cancer 2001; 93(2):236-42.
156. Mocellin S, Wang E, Marincola FM. Cytokines and Immune Response in the Tumor Microenvironment. J Immunother 2001; 24(5):392-407.
157. Tian Q et al. A polyadenylate binding protein localized to the granules of cytolytic lymphocytes induces DNA fragmentation in target cells. Cell 1991; 67(3):629-39.
158. Mocellin S et al. Dissecting tumor responsiveness to immunotherapy: The experience of peptide-based melanoma vaccines. Biochim Biophys Acta Rev Cancer 2003 in press.
159. Blazar B et al. Interleukin-10 dose dependent regulation of CD4+ and CD8+ T cell-mediated graft versus host disease. Transplantation 1998; 66:1220-1229.
160. Boehler A. The role of interleukin-10 in lung transplantation. Transpl Immunol 2002; 9(2-4):121-124.
161. Zou X et al. Downregulation of cytokine induced neutrophil chemoattractant and prolongation of rat liver allograft survival by interleukin-10. Surg Today 1998; 28:184-191.
162. Shinozaki K et al. Allograft transduction of IL-10 prolongs survival following orthtopic liver transplantation. Gene Ther 1999; 6:816-822.
163. Li W et al. Differential effects of exogenous interleukin-10 on cardiac allograft survival: Inhibition of rejection by recipient pretreatment reflects impaired host accessory cell function. Transplantation 1999; 68:1402-1409.
164. Qian S et al. Systemic administration of cellular interleukin-10 can exacerbate cardiac allograft rejection in mice. Transplantation 1996; 62:1709-1714.
165. Furukawa Y et al. Interleukin-10 (IL-10) augments allograft arterial disease. Am J Pathol 1999; 155(6):1929-1939.
166. Zheng X et al. Administration of noncytolytic IL-10/Fc in murine models of lipopolysaccharide-induced septic shock and allogeneic islet transplantation. J Immunol 1995; 154:5590-5600.

CHAPTER 2

Experimental Models for the Analysis of IL-10 Function

Carl G. Feng, Julio Aliberti, Karl F. Hoffmann, Dragana Jankovic, Marika C. Kullberg, Alan Sher and Thomas A. Wynn

Abstract

This review focuses on the regulatory functions of IL-10 in the response to parasitic and bacterial infection revealed through knockout, cytokine/receptor blocking, and transgenic mouse studies. The various mechanisms that control the production and activity of IL-10 are also discussed. Studies performed over the past few years illustrate a complex and pleiotropic nature for IL-10 in host immunity. The observations that nearly every cell in the body can respond to IL-10 and multiple cells produce the cytokine likely explain this multifaceted activity. Studies conducted in experimental infectious and inflammatory disease models have been particularly useful in defining the various regulatory activities of IL-10. Although these studies have identified many common themes for IL-10 in host immunity, they also nicely illustrate how IL-10 fine-tunes the response to individual pathogens and prevents inflammation.

Introduction

$CD4^+$ T helper (Th) cells can be divided into three major subsets, Type-1, Type-2 and Type-3 T regulatory (Treg), based upon the specific cytokines produced and the functional activities exhibited by each cell type. Type-1 Th cells produce interferon-γ (IFN-γ) and lymphotoxin (LT), which promote macrophage activation and the generation of cell-mediated immunity. Type-2 Th cells produce a variety of cytokines including IL-4, IL-5, and IL-13, and provide help for the maturation of B cells to immunoglobulin-secreting cells, thereby activating humoral defense mechanisms. In contrast to Th1 and Th2 cells, however, T regulatory cells represent a unique and more heterogeneous population, which can express a variety of immune suppressive factors including cytotoxic T-lymphocyte antigen-4 (CTLA-4), transforming growth factor (TGF)-β, and/or IL-10.

Central to the concept of T helper subset generation is the tendency for an immune response to become polarized. Thus, a Type-1 or Type-2 cytokine-producing profile will often dominate quickly during an immune response by preferentially amplifying one Th subset while down regulating the opposing response. This polarized response appears to be critical for host defense against many pathogenic organisms. Resistance to intracellular pathogens often requires a predominantly Type-1 response, while Type-2 responses are typically needed to fight extracellular parasites. A primary goal of immunological research over the past decade has been to understand the various mechanisms that influence the polarization of the immune response following infection and to exploit those mechanisms in vaccine design. Whereas a polarized response is often required to control infections, there is also a need to balance the response. The various effector molecules, particularly those associated with the Th1 pathway, are nonspecific

Interleukin-10, edited by Francesco M. Marincola. ©2006 Eurekah.com.

in their action and can be detrimental if produced for too long, in excess, or in the wrong location. The potentially harmful molecules include nitric oxide (NO), reactive oxygen intermediates (ROI), IL-1, IFN-γ, and TNF, and these factors often operate in a synergistic fashion. Therefore, it is important to produce a sufficiently potent Type 1 response to provide efficient protection from infection, while at the same time producing a regulatory Type 2 or immunosuppressive Treg cell response to prevent the protective response from causing damage to host tissues. Conversely, excessive Th2 responses must also be dampened to prevent acute anaphylactic inflammation. The sections that follow illustrate how IL-10 regulates Th1 and Th2 responses to infection.

IL-10 and Th1/Th2 Effector Choice

IL-10 was initially characterized as a Th2-specific cytokine that inhibits IFN-γ secretion by Th1 cells.[1] Because IL-10 can also be produced by activated antigen presenting cells (APC) (macrophages, dendritic cells (DC) and B lymphocytes[2-4]) it was regarded as a candidate factor that could positively influence the development of Th2 cells and negatively regulate differentiation of Th1 cells. However, experimental data have failed to support this simplistic view of IL-10's effect on Th1/T2 polarization. As anticipated, when primed with model antigens (Ag) or pathogens known to induce Th1-type responses, IL-10$^{-/-}$ animals display highly augmented immune responses frequently associated with detrimental Th1-mediated pathology. For example, IL-10$^{-/-}$ mice infected with *Toxoplasma gondii*,[5] *Plasmodium chaubudi*,[6] or certain strains of *Trypanozoma cruzi*,[7] have greatly elevated levels of IFN-γ, IL-12 and TNF-α and reduced parasitemia, but substantially increased risk of death from a toxic shock-like syndrome compared to wild-type (WT) controls. Unexpectedly, however, IL-10$^{-/-}$ mice also display enhanced Th2 responses when either challenged with allergens or exposed to Th2-type pathogens.[8-10] Together these findings show that IL-10 acts as a general negative regulator of CD4-dependent immune responses rather than a polarizing cytokine that influences Th1/Th2 commitment.

The inhibitory effect of IL-10 stems from its ability to down-regulate antigen-presenting functions of both macrophages and DC, the primary sources of Ag/MHC complexes during T cell priming.[11,12] The indirect influence of IL-10 on Th cells has been further supported by the analysis of IL-10R expression. IL-10R is expressed by most hematopoietic cells.[13] However, while its expression is down-regulated on activated CD4$^+$ T lymphocytes,[14] activation of monocytes is associated with an increase in IL-10R levels,[15] providing the molecular basis for the IL-10 responsiveness of the latter but not the former cell population.

In the context of Th effector choice, an important aspect of IL-10 effects on APC is its ability to inhibit not only the expression of MHC class II and costimulatory molecules but also the secretion of cytokines and chemokines.[12,16] Although the latter effect of IL-10 is not selective and affects most of the soluble mediators produced by activated macrophages and DC, its primary consequence is down-regulation of the Th1 development, because many of the monokines (e.g., IL-12, IL-18, IL-23 and IL-27) are IFN-γ-inducible cytokines required for optimal Th1 differentiation.[17] For the same reason, IL-10-treated macrophages or DC appear to be promoting Th2 development.[18] In contrast to this differential effect on Th1/Th2 differentiation, the accumulation of mature Th1 and Th2 effectors at the site of inflammation can be equally affected by IL-10 since it down-regulates the production of both CC and CXC chemokines.[19,20] In addition to inhibiting the production of cytokines and chemokines, IL-10 also enhances the expression of their natural antagonists by increasing the expression of either decoy (e.g., IL-1RA and chemokine receptors)[21,22] or soluble (e.g., p55 and p75 TNFR) receptors[23,24] that in turn potentiate IL-10's down-modulatory effects on APC functions.

Different IL-10-producing DC populations (e.g., from Peyer's patches[25] and liver[26]) have been associated with the development of Th2 responses. Recently, these observations have been extended by the findings that IL-10 is required for optimal development of Th2 cells by the CD8$^-$CD11c$^+$ subset of splenic DC.[27] However, since IL-10 may selectively induce apoptosis of CD8α$^+$ CD11c$^+$ cells,[27] this Th2 priming by IL-10 appears to be a result of a loss of

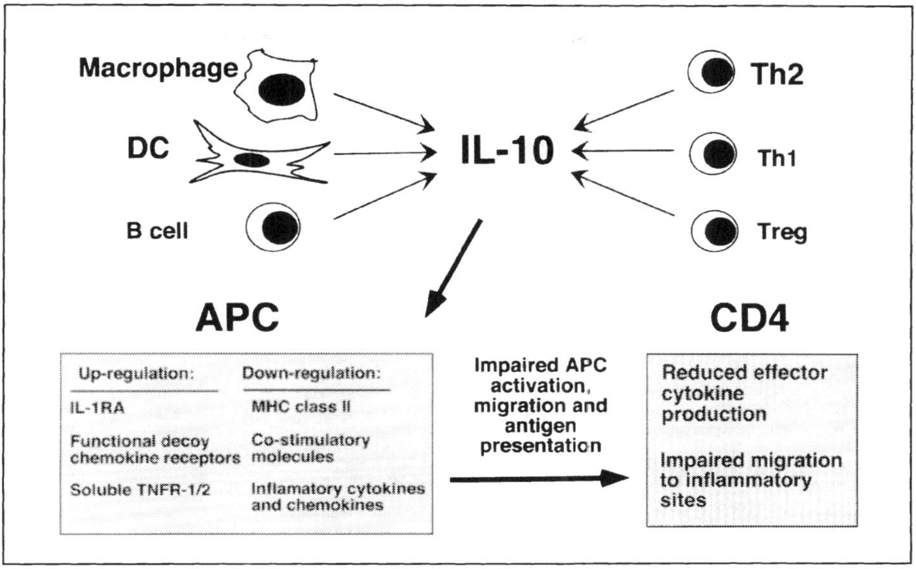

Figure 1. IL-10 as modulator of innate and adaptive immunity.

IL-12-producing DC and a subsequent lack of Th1 differentiation. In addition, while the particular DC subsets were not analyzed, naïve and *Trichinella spiralis*-infected IL-10 knockout (KO) mice display higher number of CD11c$^+$ DC in mesenteric lymph nodes when compared to WT animals.[28] Autocrine IL-10 has been shown to prevent spontaneous maturation of human DC in vitro and to limit LPS and CD40-induced maturation.[29]

While initially specifically associated with Th2 cells, the expression of IL-10 is now found in other Th subsets as well. When cultured in the presence of IL-10, murine bone marrow-derived DC promote development of IL-10$^+$ CD4$^+$ Treg lymphocytes.[30] Moreover, similar to human Th1 cells, [31] murine Th1 lymphocytes may also coexpress IL-10. For example, "classical" murine Th1 immune responses following infection with different intracellular pathogens (e.g., *Brucella abortus, Borrelia burgdorferi, Leishmania major, T. gondii*) include not only IFN-γ^+ CD4$^+$ cells but also "nonclassical" Th1 lymphocytes that concomitantly produce IFN-γ and IL-10 [32-35]

Thus, although the effect of IL-10 on Th1/Th2 effector choice is indirect and very complex (Fig. 1), IL-10 and IL-10R still represent attractive therapeutic targets for the manipulation of APC function aimed at both promoting or/and suppressing development of different types of CD4-dependent immune responses.[36,37]

IL-10 in Schistosome Infection

Like most host/helminth relationships, schistosome worms and their definitive hosts have coevolved survival strategies that maximize the transmission of parasite gametes (enclosed in the developing egg) and minimize the development of pathology in the host. While these strategies work well for the vast majority of the 200 million people currently infected with this pathogen, a small proportion of those affected will go on to develop life-threatening or severely debilitating illnesses. [38] Although there are many confounding factors that influence the schistosome/host equilibrium and clinical outcome, the induction of IL-10 during infection is a vital and indispensable process that limits host pathology and facilitates long-term survival of the parasite and host.

Schistosome cercariae release proteases as they penetrate the skin of their definitive host – a process that leads to damage of surrounding tissues and the generation of robust innate immune defense mechanisms. However, greater than 90% of infective stage cercariae survive this process and ultimately reach the lungs.[39] Prostaglandins induced and released by the cercariae [40,41] are believed to be indirectly responsible for increased schistosome survival during this critical period of infection via their effect on the host's immune system. Specifically, one prostaglandin, prostaglandin E_2 (PGE_2), up-regulates the production of keratinocyte-derived IL-10, which in turn limits the induction of anti-parasite inflammatory reactions in the skin of experimentally infected animals.[40] The production of IL-10 in skin seems to occur regardless of the parasite species used during the infection[42] and is also observed in lymph nodes draining the skin.[43] Moreover, studies in vaccinated IL-$10^{-/-}$ mice demonstrated that IL-10 dampens nearly all known anti-parasite effector mechanisms that operate during polarized Th1 and Th2 responses.[44] Finally, a recent study of *S. haematobium* infected children identified IL-10 as a major risk factor for reinfection after chemotherapy.[45] Together, the results of these studies suggest that schistosome parasites have evolved an IL-10-dependent mechanism that down-regulates the host's immune response early during infection, which maximizes their survival. However, it is also clear that IL-10 is critical to the survival of the infected host, by limiting egg-induced liver damage as infection becomes chronic.

Deposition of schistosome eggs into the intestines and liver of infected hosts induces a vigorous Th2 mediated, circumoval granulomatous response that, if not properly controlled, can lead to severe immuno-pathology.[38] Glycoconjugates and lipids derived from schistosome eggs[46-48] drive IL-10 production from B cells,[49] other APCs,[48] and Treg cells (Ref. 47 and Hesse M, Piccirillo CA, Belkaid Y, Prufer J, Mentink-Kane M, Leusink M, Cheever AW, Shevach EM and Wynn TA. The pathogenesis of schistosomiasis is controlled by cooperating IL-10-producing innate effector and regulatory T cells, submitted) possibly through a p38 protein kinase dependent signaling cascade.[50] IL-10 levels remain high even weeks after the egg-induced process is initiated and associate with global T cell hyporesponsiveness,[51-54] counter regulation of inflammatory Th1 cell populations,[51,55] decreased proliferative capacity of host cells,[53,56] and control of circumoval granulomatous responses.[55,57,58] Together, these IL-10 dependent activities create an environment that prevents the formation of an over exuberant and potentially dangerous anti-egg inflammatory response.

Further insight into the regulatory role of IL-10 during schistosome infection has recently been uncovered through studies involving experimentally infected IL-10 deficient mice and analysis of data obtained from human immuno-epidemiological field investigations. As IL-10 is associated with the control of the granulomatous response and host cell proliferation during schistosome infection, it was suggested that this cytokine might be important for down-modulation of host circumoval immune responses during chronic infection. Nevertheless, a longitudinal study using schistosome-infected IL-10 deficient animals demonstrated that the magnitude of the granulomatous response decreases substantially between wk 8 and 16 of infection,[10] suggesting that IL-10 plays only a minor role in the process of immune down-modulation. Nevertheless, further examination of the immune responses in these animals as well as double gene deficient mice (IL-10/IL-12- and IL-10/IL-4- KO) demonstrated that IL-10 critically controls Th1 and Th2 cytokine and antibody responses as well as immunopathology, especially during the acute phase of disease.[10,59] Deficiencies in IL-10 are also associated with increased pathology in infected CBA/J mice,[60] IL-4 deficient mice,[61] mice made tolerant to egg antigens,[62] $CD4^+$ T cell-depleted mice, [63] mice coinfected with *S. mansoni* and *T. gondii*,[64] and in mice immunized with egg antigens and complete Freund's adjuvant.[65] Furthermore, IL-10 also plays an important role in the development of egg-induced hepatic fibrosis by regulating IL-13Rα2 expression (decoy receptor for the collagen inducing cytokine IL-13).[66] Together these studies indicate that IL-10 production during experimental schistosomiasis is important for several infection-related pathologies.[67]

Can IL-10 contribute to the control of severe morbidity in human populations? To begin to answer this question, one recent study has elegantly confirmed the role of IL-10 in urinary tract morbidity during *S. haematobium* infection of children and adolescents in Kenya.[68] Here, the authors demonstrated that a low ratio of IL-10/TNF-α positively correlated with severe bladder wall pathology in the age- and infection intensity- matched case population. In another study performed on the shores of Lake Albert in Uganda, low levels of IL-10 were positively associated with increased fibrosis in children infected with *S. mansoni.*[69] Additional studies of this type will contribute to our understanding of the role of IL-10 in human schistosomiasis and other helminth infections.[28,70,71] Interestingly, a beneficial side effect of prolonged helminth-induced IL-10 production in chronically infected individuals is the ability of this cytokine to suppress atopy.[72] Given the many critical functions exhibited by IL-10 in this disease, It is clear that interest on IL-10 and other IL-10 related family members[73] will continue to grow in the coming years.

IL-10 in Intracellular Protozoan Infection

Due to their capacity to induce vigorous pro-inflammatory cytokine production, protozoan pathogens such as *L. major,*[74] *T. cruzi*[75] and *T. gondii*[76,77] rapidly stimulate IL-10 responses. This response quickly establishes an important equilibrium that limits damage to the host but at the same time prevents complete clearance of the organism so that transmission to new hosts can occur. Surprisingly, however, little is know about the stimuli that trigger IL-10 production in these infections or which cell types produce the cytokine. It is now widely believed that $CD4^+CD25^+$ Treg cells represent a major source of the cytokine during infection with *L. major.*[34,78] While the mechanisms that drive APC to produce IL-10 during *L. major* infection are not completely clear, it was found that IgG bound to amastigote forms by means of Fc receptor ligation can stimulate IL-10 production.[79] In the case of *T. cruzi*, some parasite membrane-derived glycoinositolphopholipids possess anti-inflammatory activity on macrophages and DC in vitro, but this effect does not appear to be due to induction of IL-10.[80] Interestingly, while DC fail to secrete IL-10 in response to *T. gondii* stimulation, T cells, macrophages and glial cells produce significant levels of the cytokine during in vivo infections.

IL-10 was originally thought to regulate resistance to protozoan infection mostly through effector cell deactivation such as by inhibiting NO expression by macrophages[81,82] or by immune deviation of T cell responses towards a type 2 cytokine profile.[83] However, when $IL\text{-}10^{-/-}$ mice became available, this paradigm had to be modified to accommodate a wider range of effects of this cytokine during infection. *T. gondii* and *T. cruzi* infection in $IL\text{-}10^{-/-}$ mice resulted in an enhanced Type 1 response and lower parasite burdens as expected, but also revealed a much more unpredicted outcome of excessive inflammation, which was associated with tissue destruction and a lethal shock-like syndrome characterized by over-production of IL-12, IFN-γ and TNF.[5,7,84,85] An additional mechanism by which IL-10 can control inflammation is through direct inhibition of chemokine expression induced by the parasite.[86]

L. major infection in $IL\text{-}10^{-/-}$ mice results in complete clearance of the parasites from skin lesions, suggesting a role for this cytokine in the induction of parasite persistence.[34,79] Belkaid and colleagues reported that $CD4^+CD25^+$ Treg cells are the major cell population secreting IL-10 and, therefore, regulating chronic persistence of leishmania parasites.[78] They hypothesized that direct inhibition of microbicidal activity by the IL-10 produced by this T cell population led to the persistence of the parasite. Nevertheless, a role for TGF-β, another immunomodulatory cytokine produced by Treg cells, has not been formally excluded. TGF-β has been shown to be associated with macrophage deactivation, inhibition of microbicidal function and proinflammatory mediators release in several models of protozoan infection.[87-90]

Metabolites of the arachidonic acid also constitute another group of anti-inflammatory mediators that can regulate immunity against protozoa parasite infections. PGE_2 production was reported in mice infected with *L. major* and *T. cruzi.*[91-94] A more direct correlation between production of PGE_2 and susceptibility to infection was observed after in vivo inhibition of PGE_2 synthesis by treatment with cyclooxygenase inhibitors.[91,92] Nevertheless, the most common cyclooxygenase inhibitors, such as Indomethacin were also shown to inhibit lipoxygenases,[95] a second class of enzymes that trigger the release of other immunomodulatory mediators. Leukotriene B_4 (LTB_4) is one of the products of the lipoxygenase metabolism of the arachidonic acid, its production had been reported during infection with *L. major*[93] that appears to have an enhancing effect over cytokine production, independently of their anti- or pro-inflammatory profile. IL-10-independent regulation of IL-12 and IFN-γ production was also reported after stimulation of mice with an extract of tachyzoites of *T. gondii*,[96] a phenomenon called "dendritic cell paralysis". It was found later that in vivo stimulation with this parasite extract induced the release of a 5-lipoxygenase-derived eicosanoid, lipoxin A_4 (LXA_4) and that 5-lipoxygenase deficient mice can not secrete LXA_4 or undergo dendritic cell paralysis.[97] The in vivo relevance of LXA_4-mediated control of IL-12 production was studied during infection of 5-lipoxygenase deficient mice with *T. gondii*.[98] These animals succumbed to infection around 30 days post-inoculation with a lower parasite burden, higher serum IL-12 levels and intense inflammation in the brain with elevated IL-12 production in situ.[98] However, when analyzed in parallel in an in vitro study, IL-10 but not LXA_4, was effective in blocking macrophage microbicidal function, suggesting that these mediators have related but not redundant effector pathways.

IL-10 in Mycobacterial Infection

Mycobacteria are slow-growing, facultative intracellular bacilli that primarily reside in phagocytes. The immune response to mycobacteria has been analyzed extensively in mouse models of *Mycobacterium tuberculosis, M. bovis* Bacillus Calmette-Guérin (BCG) and *M. avium* infections. Activation of infected macrophages and control of mycobacterial replication is critically dependent on IFN-γ produced by T lymphocytes.[99] Some bacilli, however, resist killing and survive within macrophages in the face of strong T cell responses. Although it is unclear how this latent infection is maintained, mechanisms that alter host immune responses, such as the induction of down-regulatory cytokines like IL-10 and TGF-β are thought to contribute to the persistence of mycobacterial infection. Production of IL-10 is of special interest as a possible evasion strategy because of its suppressive effects on many known immune functions required for inhibiting mycobacterial growth, including synthesis of pro-inflammatory cytokines/mediators, expression of MHC class II and costimulatory molecules.[15]

IL-10 is strongly induced at the sites of mycobacterial infection.[100-102] APC, such as macrophages and DC,[103-105] as well as T lymphocytes[106,107] are capable of producing IL-10 in response to mycobacterial infection. Interestingly, although originally described as a Th2 cytokine, IL-10 also appears to be produced in large quantities by Th1 IFN-γ-producing $CD4^+$ lymphocytes during mycobacterial infection.[108,109]

IL-10 inhibits cellular responses induced by mycobacterial infection at multiple levels. After activation with IFN-γ murine macrophages release pro-inflammatory cytokines and NO to control the intracellular growth of *M. tuberculosis* and *M. bovis*.[110-113] This IFN-γ-mediated bactericidal effect, however, is inhibited in the presence of IL-10.[114] Moreover, IL-10 prevents TNF-dependent apoptosis of *M. tuberculosis*-infected macrophages by inhibiting TNF production[115] or by inducing the release of TNF receptor 2 that could form nonactive TNF-TNFR2 complexes.[116,117] The induction of macrophage apoptosis may restrict mycobacterial spreading[118] as well as facilitate antigen presentation to T cells[119] thereby contributing to host control of the infection. Although IL-10 does not exhibit a direct suppressive effect on Th1 cells (see previous section), the cytokine may influence the T cell response to mycobacterial infection by modulating APC functions. Mycobacterium-induced IL-10 inhibits IL-12 production by DC in vitro and in vivo.[104,105] In addition, BCG-infected, IL-10-deficient DC have been shown to migrate more efficiently to draining lymph nodes compared to cells from WT mice, suggesting that autocrine IL-10 regulates DC migration in response to BCG infection in vivo.[104]

Table 1. Effects of manipulation of IL-10 level on host resistance to mycobacterial infection

Mycobacterium Spp.	Methods	Bacterial Burdens[a]	References
M. avium	Anti-IL-10[b]	Reduced	123, 124
	IL-10 KO[c]	Reduced	125
	IL-10 Tg[d]	Increased	107
BCG	IL-10 KO	Unchanged	128
	IL-10 KO	Transiently reduced	126, 127
	IL-10 Tg	Increased	130, 131
M. tuberculosis	IL-10 KO	Unchanged	102, 129
	IL-10 KO	Transiently reduced	125
	IL-10 Tg	Increased	101

[a] Compared to those in WT mice; [b] Treated with antibody specific for IL-10; [c] IL-10 deficient mice; [d] IL-10 transgenic mice

As noted above, since IL-10 has a major down-regulatory effect on cell-mediated immunity, it has been hypothesized that the production of this cytokine promotes the long-term survival of mycobacteria in infected hosts.[120-122] Initial studies,[123,124] which used neutralizing antibody to block IL-10 function in vivo, in general supported this concept. However, more recent studies employing IL-$10^{-/-}$ mice have yielded conflicting results (Table 1). For example, IL-$10^{-/-}$ mice show increased resistance to *M. avium*[125] and in some[125-127] but not all[102,128,129] studies display transiently enhanced control of *M. tuberculosis* and BCG infection. The discrepancy between these studies possibly results from variation in the virulence of the mycobacteria, the time-points analysed and most importantly, the route of infection.

Although IL-$10^{-/-}$ mice display only minimally enhanced resistance to mycobacterial infection, such observation does not rule out a role for IL-10 as one of several redundant mechanisms regulating host resistance to these microorganisms. It has been demonstrated that the over-expression of IL-10 in transgenic mice results in significantly impaired host resistance to *M. tuberculosis,*[101] BCG[130,131] and *M. avium*[107] infection. Because the expression of transgenic IL-10 can be controlled by cell lineage-specific promoters, the relative effect of T cell- vs APC-derived IL-10 on the host immune response to mycobacterial infection was investigated. Over-production of IL-10 by T cells,[101,130] macrophages[131] or MHC class II expressing cells [107] lead to dramatically elevated bacterial burdens and impaired macrophage functions. IFN-γ responses, however, were not markedly decreased in these infected transgenic animals, suggesting normal development of Th1 effector cells. Together, these observations are consistent with the in vitro findings that IL-10 can over-ride the macrophage activation effects of IFN-γ.[114]

In conclusion, both in vivo and in vitro studies demonstrated that excessive IL-10 production can promote intracellular pathogen growth in macrophages and argue that IL-10-mediated immune down-regulation may contribute to the maintenance of latency in chronic mycobacterial infection, possibly as one of several redundant mechanisms.

The Role of IL-10 in the Regulation of Inflammatory Bowel Disease

Inflammatory bowel disease (IBD), which includes Crohn's disease and ulcerative colitis, is the major chronic inflammatory disease of the intestinal tract. Although the etiology of IBD is unknown, the intestinal flora is believed to play an important role in its pathogenesis. This is perhaps best illustrated in experimental models of the disease—e.g., the IL-$10^{-/-}$ mouse model—

in which various immunodeficient animals develop intestinal inflammation when housed in conventional animal facilities, but not when reared under specific pathogen-free or germ-free conditions.[132]

The immune mechanisms that regulate intestinal inflammation have been extensively studied over the years, and data from the severe combined immunodeficiency (SCID) transfer model have been particularly useful in defining both pathogenic and disease-protective $CD4^+$ T cell responses in IBD. In the SCID transfer model, colitis is induced in T cell-deficient SCID or $Rag^{-/-}$ recipients by transfer of naïve $CD4^+$ $CD45RB^{hi}$ T cells.[133,134] Cotransfer of the $CD45RB^{low}$ memory T cell subset prevents the development of inflammation, defining a population of $CD4^+$ Treg cells with disease-suppressive function.[134-136] Subsequent studies have demonstrated that IL-10 plays a disease-protective role in this model as (1) systemic administration of recombinant IL-10 prevents development of $CD45RB^{hi}$-induced colitis, (2) anti-IL-10R treatment reverses the disease suppression mediated by the $CD45RB^{low}$ cells, and (3) $CD45RB^{low}$ cells from IL-$10^{-/-}$ animals fail to protect from disease.[137,138] Furthermore, $CD45RB^{hi}$ cells isolated from IL-10 transgenic mice do not induce colitis in SCID recipients and these transgenic IL-10-secreting $CD45RB^{hi}$ cells are even able to protect from colitis induced by $CD45RB^{hi}$ cells from WT mice.[139]

IL-10 clearly controls intestinal inflammation also in other models of colitis. For example, IL-10 therapy has been shown beneficial in preventing and/or partially reversing disease in the IL-$10^{-/-}$ and trinitrobenzene sulphonic acid (TNBS) colitis models.[140,141] Moreover, Treg cell suppression of T-cell dependent as well as T-cell independent *Helicobacter hepaticus*-triggered intestinal inflammation in $Rag^{-/-}$ mice is reversed by anti-IL-10R treatment.[142,143] Interestingly, administration of anti-IL-10R mAb to normal BALB/c mice leads to the induction of colitis,[144] arguing that IL-10 is required also in intact immunocompetent animals to maintain intestinal homeostasis. Studies from the *H. hepaticus* colitis model have further demonstrated that whereas infected IL-$10^{-/-}$ animals develop a pathogenic Th1 type response, infected WT mice that are disease free mount an IL-10-dominated immune response against the bacterium.[145] These studies support the hypothesis that in immunocompetent hosts, intestinal flora induces IL-10-secreting $CD4^+$ T cells that prevent pathologic immune responses towards intestinal antigens. The cellular source of the disease-protective IL-10 in most of the colitis models are indeed believed to be $CD4^+$ Treg cells,[138,142,143,145] although B cell-derived IL-10 has been reported to suppress intestinal inflammation in TCRα-deficient mice.[146]

There are likely multiple mechanisms by which IL-10 exerts its disease-suppressive effect in IBD. Treg cells, through their production of IL-10, are known to control the expansion of colitogentic $CD4^+$ T cells.[138,142,143,147,148] Moreover, in addition to its down-regulatory effects on APC populations,[11,12,149] IL-10 has been shown to promote the development of IL-10-secreting $CD4^+$ Treg cells in vitro[150] and to enhance the differentiation of DC that prime such Treg cells.[30] Evidence that IL-10 may prevent intestinal inflammation by acting on the innate arm of the immune response comes from a report describing the development of enterocolitis in mice whose macrophages and neutrophils are rendered IL-10 unresponsive by specific disruption of the Stat3 gene.[151] Likewise, as mentioned above, Treg cells are able to in an IL-10-dependent fashion suppress the colitis that develops in *H. hepaticus*-infected $Rag^{-/-}$ mice on the 129SvEv background, suggesting that cells of nonT lymphocyte compartments are the targets of IL-10 activity.[143] Besides IL-10, TGF-β plays an important role in protection against colitis.[152-154] The relation between IL-10 and TGF-β in disease suppression is not yet clear, however studies in the TNBS colitis model suggest that IL-10 acts by down-regulating the Th1 response, thereby facilitating TGF-β secretion in the host.[155] IL-10 may also enhance TGF-β receptor type II expression and restore TGF-β responsiveness of activated T cells.[156]

Similar to the findings in experimental models, the gut flora has also been implicated in the development of IBD in humans.[157-159] Moreover, while normal individuals display peripheral tolerance against resident autologous flora mediated by $CD4^+$ T cells secreting IL-10 and TGF-β[160] this state is broken in active IBD.[161] In contrast to rodent models, however, systemic treatment of IBD patients with recombinant human IL-10 has thus far not been very effective, and other approaches are therefore being developed for use in humans.[162] Encouraging results have been obtained from experimental models using IL-10-secreting *Lactococcus lactis* to treat IL-$10^{-/-}$ mice as well as mice exposed to dextran sodium sulfate,[163] and a phase I clinical trial using IL-10-secreting bacteria in patients with Crohn's disease is currently underway.[162] Taken together, IL-10 clearly has suppressive effects on inflammatory responses in the intestine and with improved methods for delivery this cytokine may prove beneficial as a treatment for humans with IBD.

References

1. Fiorentino DF, Bond MW, Mosmann TR. Two types of mouse T helper cell. IV. Th2 clones secrete a factor that inhibits cytokine production by Th1 clones. J Exp Med 1989; 170:2081-2095.
2. Shnyra A, Brewington R, Alipio A et al. Reprogramming of lipopolysaccharide-primed macrophages is controlled by a counterbalanced production of IL-10 and IL-12. J Immunol 1998; 160:3729-3736.
3. Akbari O, DeKruyff RH, Umetsu DT. Pulmonary dendritic cells producing IL-10 mediate tolerance induced by respiratory exposure to antigen. Nat Immunol 2001; 2:725-731.
4. Mauri C, Gray D, Mushtaq N et al. Prevention of arthritis by interleukin 10-producing B cells. J Exp Med 2003; 197:489-501.
5. Gazzinelli RT, Wysocka M, Hieny S et al. In the absence of endogenous IL-10, mice acutely infected with Toxoplasma gondii succumb to a lethal immune response dependent on CD4+ T cells and accompanied by overproduction of IL-12, IFN-γ and TNF-α. J Immunol 1996; 157:798-805.
6. Li C, Corraliza I, Langhorne J. A defect in interleukin-10 leads to enhanced malarial disease in Plasmodium chabaudi chabaudi infection in mice. Infect Immun 1999; 67:4435-4442.
7. Hunter CA, Ellis-Neyes LA, Slifer T et al. IL-10 is required to prevent immune hyperactivity during infection with Trypanosoma cruzi. J Immunol 1997; 158:3311-3316.
8. Grunig G, Corry DB, Leach MW et al. Interleukin-10 is a natural suppressor of cytokine production and inflammation in a murine model of allergic bronchopulmonary aspergillosis. J Exp Med 1997; 185:1089-1099.
9. Wynn TA, Morawetz R, Scharton-Kersten T et al. Analysis of granuloma formation in double cytokine-deficient mice reveals a central role for IL-10 in polarizing both T helper cell 1- and T helper cell 2-type cytokine responses in vivo. J Immunol 1997; 159:5014-5023.
10. Wynn TA, Cheever AW, Williams ME et al. IL-10 regulates liver pathology in acute murine Schistosomiasis mansoni but is not required for immune down-modulation of chronic disease. J Immunol 1998; 160:4473-4480.
11. de Waal Malefyt R, Haanen J, Spits H et al. Interleukin 10 (IL-10) and viral IL-10 strongly reduce antigen-specific human T cell proliferation by diminishing the antigen-presenting capacity of monocytes via downregulation of class II major histocompatibility complex expression. J Exp Med 1991; 174:915-924.
12. Fiorentino DF, Zlotnik A, Vieira P et al. IL-10 acts on the antigen-presenting cell to inhibit cytokine production by Th1 cells. J Immunol 1991; 146:3444-3451.
13. Tan JC, Braun S, Rong H et al. Characterization of recombinant extracellular domain of human interleukin-10 receptor. J Biol Chem 1995; 270:12906-12911.
14. Liu Y, Wei SH, Ho AS et al. Expression cloning and characterization of a human IL-10 receptor. J Immunol 1994; 152:1821-1829.
15. Moore KW, de Waal Malefyt R, Coffman RL et al. Interleukin-10 and the interleukin-10 receptor. Annu Rev Immunol 2001; 19:683-765.
16. de Waal Malefyt R, Figdor CG, Huijbens R et al. Effects of IL-13 on phenotype, cytokine production, and cytotoxic function of human monocytes. Comparison with IL-4 and modulation by IFN-γ or IL-10. J Immunol 1993; 151:6370-6381.
17. Robinson DS, O'Garra A. Further checkpoints in Th1 development. Immunity 2002; 16:755-758.
18. De Smedt T, Van Mechelen M, De Becker G et al. Effect of interleukin-10 on dendritic cell maturation and function. Eur J Immunol 1997; 27:1229-1235.
19. Berkman N, John M, Roesems G et al. Inhibition of macrophage inflammatory protein-1 alpha expression by IL-10. Differential sensitivities in human blood monocytes and alveolar macrophages. J Immunol 1995; 155:4412-4418.
20. Kopydlowski KM, Salkowski CA, Cody MJ et al. Regulation of macrophage chemokine expression by lipopolysaccharide in vitro and in vivo. J Immunol 1999; 163:1537-1544.

21. Jenkins JK, Malyak M, Arend WP. The effects of interleukin-10 on interleukin-1 receptor antagonist and interleukin-1 beta production in human monocytes and neutrophils. Lymphokine Cytokine Res 1994; 13:47-54.
22. D'Amico G, Frascaroli G, Bianchi G et al. Uncoupling of inflammatory chemokine receptors by IL-10: Generation of functional decoys. Nat Immunol 2000; 1:387-391.
23. Hart PH, Hunt EK, Bonder CS et al. Regulation of surface and soluble TNF receptor expression on human monocytes and synovial fluid macrophages by IL-4 and IL-10. J Immunol 1996; 157:3672-3680.
24. Joyce DA, Steer JH. IL-4, IL-10 and IFN-γ have distinct, but interacting, effects on differentiation-induced changes in TNF-α and TNF receptor release by cultured human monocytes. Cytokine 1996; 8:49-57.
25. Iwasaki A, Kelsall BL. Freshly isolated Peyer's patch, but not spleen, dendritic cells produce interleukin 10 and induce the differentiation of T helper type 2 cells. J Exp Med 1999; 190:229-239.
26. Khanna A, Morelli AE, Zhong C et al. Effects of liver-derived dendritic cell progenitors on Th1- and Th2-like cytokine responses in vitro and in vivo. J Immunol 2000; 164:1346-1354.
27. Maldonado-Lopez R, Maliszewski C, Urbain J et al. Cytokines regulate the capacity of CD8α^+ and CD8α^- dendritic cells to prime Th1/Th2 cells in vivo. J Immunol 2001; 167:4345-4350.
28. Bliss SK, Alcaraz A, Appleton JA. IL-10 prevents liver necrosis during murine infection with Trichinella spiralis. J Immunol 2003; 171:3142-3147.
29. Corinti S, Albanesi C, la Sala A et al. Regulatory activity of autocrine IL-10 on dendritic cell functions. J Immunol 2001; 166:4312-4318.
30. Wakkach A, Fournier N, Brun V et al. Characterization of dendritic cells that induce tolerance and T regulatory 1 cell differentiation in vivo. Immunity 2003; 18:605-617.
31. Del Prete G, De Carli M, Almerigogna F et al. Human IL-10 is produced by both type 1 helper (Th1) and type 2 helper (Th2) T cell clones and inhibits their antigen-specific proliferation and cytokine production. J Immunol 1993; 150:353-360.
32. Svetic A, Jian YC, Lu P et al. Brucella abortus induces a novel cytokine gene expression pattern characterized by elevated IL-10 and IFN-γ in CD4$^+$ T cells. Int Immunol 1993; 5:877-883.
33. Pohl-Koppe A, Balashov KE, Steere AC et al. Identification of a T cell subset capable of both IFN-γ and IL-10 secretion in patients with chronic Borrelia burgdorferi infection. J Immunol 1998; 160:1804-1810.
34. Belkaid Y, Hoffmann KF, Mendez S et al. The role of interleukin (IL)-10 in the persistence of Leishmania major in the skin after healing and the therapeutic potential of anti-IL-10 receptor antibody for sterile cure. J Exp Med 2001; 194:1497-1506.
35. Jankovic D, Kullberg MC, Hieny S et al. In the absence of IL-12, CD4$^+$ T cell responses to intracellular pathogens fail to default to a Th2 pattern and are host protective in an IL-10$^{-/-}$ setting. Immunity 2002; 16:429-439.
36. Castro AG, Neighbors M, Hurst SD et al. Anti-interleukin 10 receptor monoclonal antibody is an adjuvant for T helper cell type 1 responses to soluble antigen only in the presence of lipopolysaccharide. J Exp Med 2000; 192:1529-1534.
37. Igietseme JU, Ananaba GA, Bolier J et al. Suppression of endogenous IL-10 gene expression in dendritic cells enhances antigen presentation for specific Th1 induction: Potential for cellular vaccine development. J Immunol 2000; 164:4212-4219.
38. Hoffmann KF, Wynn TA, Dunne DW. Cytokine-mediated host responses during schistosome infections; walking the fine line between immunological control and immunopathology. Adv Parasitol 2002; 52:265-307.
39. Wilson RA, Coulson PS, Dixon B. Migration of the schistosomula of Schistosoma mansoni in mice vaccinated with radiation-attenuated cercariae, and normal mice: An attempt to identify the timing and site of parasite death. Parasitology 1986; 92(Pt 1):101-116.
40. Ramaswamy K, Kumar P, He YX. A role for parasite-induced PGE2 in IL-10-mediated host immunoregulation by skin stage schistosomula of Schistosoma mansoni. J Immunol 2000; 165:4567-4574.
41. Angeli V, Faveeuw C, Roye O et al. Role of the parasite-derived prostaglandin D2 in the inhibition of epidermal Langerhans cell migration during schistosomiasis infection. J Exp Med 2001; 193:1135-1147.
42. He YX, Chen L, Ramaswamy K. Schistosoma mansoni, S. haematobium, and S. japonicum: Early events associated with penetration and migration of schistosomula through human skin. Exp Parasitol 2002; 102:99-108.
43. Betts CJ, Wilson RA. Th1 cytokine mRNA expression dominates in the skin-draining lymph nodes of C57BL/6 mice following vaccination with irradiated Schistosoma mansoni cercariae, but is down-regulated upon challenge infection. Immunology 1998; 93:49-54.

44. Hoffmann KF, James SL, Cheever AW et al. Studies with double cytokine-deficient mice reveal that highly polarized Th1- and Th2-type cytokine and antibody responses contribute equally to vaccine-induced immunity to Schistosoma mansoni. J Immunol 1999; 163:927-938.
45. van den Biggelaar AH, Borrmann S, Kremsner P et al. Immune responses induced by repeated treatment do not result in protective immunity to Schistosoma haematobium: Interleukin (IL)-5 and IL-10 responses. J Infect Dis 2002; 186:1474-1482.
46. Okano M, Satoskar AR, Nishizaki K et al. Lacto-N-fucopentaose III found on Schistosoma mansoni egg antigens functions as adjuvant for proteins by inducing Th2-type response. J Immunol 2001; 167:442-450.
47. van der Kleij D, Latz E, Brouwers JF et al. A novel host-parasite lipid cross-talk. Schistosomal lyso-phosphatidylserine activates toll-like receptor 2 and affects immune polarization. J Biol Chem 2002; 277:48122-48129.
48. Van der Kleij D, Van Remoortere A, Schuitemaker JH et al. Triggering of innate immune responses by schistosome egg glycolipids and their carbohydrate epitope GalNAc beta 1-4(Fuc alpha 1-2Fuc alpha 1- 3)GlcNAc. J Infect Dis 2002; 185:531-539.
49. Velupillai P, Harn DA. Oligosaccharide-specific induction of interleukin 10 production by B220$^+$ cells from schistosome-infected mice: A mechanism for regulation of CD4$^+$ T-cell subsets. Proc Natl Acad Sci USA 1994; 91:18-22.
50. Carneiro-Santos P, Alves-Oliveira LF, Correa-Oliveira R et al. P38 mitogen-activated protein kinase influence on the production of IL-10 in human schistosomiasis mansoni. Parasite Immunol 2002; 24:493-497.
51. Sher A, Fiorentino D, Caspar P et al. Production of IL-10 by CD4$^+$ T lymphocytes correlates with down- regulation of Th1 cytokine synthesis in helminth infection. J Immunol 1991; 147:2713-2716.
52. Malaquias LC, Falcao PL, Silveira AM et al. Cytokine regulation of human immune response to Schistosoma mansoni: Analysis of the role of IL-4, IL-5 and IL-10 on peripheral blood mononuclear cell responses. Scand J Immunol 1997; 46:393-398.
53. King CL, Medhat A, Malhotra I et al. Cytokine control of parasite-specific anergy in human urinary schistosomiasis. IL-10 modulates lymphocyte reactivity. J Immunol 1996; 156:4715-4721.
54. Montenegro SM, Miranda P, Mahanty S et al. Cytokine production in acute versus chronic human Schistosomiasis mansoni: The cross-regulatory role of interferon-gamma and interleukin-10 in the responses of peripheral blood mononuclear cells and splenocytes to parasite antigens. J Infect Dis 1999; 179:1502-1514.
55. Flores-Villanueva PO, Zheng XX, Strom TB et al. Recombinant IL-10 and IL-10/Fc treatment down-regulate egg antigen- specific delayed hypersensitivity reactions and egg granuloma formation in schistosomiasis. J Immunol 1996; 156:3315-3320.
56. Araujo MI, de Jesus AR, Bacellar O et al. Evidence of a T helper type 2 activation in human schistosomiasis. Eur J Immunol 1996; 26:1399-1403.
57. Falcao PL, Malaquias LC, Martins-Filho OA et al. Human Schistosomiasis mansoni: IL-10 modulates the in vitro granuloma formation. Parasite Immunol 1998; 20:447-454.
58. Rezende SA, Silva-Teixeira DN, Drummond SC et al. IL-10 plays a role in the modulation of human granulomatous hypersensitivity against Schistosoma mansoni eggs induced by immune complexes. Scand J Immunol 1997; 46:96-102.
59. Hoffmann KF, Cheever AW, Wynn TA. IL-10 and the dangers of immune polarization: Excessive type 1 and type 2 cytokine responses induce distinct forms of lethal immunopathology in murine schistosomiasis. J Immunol 2000; 164:6406-6416.
60. Bosshardt SC, Freeman Jr GL, Secor WE et al. IL-10 deficit correlates with chronic, hypersplenomegaly syndrome in male CBA/J mice infected with Schistosoma mansoni. Parasite Immunol 1997; 19:347-353.
61. Brunet LR, Finkelman FD, Cheever AW et al. IL-4 protects against TNF-α-mediated cachexia and death during acute schistosomiasis. J Immunol 1997; 159:777-785.
62. Fallon PG, Dunne DW. Tolerization of mice to Schistosoma mansoni egg antigens causes elevated type 1 and diminished type 2 cytokine responses and increased mortality in acute infection. J Immunol 1999; 162:4122-4132.
63. Fallon PG, Richardson EJ, Smith P et al. Elevated type 1, diminished type 2 cytokines and impaired antibody response are associated with hepatotoxicity and mortalities during Schistosoma mansoni infection of CD4-depleted mice. Eur J Immunol 2000; 30:470-480.
64. Marshall AJ, Brunet LR, van Gessel Y et al. Toxoplasma gondii and Schistosoma mansoni synergize to promote hepatocyte dysfunction associated with high levels of plasma TNF-α and early death in C57BL/6 mice. J Immunol 1999; 163:2089-2097.
65. Rutitzky LI, Hernandez HJ, Stadecker MJ. Th1-polarizing immunization with egg antigens correlates with severe exacerbation of immunopathology and death in schistosome infection. Proc Natl Acad Sci USA 2001; 98:13243-13248.

66. Chiaramonte MG, Mentink-Kane M, Jacobson BA et al. Regulation and function of the interleukin 13 receptor α 2 during a T helper cell type 2-dominant immune response. J Exp Med 2003; 197:687-701.
67. Maizels RM, Yazdanbakhsh M. Immune regulation by helminth parasites: Cellular and molecular mechanisms. Nat Rev Immunol 2003; 3:733-744.
68. King CL, Malhotra I, Mungai P et al. Schistosoma haematobium-induced urinary tract morbidity correlates with increased tumor necrosis factor-α and diminished interleukin-10 production. J Infect Dis 2001; 184:1176-1182.
69. Booth M, Mwatha JK, Joseph S et al. Peri-portal fibrosis in human Schistosoma mansoni infection is associated with low IL-10, low IFN γ, high TNF-alpha or low RANTES, depending on age and gender. In press.
70. Schopf LR, Hoffmann KF, Cheever AW et al. IL-10 is critical for host resistance and survival during gastrointestinal helminth infection. J Immunol 2002; 168:2383-2392.
71. Helmby H, Grencis RK. Contrasting roles for IL-10 in protective immunity to different life cycle stages of intestinal nematode parasites. Eur J Immunol 2003; 33:2382-2390.
72. van den Biggelaar AH, van Ree R, Rodrigues LC et al. Decreased atopy in children infected with Schistosoma haematobium: A role for parasite-induced interleukin-10. Lancet 2000; 355:1723-1727.
73. Fickenscher H, Hor S, Kupers H et al. The interleukin-10 family of cytokines. Trends Immunol 2002; 23:89-96.
74. Heinzel FP, Sadick MD, Mutha SS et al. Production of interferon gamma, interleukin 2, interleukin 4, and interleukin 10 by CD4+ lymphocytes in vivo during healing and progressive murine leishmaniasis. Proc Natl Acad Sci USA 1991; 88:7011-7015.
75. Silva JS, Morrissey PJ, Grabstein KH et al. Interleukin 10 and interferon γ regulation of experimental Trypanosoma cruzi infection. J Exp Med 1992; 175:169-174.
76. Gazzinelli RT, Eltoum I, Wynn TA et al. Acute cerebral toxoplasmosis is induced by in vivo neutralization of TNF-α and correlates with the down-regulated expression of inducible nitric oxide synthase and other markers of macrophage activation. J Immunol 1993; 151:3672-3681.
77. Hunter CA, Abrams JS, Beaman MH et al. Cytokine mRNA in the central nervous system of SCID mice infected with Toxoplasma gondii: Importance of T-cell-independent regulation of resistance to T. gondii. Infect Immun 1993; 61:4038-4044.
78. Belkaid Y, Piccirillo CA, Mendez S et al. CD4+CD25+ regulatory T cells control Leishmania major persistence and immunity. Nature 2002; 420:502-507.
79. Kane MM, Mosser DM. The role of IL-10 in promoting disease progression in leishmaniasis. J Immunol 2001; 166:1141-1147.
80. Brodskyn C, Patricio J, Oliveira R et al. Glycoinositolphospholipids from Trypanosoma cruzi interfere with macrophages and dendritic cell responses. Infect Immun 2002; 70:3736-3743.
81. Vieth M, Will A, Schroppel K et al. Interleukin-10 inhibits antimicrobial activity against Leishmania major in murine macrophages. Scand J Immunol 1994; 40:403-409.
82. Gazzinelli RT, Oswald IP, James SL et al. IL-10 inhibits parasite killing and nitrogen oxide production by IFN-γ-activated macrophages. J Immunol 1992; 148:1792-1796.
83. Chakkalath HR, Titus RG. Leishmania major-parasitized macrophages augment Th2-type T cell activation. J Immunol 1994; 153:4378-4387.
84. Holscher C, Mohrs M, Dai WJ et al. Tumor necrosis factor α-mediated toxic shock in Trypanosoma cruzi-infected interleukin 10-deficient mice. Infect Immun 2000; 68:4075-4083.
85. Suzuki Y, Sher A, Yap G et al. IL-10 is required for prevention of necrosis in the small intestine and mortality in both genetically resistant BALB/c and susceptible C57BL/6 mice following peroral infection with Toxoplasma gondii. J Immunol 2000; 164:5375-5382.
86. Aliberti JC, Souto JT, Marino AP et al. Modulation of chemokine production and inflammatory responses in interferon-γ- and tumor necrosis factor-R1-deficient mice during Trypanosoma cruzi infection. Am J Pathol 2001; 158:1433-1440.
87. Silva JS, Twardzik DR, Reed SG. Regulation of Trypanosoma cruzi infections in vitro and in vivo by transforming growth factor beta (TGF-β). J Exp Med 1991; 174:539-545.
88. Gazzinelli RT, Oswald IP, Hieny S et al. The microbicidal activity of interferon-γ-treated macrophages against Trypanosoma cruzi involves an L-arginine-dependent, nitrogen oxide-mediated mechanism inhibitable by interleukin-10 and transforming growth factor-β. Eur J Immunol 1992; 22:2501-2506.
89. Li J, Hunter CA, Farrell JP. Anti-TGF-β treatment promotes rapid healing of Leishmania major infection in mice by enhancing in vivo nitric oxide production. J Immunol 1999; 162:974-979.
90. Langermans JA, Nibbering PH, Van Vuren et al. Transforming growth factor-β suppresses interferon-γ-induced toxoplasmastatic activity in murine macrophages by inhibition of tumour necrosis factor-α production. Parasite Immunol 2001; 23:169-175.

91. Farrell JP, Kirkpatrick CE. Experimental cutaneous leishmaniasis. II ed. A possible role for prostaglandins in exacerbation of disease in Leishmania major-infected BALB/c mice. J Immunol 1987; 138:902-907.
92. Celentano AM, Gorelik G, Solana ME et al. PGE2 involvement in experimental infection with Trypanosoma cruzi subpopulations. Prostaglandins 1995; 49 141-153.
93. Milano S, Arcoleo F, Dieli M et al. Ex vivo evidence for PGE2 and LTB4 involvement in cutaneous leishmaniasis: Relation with infection status and cytokine production. Parasitology 1996; 112(Pt 1):13-19.
94. Pinge-Filho P, Tadokoro CE, Abrahamsohn IA. Prostaglandins mediate suppression of lymphocyte proliferation and cytokine synthesis in acute Trypanosoma cruzi infection. Cell Immunol 1999; 193:90-98.
95. Burka JF, Flower RJ. Effects of modulators of arachidonic acid metabolism on the synthesis and release of slow-reacting substance of anaphylaxis. Br J Pharmacol 1979; 65:35-41.
96. Reis e Sousa C, Yap G, Schulz O et al. Paralysis of dendritic cell IL-12 production by microbial products prevents infection-induced immunopathology. Immunity 1999; 11:637-647.
97. Aliberti J, Hieny S, Reis e Sousa C et al. Lipoxin-mediated inhibition of IL-12 production by DCs: A mechanism for regulation of microbial immunity. Nat Immunol 2002; 3:76-82.
98. Aliberti J, Serhan C, Sher A. Parasite-induced lipoxin A4 is an endogenous regulator of IL-12 production and immunopathology in Toxoplasma gondii infection. J Exp Med 2002; 196:1253-1262.
99. Flynn JL, Chan J. Immunology of tuberculosis. Annu Rev Immunol 2001; 19:93-129.
100. Manca C, Tsenova L, Barry 3rd CE et al. Mycobacterium tuberculosis CDC1551 induces a more vigorous host response in vivo and in vitro, but is not more virulent than other clinical isolates. J Immunol 1999; 162:6740-6746.
101. Turner J, Gonzalez-Juarrero M, Ellis DL et al. In vivo IL-10 production reactivates chronic pulmonary tuberculosis in C57BL/6 mice. J Immunol 2002; 169:6343-6351.
102. Jung YJ, Ryan L, LaCourse R et al. Increased interleukin-10 expression is not responsible for failure of T helper 1 immunity to resolve airborne Mycobacterium tuberculosis infection in mice. Immunology 2003; 109:295-299.
103. Flesch IE, Kaufmann SH. Role of macrophages and alpha beta T lymphocytes in early interleukin 10 production during Listeria monocytogenes infection. Int Immunol 1994; 6:463-468.
104. Demangel C, Bertolino P, Britton WJ. Autocrine IL-10 impairs dendritic cell (DC)-derived immune responses to mycobacterial infection by suppressing DC trafficking to draining lymph nodes and local IL-12 production. Eur J Immunol 2002; 32:994-1002.
105. Hickman SP, Chan J, Salgame P. Mycobacterium tuberculosis induces differential cytokine production from dendritic cells and macrophages with divergent effects on naive T cell polarization. J Immunol 2002; 168:4636-4642.
106. Lyadova IV, Eruslanov EB, Khaidukov SV et al. Comparative analysis of T lymphocytes recovered from the lungs of mice genetically susceptible, resistant, and hyperresistant to Mycobacterium tuberculosis-triggered disease. J Immunol 2000; 165:5921-5931.
107. Feng CG, Kullberg MC, Jankovic D et al. Transgenic mice expressing human interleukin-10 in the antigen-presenting cell compartment show increased susceptibility to infection with Mycobacterium avium associated with decreased macrophage effector function and apoptosis. Infect Immun 2002; 70:6672-6679.
108. Gerosa F, Nisii C, Righetti S et al. $CD4^+$ T cell clones producing both interferon-γ and interleukin-10 predominate in bronchoalveolar lavages of active pulmonary tuberculosis patients. Clin Immunol 1999; 92:224-234.
109. Lyadova I, Yeremeev V, Majorov K et al. An ex vivo study of T lymphocytes recovered from the lungs of I/St mice infected with and susceptible to Mycobacterium tuberculosis. Infect Immun 1998; 66:4981-4988.
110. Rook GA, Steele J, Ainsworth M et al. Activation of macrophages to inhibit proliferation of Mycobacterium tuberculosis: Comparison of the effects of recombinant gamma-interferon on human monocytes and murine peritoneal macrophages. Immunology 1986; 59:333-338.
111. Flesch I, Kaufmann SH. Mycobacterial growth inhibition by interferon-γ-activated bone marrow macrophages and differential susceptibility among strains of Mycobacterium tuberculosis. J Immunol 1987; 138:4408-4413.
112. Flesch IE, Kaufmann SH. Activation of tuberculostatic macrophage functions by gamma interferon, interleukin-4, and tumor necrosis factor. Infect Immun 1990; 58:2675-2677.
113. Ding AH, Nathan CF, Stuehr DJ. Release of reactive nitrogen intermediates and reactive oxygen intermediates from mouse peritoneal macrophages. Comparison of activating cytokines and evidence for independent production. J Immunol 1988; 141:2407-2412.

114. Flesch IE, Hess JH, Oswald IP et al. Growth inhibition of Mycobacterium bovis by IFN-γ stimulated macrophages: Regulation by endogenous tumor necrosis factor-α and by IL-10. Int Immunol 1994; 6:693-700.
115. Rojas M, Olivier M, Gros P et al. TNF-α and IL-10 modulate the induction of apoptosis by virulent Mycobacterium tuberculosis in murine macrophages. J Immunol 1999; 162:6122-6131.
116. Balcewicz-Sablinska MK, Keane J, Kornfeld H et al. Pathogenic Mycobacterium tuberculosis evades apoptosis of host macrophages by release of TNF-R2, resulting in inactivation of TNF- α. J Immunol 1998; 161:2636-2641.
117. Balcewicz-Sablinska MK, Gan H, Remold HG. Interleukin 10 produced by macrophages inoculated with Mycobacterium avium attenuates mycobacteria-induced apoptosis by reduction of TNF-α activity. J Infect Dis 1999; 180:1230-1237.
118. Fratazzi C, Arbeit RD, Carini C et al. Macrophage apoptosis in mycobacterial infections. J Leukocyte Biol 1999; 66:763-764.
119. Schaible UE, Winau F, Sieling PA et al. Apoptosis facilitates antigen presentation to T lymphocytes through MHC-I and CD1 in tuberculosis. Nat Med 2003; 9:1039-1046.
120. Murray PJ. Defining the requirements for immunological control of mycobacterial infections. Trends Microbiol 1999; 7:366-372.
121. Redpath S, Ghazal P, Gascoigne NR. Hijacking and exploitation of IL-10 by intracellular pathogens. Trends Microbiol 2001; 9:86-92.
122. Trinchieri G. Regulatory role of T cells producing both interferon γ and interleukin 10 in persistent infection. J Exp Med 2001; 194:F53-57.
123. Bermudez LE, Champsi J. Infection with Mycobacterium avium induces production of interleukin-10 (IL-10), and administration of anti-IL-10 antibody is associated with enhanced resistance to infection in mice. Infect Immun 1993; 61:3093-3097.
124. Denis M, Ghadirian E. IL-10 neutralization augments mouse resistance to systemic Mycobacterium avium infections. J Immunol 1993; 151:5425-5430.
125. Roach DR, Martin E, Bean AG et al. Endogenous inhibition of antimycobacterial immunity by IL-10 varies between mycobacterial species. Scand J Immunol 2001; 54:163-170.
126. Murray PJ, Young RA. Increased antimycobacterial immunity in interleukin-10-deficient mice. Infect Immun 1999; 67:3087-3095.
127. Jacobs M, Brown N, Allie N et al. Increased resistance to mycobacterial infection in the absence of interleukin-10. Immunology 2000; 100:494-501.
128. Erb KJ, Kirman J, Delahunt B et al. IL-4, IL-5 and IL-10 are not required for the control of M. bovis-BCG infection in mice. Immunol Cell Biol 1998; 76:41-46.
129. North RJ. Mice incapable of making IL-4 or IL-10 display normal resistance to infection with Mycobacterium tuberculosis. Clin Exp Immunol 1998; 113:55-58.
130. Murray PJ, Wang L, Onufryk C et al. T cell-derived IL-10 antagonizes macrophage function in mycobacterial infection. J Immunol 1997; 158:315-321.
131. Lang R, Rutschman RL, Greaves DR et al. Autocrine deactivation of macrophages in transgenic mice constitutively overexpressing IL-10 under control of the human CD68 promoter. J Immunol 2002; 168:3402-3411.
132. Strober W, Fuss IJ, Blumberg RS. The immunology of mucosal models of inflammation. Annu Rev Immunol 2002; 20:495-549.
133. Morrissey PJ, Charrier K, Braddy S et al. $CD4^+$ T cells that express high levels of CD45RB induce wasting disease when transferred into congenic severe combined immunodeficient mice. Disease development is prevented by cotransfer of purified $CD4^+$ T cells. J Exp Med 1993; 178:237-244.
134. Powrie F, Leach MW, Mauze S et al. Phenotypically distinct subsets of $CD4^+$ T cells induce or protect from chronic intestinal inflammation in C B-17 scid mice. Int Immunol 1993; 5:1461-1471.
135. Powrie F, Correa-Oliveira R, Mauze S et al. Regulatory interactions between $CD45RB^{high}$ and $CD45RB^{low}$ $CD4^+$ T cells are important for the balance between protective and pathogenic cell-mediated immunity. J Exp Med 1994; 179:589-600.
136. Read S, Malmström V, Powrie F. Cytotoxic T lymphocyte-associated antigen 4 plays an essential role in the function of $CD25^+CD4^+$ regulatory cells that control intestinal inflammation. J Exp Med 2000; 192:295-302.
137. Powrie F, Leach MW, Mauze S et al. Inhibition of Th1 responses prevents inflammatory bowel disease in scid mice reconstituted with $CD45RB^{hi}$ $CD4^+$ T cells. Immunity 1994; 1:553-562.
138. Asseman C, Mauze S, Leach MW et al. An essential role for interleukin 10 in the function of regulatory T cells that inhibit intestinal inflammation. J Exp Med 1999; 190:995-1004.
139. Hagenbaugh A, Sharma S, Dubinett SM et al. Altered immune responses in interleukin 10 transgenic mice. J Exp Med 1997; 185:2101-2110.

140. Berg DJ, Davidson N, Kühn R et al. Enterocolitis and colon cancer in interleukin-10-deficient mice are associated with aberrant cytokine production and CD4$^+$ TH1-like responses. J Clin Invest 1996; 98:1010-1020.
141. Duchmann R, Schmitt E, Knolle P et al. Tolerance towards resident intestinal flora in mice is abrogated in experimental colitis and restored by treatment with interleukin-10 or antibodies to interleukin-12. Eur J Immunol 1996; 26:934-938.
142. Kullberg MC, Jankovic D, Gorelick PL et al. Bacteria-triggered CD4$^+$ T regulatory cells suppress Helicobacter hepaticus-induced colitis. J Exp Med 2002; 196:505-515.
143. Maloy KJ, Salaun L, Cahill R et al. CD4$^+$CD25$^+$ T_R cells suppress innate immune pathology through cytokine-dependent mechanisms. J Exp Med 2003; 197:111-119.
144. Asseman C, Read S, Powrie F. Colitogenic Th1 Cells are present in the antigen-experienced T cell pool in normal mice: Control by CD4$^+$ regulatory T cells and IL-10. J Immunol 2003; 171:971-978.
145. Kullberg MC, Ward JM, Gorelick PL et al. Helicobacter hepaticus triggers colitis in specific-pathogen-free interleukin-10 (IL-10)-deficient mice through an IL-12- and gamma interferon-dependent mechanism. Infect Immun 1998; 66:5157-5166.
146. Mizoguchi A, Mizoguchi E, Takedatsu H et al. Chronic intestinal inflammatory condition generates IL-10-producing regulatory B cell subset characterized by CD1d upregulation. Immunity 2002; 16:219-230.
147. Annacker O, Burlen-Defranoux O, Pimenta-Araujo R et al. Regulatory CD4 T cells control the size of the peripheral activated/memory CD4 T cell compartment. J Immunol 2000; 164:3573-3580.
148. Annacker O, Pimenta-Araujo R, Burlen-Defranoux O et al. CD25$^+$ CD4$^+$ T cells regulate the expansion of peripheral CD4 T cells through the production of IL-10. J Immunol 2001; 166:3008-3018.
149. Willems F, Marchant A, Delville JP et al. Interleukin-10 inhibits B7 and intercellular adhesion molecule-1 expression on human monocytes. Eur J Immunol 1994; 24:1007-1009.
150. Groux H, O'Garra A, Bigler M et al. A CD4$^+$ T-cell subset inhibits antigen-specific T-cell responses and prevents colitis. Nature 1997; 389:737-742.
151. Takeda K, Clausen BE, Kaisho T et al. Enhanced Th1 activity and development of chronic enterocolitis in mice devoid of Stat3 in macrophages and neutrophils. Immunity 1999; 10:39-49.
152. Powrie F, Carlino J, Leach MW et al. A critical role for transforming growth factor-β but not interleukin 4 in the suppression of T helper type 1-mediated colitis by CD45RB^{low} CD4$^+$ T cells. J Exp Med 1996; 183:2669-2674.
153. Neurath MF, Fuss I, Kelsall BL et al. Experimental granulomatous colitis in mice is abrogated by induction of TGF-beta-mediated oral tolerance. J Exp Med 1996; 183:2605-2616.
154. Kitani A, Fuss IJ, Nakamura K et al. Treatment of experimental (Trinitrobenzene sulfonic acid) colitis by intranasal administration of transforming growth factor (TGF)-β1 plasmid: TGF-β1-mediated suppression of T helper cell type 1 response occurs by interleukin (IL)-10 induction and IL-12 receptor beta2 chain downregulation. J Exp Med 2000; 192:41-52.
155. Fuss IJ, Boirivant M, Lacy B et al. The interrelated roles of TGF-β and IL-10 in the regulation of experimental colitis. J Immunol 2002; 168:900-908.
156. Cottrez F, Groux H. Regulation of TGF-beta response during T cell activation is modulated by IL-10. J Immunol 2001; 167:773-778.
157. Sartor RB. Pathogenesis and immune mechanisms of chronic inflammatory bowel diseases. Am J Gastroenterol 1997; 92(12 Suppl):5S-11S.
158. Gionchetti P, Rizzello F, Venturi A et al. Review—Antibiotic treatment in inflammatory bowel disease: Rifaximin, a new possible approach. Eur Rev Med Pharmacol Sci 1999; 3:27-30.
159. Campieri M, Gionchetti P. Probiotics in inflammatory bowel disease: New insight to pathogenesis or a possible therapeutic alternative? Gastroenterology 1999; 116:1246-1249.
160. Khoo UY, Proctor IE, Macpherson AJ. CD4$^+$ T cell down-regulation in human intestinal mucosa: Evidence for intestinal tolerance to luminal bacterial antigens. J Immunol 1997; 158:3626-3634.
161. Duchmann R, Kaiser I, Hermann E et al. Tolerance exists towards resident intestinal flora but is broken in active inflammatory bowel disease (IBD). Clin Exp Immunol 1995; 102:448-455.
162. Braat H, Peppelenbosch MP, Hommes DW. Interleukin-10-based therapy for inflammatory bowel disease. Expert Opin Biol Ther 2003; 3:725-731.
163. Steidler L, Hans W, Schotte L et al. Treatment of murine colitis by Lactococcus lactis secreting interleukin-10. Science 2000; 289:1352-1355.

Chapter 3A

Role of IL-10 in Autoimmune Pathology

Lionel Almeras and Lionel Prin

Abstract

The mechanisms involved in initiating and maintaining the inflammatory response associated with autoimmune diseases is yet to be defined. Nevertheless, cytokines with pro-inflammatory or anti-inflammatory activity play a crucial role in the development of auto-aggressive responses, as demonstrated in animal models and human clinical trials. Studies on interleukin (IL)-10—a cytokine that is able to regulate inflammatory and immune responses—can provide insights into the manner in which homeostatic networks modulate disease pathogenesis involving immune dysfonction. Numerous polymorphisms have been detected in the IL-10 promoter, and some of these have been associated with levels of IL-10 production. Here, we review a number of studies that have sought to explore the relationships between IL-10 promoter polymorphisms and the initiation/progression of some autoimmune diseases. As expected, with regard to the pleiotropic role of IL-10 and to the complexity of genetic interactions in autoimmune diseases, the findings are not yet conclusive, but might have indicative value when associated with other markers.

Introduction

In developed countries, approximately 4-5% of the population suffers from some form of autoimmune disease, making the latter a major health issue. Cytokines play a crucial role in the regulation of immune and inflammatory responses and therefore allow the targetting of potential candidate genes for diseases involving immune dysfonction. The broad spectrum of IL-10 activity in the immune system suggests that this cytokine is a key component in protective and pathological responses. This pleiotropic cytokine is mainly produced "endogenously" by Th2 cells, monocytes/macrophages and B cells (genetic influence?).[1] In addition, "exogenous" IL-10 related cytokines can also be produced by several viruses such as herpes virus or pox virus genera (environmental influence?).[2]

IL-10 is highly regarded as being involved in anti-inflammatatory and immunosuppressive functions, in large part on the basis of results obtained using animal models.[3] IL-10 also contributes to enhancing B cell survival and proliferation, and antibody production. By studying its direct effects on individuals' immune cell populations, it has been observed that IL-10 can act at different phases of the immune response. Thus, it suppresses several cell immune functions, and therefore can prevent the proliferation of autoreactive lymphocytes. For example, IL-10 prevents the migration of dendritic cells to lymph nodes[4] and the production of proinflammatory chemokines.[5] IL-10 also inhibits the synthesis of IL-12 and tumor necrosis factor (TNF)-α, and leads to the down-regulation of major histocompatibility complex (MHC) class II antigens and costimulatory molecules in antigen-presenting cells.[5] The presence of IL-10 interferes with Th1 cell functions. It blocks the activation of the synthesis of interferon (IFN)-γ and proinflammatory cytokines.[6,7] Anergic T cells are obtained when these T cells produce IL-10.[8] In animal models of autoimmunity, recent data shows that IL-10 regulates autoreactive T cells and inhibits the progression of autoimmune diabetes[9] and of experimental

Interleukin-10, edited by Francesco M. Marincola. ©2006 Eurekah.com.

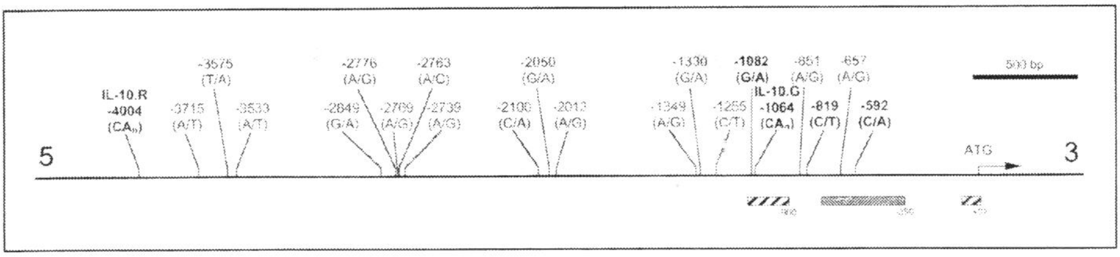

Figure 1. Schematic representation of IL-10 gene promoter polymorphisms. The locations of the various known single nucleotide polymorphisms (SNPs) are defined upstream of the transcription start site (ATG), as are the two CA-repeat microsatellites, IL-10.R and IL-10.G. Bold letters correspond to polymorphisms usually used to investigate an association with disease susceptibility. Positive and negative regulatory regions, described previously by Kube et al[27] are schematically represented by shadowy and grey boxes respectively.

autoimmune encephalomyelitis (EAE).[10] In this last model of EAE, only high levels of IL-10 were induced in the protocol used. Nevertheless, the presence of other cytokines produced endogenously within the inflammatory site may increase[11] or impair[12] such modulating effects of IL-10. Thus, the transgenic expression of IL-10 in mice can exacerbate inflammatory processes[13] and autoimmune diseases.[14]

The expression of IL-10 appears to be strictly regulated. The levels of its constitutive expression in normal cells are very low.[15] The IL-10 gene exhibits substantial polymorphism in the promoter region, which appears to be correlated with variations in transcription.[16,17] Several groups have investigated IL-10 promoter polymorphisms, which may result in an altered level of expression. Therefore, the IL-10 gene must be considered as important candidate for conferring susceptibility and severity to the development of autoimmunity. In this review, IL-10 gene structure and polymorphisms, and the links between IL-10 polymorphisms and auto-immune diseases are summarized.

Chromosomal Localization and Gene Structure of IL-10

The first gene structures of IL-10 were determined by cloning cDNAs encoding the cytokines in mouse (mIL-10) and human IL-10 (hIL-10).[18,19] The degree of homology exhibited by the nucleotide sequence of the respective cDNA clones exceeds 80%. The human IL-10 gene covers 4894 bp, comprising 5 exons, and is mapped in position q31-q32 on chromosome 1.[20,21] Westendorp et al identified a strong genetic influence in the production of cytokines.[22] Their analysis of heritability in monozygotic twins and family studies has suggested that genetic factors account for 75% of the inter-individual variation in IL-10 secretion.[22] The nature of the 5'-flanking region of the IL-10 gene may contribute to the observed differences.

IL-10 production appears to be controlled at the transcriptional level.[23] The level of transcriptional activation of a gene depends upon the binding of regulatory factors to specific recognition sequences in the promoter. In the 5' noncoding region of the IL-10 gene, the presence of sequential patterns associated with the regulation of gene transcription were identified.[21,24-26] IL-10 promoter sequences were cloned and analyzed to detect 5'-flanking regions that can activate or suppress the expression of IL-10.[27] A typical TATA box sequence is located 97bp upstream of the first ATG within a corresponding region previously shown to have promoter activity for human IL-10.[27] Three critical regions have been identified (Fig. 1): two positive regulatory sequences located around TATA-box, and between -1100/-900; and a negative regulatory elements found between -750/-350, downstream from the transcription initiation site.[27] Polymorphisms have been described in the IL-10 promoter region and many have a functional effect on IL-10 transcription. Several groups investigated the IL-10 promoter to determine whether an association existed between IL-10 production and these polymorphic elements.

Table 1. Associations between IL-10 promoter locus polymorphisms and IL-10 level production

Polymorphisms	Locations	Alleles/Haplotypes	Phenotype	References
SNPs				
	-1082	G/G	High	31
		G/A	Medium	31
		A/A	Low	31
	-1082/-819/-592	GCC	High	16,37,86
		ACC	Medium	16,37
		ATA	Low	16,37
	-3575/-2849/-2763	TGC	High	32,33,36
		AAA	Low	32,33,36
Microsatellites				
	-1064 [CA]n-repeat	R2/R2	High	35,39
		R2/R4	High	35
		R3/R3	Low	35,39
	-1064/-4004 [CA]n-repeat	R3.G7	Low	35
		R2.G14	High	35

IL-10 Polymorphisms

Single nucleotide polymorphisms (SNPs) or microsatellites, located within the promoter, affect gene transcription and cause inter-individual variations in cytokine production. Such differential cytokine production may confer a certain degree of flexibility in the immune response, with protective or aggressive effects leading to resistance or susceptibility with regard to autoimmune diseases.[28]

Several single nucleotide polymorphisms (SNPs) and two CA-repeat microsatellites have been described in the IL-10 promoter[29-34] (Fig. 1). Among the SNPs identified in the promoter region of this gene, three single base pair substitutions assigned to positions -1082, -819, and -592 have been studied in detail. The -1082 polymorphism is a G-to-A substitution occurring within a putative Ets transcription factor-binding site, and deletion of this site affects the activity of the IL-10 gene promoter.[27] The -819 polymorphism is a C-to-T substitution and lies within a putative positive regulatory region.[27] The -592 polymorphism is a C-to-A substitution and links within a putative STAT 3 binding site and a negative regulatory region.[27] These three polymorphisms exhibit strong linkage disequilibrium, and three main haplotypes (GCC, ACC, and ATA) appear to segregate in most populations.[31]

Several studies reported an association between alleles or haplotypes of IL-10 polymorphisms and IL-10 production, at least in vitro, in individual immune cell populations.[31-33,35] To investigate whether 5' flanking region polymorphisms affect IL-10 production to a known stimulus, using for example lipopolysaccharide (LPS) or concanavalin A (Con A),[16,31,32] different methods were used. These include whole blood culture,[16,36] peripheral blood mononuclear cells (PBMC) culture,[35,37] or transfection studies.[16] A G at position -1082 was shown to result in higher IL-10 production.[31] In vitro studies[16,31,37] also suggested that -1082G, -819C, and -592C/GCC genotypes are associated with higher levels of IL-10 production. In contrast, the ATA/ATA genotype is associated with low levels of IL-10 production (Table 1).

Gibson et al identified seven novel genetic SNPs in the distal part of the 5' promoter region of IL-10 [32] (Fig. 1). For three of these sites (-3575 T/A, -2849 G/A, and -2763 C/A), both alleles are commonly represented. In contrast, for the four other sites (-3715 A/T, -2776 A/G, -2100 C/A, and -2050 G/A), the AACG allele predominates while the frequency of occurrence of the other alleles is less than 5%. Few groups have studied these distal SNP promoter genotypes and haplotypes in the IL-10 locus and their association with IL-10 production. The AA carriers in position -2849 show a significantly lower level of IL-10 production compared to AG or GG carriers.[36] Moreover, the -3575T/-2849G/-2763C haplotype is associated with high IL-10 production, while the -3575A/-2849A/-2763A haplotype is associated with low IL-10 production[32,33] (Table 1).

The IL-10 promoter region also contains two microsatellite polymorphisms, IL-10.G and IL-10.R, located respectively at -1.1 and -4 kb from the transcription initiation site.[29,30] Sixteen and four alleles have been identified respectively for IL-10.G (commonly referred to as IL-10.G1 to IL-10.G16) and IL-10.R (commonly referred to as IL-10.R1 to IL-10.R4), and variations in IL-10 secretion have been associated with particular haplotypes defined by these microsatellite markers[35] (Table 1). The IL-10.R3 allele has been associated with lower IL-10 secretion than other IL-10.R alleles.[35] Haplotypes containing the IL-10.R2 allele are associated with increased IL-10 secretion.[35] The IL-10.R3/G7 haplotype was particularly associated with lower IL-10 secretion, whereas the IL-10.R2/G14 haplotype displayed the greatest enhancement in IL-10 secretion.[35] Variable associations between the proximal IL-10.G microsatellite alleles and IL-10 production, and between SNPs or SNP haplotypes in the 1.4 kb IL-10 proximal promoter region and IL-10 production have been also been reported.[35,38,39]

All of these findings indicates that IL-10 gene polymorphisms are associated with differential IL-10 production in vitro. Thus, the heritable differences in IL-10 production are strongly linked to the gene's structure. Several recent reports on IL-10 promoter genotype/haplotype frequencies in different populations have demonstrated that the distribution of these polymorphisms varies according to ethnicity.[39-44] Indeed, ethnic origin is an important component for several immune-mediated diseases.

IL-10 Polymorphisms and Autoimmune Diseases

The relationships between inflammatory processes and autoimmune responses have yet to be clearly defined. Nevertheless, the role of cytokines is crucial in such immune dysfunctions, as shown by animal models and human clinical trials. Levels of IL-10 production are therefore critical in homeostatic mechanisms involved in inflammatory reactions but also in antibody production. The differential expression of IL-10, described in a number of autoimmune disorders, including multiple sclerosis (MS),[45,46] rheumatoid arthritis (RA),[47-49] systemic lupus erythematosus (SLE) and Sjögren's syndrome,[50] emphasizes the need for more detailed analysis of IL-10 polymorphism.

IL-10 Polymorphisms and MS

Multiple sclerosis (MS) is an autoimmune/inflammatory disease of the central nervous system (CNS) of unknown aetiology resulting in polymorphic and unpredictable clinical manifestations. Current views attribute environmental factors to the triggering of the onset of MS in genetically susceptible individuals. Evidence for the role of genetic factors is shown by studies on twins and by the development of the disease in family members. The autoimmune nature of MS suggests that cytokine genes may be potent candidates, with different loci contributing to disease susceptibility and/or to disease progression.

Data concerning the relationship between IL-10 polymorphism and susceptibility to MS is summarized in Table 2. In studies on the relationship between MS and IL-10 promoter polymorphisms (-1082, -819, -592, and microsatellites), no association between the three SNPs and susceptibility to, or severity of MS could be established.[51-53] However, the study of recently described IL-10 SNPs,[32] by De Jong et al, indicated that the frequency of IL-10 -2849G polymorphism is lower in patients with primary progressive forms than in patients having relapse-onset MS and control subjects.[36]

In a Spanish population, both IL-10.G and IL-10.R microsatellites, as well as the three promoter SNPs, failed to exhibit any association between IL-10 haplotypes and susceptibility to MS.[54] Nevertheless, in this population, the IL-10.G12 allele showed a significant difference between MS patients and controls. In contrast, this association was not detected in three other

Table 2. Characteristics of tested markers and their reported associations with autoimmune disease

Disease	Location of Polymorphism	Associated Alleles, Haplotypes or Genotypes	Population	Clinical Criteria	Refs.
MS	-1082/-819 /-592	none	Caucasian (UK)	EDSS[a]	51
	-1082/-819 /-592	none	Caucasian	MS forms	53
	-1082	none	Caucasian (Germany)	PI[b]	52
	-3575/-2849/ -2763/-1082/-819	-2849 G	Caucasian	MS forms	36
	-1082/-819 /-592 and IL-10.R/G	IL-10.G12	Spanish		54
	IL-10.G	none	Swedish		56
	IL-10.G	none	Northern Irish	MS forms	57
	IL-10.G	none	French		55
	IL-10.R/G	IL-10.G genotypes[e]	French	PI	58
SLE	-1082/-819 /-592	ACC GCC	Northwest England	Ro+[c]	71
	-1082/-819 /-592	ATA	Southern Chinese	renal disease	40
	-1082/-819 /-592	none	Caucasian		72
	-1082/-819 /-592	none	Anglo-Saxon Caucasian	dsDNA-Abs renal disease	73
	-3575/-2849/-2763	-2763 C	African-American		32
	IL-10.R/G	IL-10.G9 IL-10.G13	Caucasian (UK)		38
	IL-10.G	IL-10.G10	Mexican-American		74
	IL-10.R/G	IL-10.G11	Italian		75
	IL-10.R/G	none	Taiwan		77
	IL-10.G	none	Mexican		78
	IL-10.G	none	Mexican Swedish		76
RA	-1082/-819 /-592	none	UK		80
	-1082	none	French		82
	-1082/-819 /-592	-1082 A	Dutch	Joint destruction	17
	-1082/-819 /-592 and IL-10.R/G	IL-10.G12	Spanish		81
	-1082/-819 /-592	ACC	Caucasian	IgA RF+/IgG RF-	83
	IL-10.R/G	IL-10.R2 IL-10.R3	Caucasian	Different origins	39
	IL-10.R/G	none	South African Caucasian (UK)		44
Primary	-1082/-819 /-592	GCC	Finnish		86
Sjögren's	-1082/-819 /-592	GCC	Caucasian		87
Syndrome	IL-10.G	IL-10.G9		IgA RF/anti-Ro/anti-La	88
Psoriasis	-1082	none	Caucasian	age-at-onset	84
	IL-10.R/G	none	Caucasian (German)	age-at-onset	85
Type 1	-1082/-819 /-592	ATA	Japanese	age-at-onset	90
Diabetes	-592	C	Japanese	GADab+*d*	89

[a] EDSS: Expanded Disability Status Scale;[98] [b] PI: Progression Index (PI=EDSS/duration [years]); [c] Ro+: patients with SLE positive for anti-Ro antibodies; [d] GADab+: patients with type 1 Diabetes positive for antibodies to glutamic acid decarboxylase; [e] IL-10.G9/9, G10/13, G11/13 and G13/14 genotypes were associated with MS patients with mild disease progression; IL-10.G9/10, G9/11, G9/13 and G12/13 were associated with MS patients with severe disease progression.

studies conducted on northern Irish, Swedish or French populations.[55-57] We also analyzed the distribution of both the IL-10.R and the IL-10.G alleles in a French Caucasian population[58] with a predisposition to MS. Our study corroborates previous reports that have demonstrated the absence of an association between any of the IL-10 microsatellite polymorphisms and susceptibility to MS. On the other hand, some IL-10.G genotypes were distributed differently in MS patients that were stratified according to the rate of progression of their disease (i.e., progression index).[58]

Although the genetic factor in MS is substantial—30% concordance in identical twins[59]—attempts to find associations between these IL-10 promoter polymorphisms and MS have failed. All of the data indicates that IL-10 does not seem to be a major susceptibility locus in MS. Such data is consistent with full genome analysis results, which have demonstrated that there is no single master gene for MS.[60-62]

IL-10 Polymorphisms and SLE

Systemic lupus erythematosus (SLE) is characterized by inflammatory damage in different organ systems and by the excessive production of autoantibodies, especially to a spectrum of nuclear antigens such as DNA and to other intracellular antigens involved in transcription and translation.[63,64] Genetic factors seem to play a crucial role in disease susceptibility as shown by linkage analyses in human families with SLE.[65] Genes also seem to contribute to the extreme heterogeneity in disease progression among individuals.[66] Different data has emphasized the role of the 1q21-q44 region of human chromosome 1, which contains the IL-10 gene (i.e., 1q31-q32), and which is syntenic with a segment of the murine genome containing several different susceptibility loci for lupus-like diseases.[67] IL-10, which is active in B cell lineages, can promote the production of antibodies, and might play a direct pathogenic role in SLE.[68] Abnormalities in cytokine production, and increased sera levels of IL-10[69,70] in particular, have been found in patients with SLE. In this regard, several groups have studied IL-10 promoter polymorphisms in patients with SLE compared to normal controls to determine whether associations exist between the cluster of inherited IL-10 genes and SLE or autoantibody production (Table 2).

The three SNPs in the promoter region (-1082/-819/-592) were investigated in Chinese and English populations, but no significant change was found in the allele frequency in the SLE group when compared to controls.[40,71,72] When autoantibody status and clinical features were taken into account to stratify SLE patients, some haplotypes occured more frequently. Mok et al described an overrepresentation of the ATA haplotype associated with renal involvement in SLE patients.[40] Lazarus et al described an association between anti-Ro positivity and the ACC and GCC haplotypes (i.e., previously associated with high IL-10 production).[71] In contrast, Crawley et al found no association between anti-Ro positivity and either the GCC or the ACC haplotype.[73] The SNPs in the promoter region of the IL-10 locus, identified by Gibson et al, combined with the three previously identified SNPs, form new haplotypes.[32] Some of them segregate with high IL-10 production and are associated with SLE in African-American populations.[32] These new SNPs should be studied in a large number of subjects with ethnic differences to improve upon these initial findings.

An initial study on microsatellite polymorphisms in Caucasian patients suggested an association between the IL-10.G9 and IL-10.G13 alleles and SLE.[38] A large study involving Mexican Americans has shown an association between SLE and the IL-10.G10 allele.[74] Recent findings based on a study carried out in the Italian population suggest that the IL-10.G11 microsatellite plays a role in the susceptibility to SLE.[75] In contrast, in Mexican, Swedish or Taiwanese populations, no association has been established between IL-10.G polymorphism and SLE.[76-78] These findings indicate the presence of genetic heterogeneity within and between ethnic groups in the SLE-associated IL-10.G alleles. Although levels of IL-10 production seem to play a major role in SLE, these divergent findings corroborate the fact that many others genes could affect the expression and progression of the disease.

IL-10 Polymorphisms and Rheumatoid Arthritis

Rheumatoid arthritis (RA) is a systemic inflammatory arthropathy of unknown aetiology, characterized by chronic synovitis and progressive joint destruction. It is a frequent and complex multifactorial disease where both genetic and environmental factors contribute to disease susceptibility and severity.[79] In RA, IL-10 might play a dual role as an anti-inflammatory cytokine and also as an active agent on synovial B cells, which are postulated to be important for the synthesis of autoantibody-like rheumatoid factor (RF). High levels of IL-10 in serum and in synovial fluid have been found in patients with RA.[49] Differences in severity between individuals might therefore be related to different levels of cytokine production resulting from polymorphisms in their genes.

Results on IL-10 polymorphisms as a potential marker of RA susceptibility or severity are summarized in Table 2. No significant differences in the allele or haplotype frequencies of the three polymorphisms (-1082/-819/-592) were found between controls and RA patients in different ethnic populations (i.e., French Caucasian, Spanish, UK and Dutch subjects).[17,80-82] This data is consistent with the findings of Hajeer et al,[83] which showed a similar distribution of the -1082, -819 and -592 haplotypes in RA patients and controls. Nevertheless, a significant association was observed between the ACC haplotype and the detection of IgA positive RF in RA patients. Moreover, this prevalence was greater in patients who were negative for IgG RF.[83] In addition, a recent work reports that patients with the -1082AA genotype developed less joint damage than patients with the -1082GG genotype.[17]

Discordant results were also found when IL-10 microsatellite analysis was performed in RA populations (Table 2). Eskdale et al observed, in RA patients from different locations (i.e., Oxford, Glasgow and Atlanta), an increase in IL-10.R2 and a reduction in IL-10.R3 alleles.[39] In contrast, MacKay et al were unable to confirm any association between IL-10.R alleles and RA in two ethnic groups (South African and UK Caucasian populations).[44] In both studies, IL-10.R allele distribution was different in each of the tested ethnic populations.[39,44] In a more recent work, IL-10.G12 has been associated with susceptibility to RA in a Spanish population.[81] All of these findings indicate the presence of a genetic heterogeneity within and between ethnic groups with regard to IL-10 promoter polymorphisms. As for other autoimmune diseases, such results underline the difficulty in detecting the contribution of individual loci in multiethnic populations.

IL-10 Polymorphisms and Other Autoimmune Diseases

IL-10 promoter polymorphisms have been analyzed in several auto-immune diseases. No association has been established between the different IL-10 promoter polymorphisms described above and susceptibility to psoriasis.[84,85] In contrast, a predisposition for primary Sjögren Syndrome has been described in GCC (-1082/-819/-592) patient carriers.[86,87] Moreover, this haplotype is associated with high levels of plasma IL-10 and seems to be related to the disease progression.[86] A relationship between IL-10 microsatellite polymorphisms and biological markers of disease activity was also found in primary Sjögren Syndrome.[88] In autoimmune diabetes, a study performed in one ethnic group showed that IL-10 promoter (-592A/C) polymorphism was a weak determinant for susceptibility to the disease in Japanese individuals. On the other hand, IL-10 (-592) polymorphism is associated with clinical heterogeneity.[89] Additionally, within the same ethnic population, the frequency of the ATA/ATA (-1082/-819/-592) genotype was significantly higher in individuals with type 1 diabetes, stratified according to the age at disease onset.[90] All of these findings also underscore the difficulty in comparing different studies in which various clinical and biological parameters are examined in individuals selected from different ethnic groups.

Concluding Remarks

Interest in IL-10 genetic analysis in this last decade is attributable to the fact that protein production could be altered by IL-10 promoter polymorphisms. Thus, it has been hypothesized that potential variations in IL-10 protein levels might influence disease severity by upsetting the balance between pro-inflammatory and anti-inflammatory processes. The data summarized in this review shows that the genetic influences of IL-10 in the initiation and/or progression of autoimmune disease is not conclusive. There are several possible reasons for these divergent findings. Firstly, recent works have demonstrated that IL-10, traditionally classified as an anti-inflammatory and immuno-suppressive cytokine, can play a dual role,[91] possibly with opposing effects depending on the context, as shown by global effects observed in animal models.[92] Other cytokines, similar to IL-10 and having potent redundant activities, can also modulate the inflammatory and immune response.[11] Secondly, screening of the whole human genome, performed by several groups, has indicated that multiple loci contribute to genetic susceptibility in autoimmune diseases.[93-97] Thus, analysis of the impact of individual loci is too restrictive in complex diseases resulting from the involvement of multiple genes having variable effects, which, moreover, can vary according to ethnic group.

Acknowledgements

L. Almeras was supported by a grant from the Conseil Regional du Nord Pas de Calais and CHU de Lille and Biogen. The authors also wish to thank Drs. S. Dubucquoi and D. Lefranc for critically reviewing the manuscript.

References

1. Mosmann TR. Properties and functions of interleukin-10. Adv Immunol 1994; 56:1-26.
2. Fickenscher H, Hor S, Kupers H et al. The interleukin-10 family of cytokines. Trends Immunol 2002; 23:89-96.
3. Kuhn R, Lohler J, Rennick D et al. Interleukin-10-deficient mice develop chronic enterocolitis. Cell 1993; 75:263-74.
4. D'Amico G, Frascaroli G, Bianchi G et al. Uncoupling of inflammatory chemokine receptors by IL-10: Generation of functional decoys. Nat Immunol 2000; 1:387-91.
5. Moore KW, de Waal Malefyt R, Coffman RL et al. Interleukin-10 and the interleukin-10 receptor. Annu Rev Immunol 2001; 19:683-765.
6. Moore KW, O'Garra A, de Waal Malefyt R et al. Interleukin-10. Annu Rev Immunol 1993; 11:165-90.
7. D'Andrea A, Aste-Amezaga M, Valiante NM et al. Interleukin 10 (IL-10) inhibits human lymphocyte interferon gamma-production by suppressing natural killer cell stimulatory factor/IL-12 synthesis in accessory cells. J Exp Med 1993; 178:1041-8.
8. Gilliet M, Liu YJ. Generation of human CD8 T regulatory cells by CD40 ligand-activated plasmacytoid dendritic cells. J Exp Med 2002; 195:695-704.
9. Gonzalez A, AndreSchmutz I, Carnaud C et al. Damage control, rather than unresponsiveness, effected by protective DX5+ T cells in autoimmune diabetes. Nat Immunol 2001; 2:1117-25.
10. Barrat FJ, Cua DJ, Boonstra A et al. In vitro generation of interleukin 10-producing regulatory CD4(+) T cells is induced by immunosuppressive drugs and inhibited by T helper type 1 (Th1)- and Th2-inducing cytokines. J Exp Med 2002; 195:603-16.
11. Kotenko SV. The family of IL-10-related cytokines and their receptors: Related, but to what extent? Cytokine Growth Factor Rev 2002; 13:223-40.
12. Cai G, Kastelein RA, Hunter CA. IL-10 enhances NK cell proliferation, cytotoxicity and production of IFN-gamma when combined with IL-18. Eur J Immunol 1999; 29:2658-65.
13. Lee CG, Homer RJ, Cohn L et al. Transgenic overexpression of interleukin (IL)-10 in the lung causes mucus metaplasia, tissue inflammation, and airway remodeling via IL-13-dependent and -independent pathways. J Biol Chem 2002; 277:35466-74.
14. Balasa B, Sarvetnick N. The paradoxical effects of interleukin 10 in the immunoregulation of autoimmune diabetes. J Autoimmun 1996; 9:283-6.
15. Tone M, Powell MJ, Tone Y et al. IL-10 gene expression is controlled by the transcription factors Sp1 and Sp3. J Immunol 2000; 165:286-91.
16. Crawley E, Kay R, Sillibourne J et al. Polymorphic haplotypes of the interleukin-10 5' flanking region determine variable interleukin-10 transcription and are associated with particular phenotypes of juvenile rheumatoid arthritis. Arthritis Rheum 1999; 42:1101-8.

17. Huizinga TW, Keijsers V, Yanni G et al. Are differences in interleukin 10 production associated with joint damage? Rheumatology (Oxford) 2000; 39:1180-8.
18. Moore KW, Vieira P, Fiorentino DF et al.Homology of cytokine synthesis inhibitory factor (IL-10) to the Epstein-Barr virus gene BCRFI. Science 1990; 248:1230-4.
19. Vieira P, de Waal-Malefyt R, Dang MN et al. Isolation and expression of human cytokine synthesis inhibitory factor cDNA clones: Homology to Epstein-Barr virus open reading frame BCRFI. Proc Natl Acad Sci USA 1991; 88:1172-6.
20. Kim JM, Brannan CI, Copeland NG et al. Structure of the mouse IL-10 gene and chromosomal localization of the mouse and human genes. J Immunol 1992; 148:3618-23.
21. Eskdale J, Kube D, Tesch H et al. Mapping of the human IL10 gene and further characterization of the 5' flanking sequence. Immunogenetics 1997; 46:120-8.
22. Westendorp RG, Langermans JA, Huizinga TW et al. Genetic influence on cytokine production and fatal meningococcal disease. Lancet 1997; 349:170-3.
23. Bienvenu J, Doche C, Gutowski MC et al. Production of proinflammatory cytokines and cytokines involved in the TH1/TH2 balance is modulated by pentoxifylline. J Cardiovasc Pharmacol 1995; 25(Suppl 2):S80-4.
24. Platzer C, Fritsch E, Elsner T et al. Cyclic adenosine monophosphate-responsive elements are involved in the transcriptional activation of the human IL-10 gene in monocytic cells. Eur J Immunol 1999; 29:3098-104.
25. Meisel C, Vogt K, Platzer C et al. Differential regulation of monocytic tumor necrosis factor-alpha and interleukin-10 expression. Eur J Immunol 1996; 26:1580-6.
26. Mori N, Prager D. Activation of the interleukin-10 gene in the human T lymphoma line HuT 78: Identification and characterization of NF-kappa B binding sites in the regulatory region of the interleukin-10 gene. Eur J Haematol 1997; 59:162-70.
27. Kube D, Platzer C, von Knethen A et al. Isolation of the human interleukin 10 promoter. Characterization of the promoter activity in Burkitt's lymphoma cell lines. Cytokine 1995; 7:1-7.
28. Bidwell J, Keen L, Gallagher G et al. Cytokine gene polymorphism in human disease: On-line databases. Genes Immun 1999; 1:3-19.
29. Eskdale J, Gallagher G. A polymorphic dinucleotide repeat in the human IL-10 promoter. Immunogenetics 1995; 42:444-5.
30. Eskdale J, Kube D, Gallagher G. A second polymorphic dinucleotide repeat in the 5' flanking region of the human IL10 gene. Immunogenetics 1996; 45:82-3.
31. Turner DM, Williams DM, Sankaran D et al. An investigation of polymorphism in the interleukin-10 gene promoter. Eur J Immunogenet 1997; 24:1-8.
32. Gibson AW, Edberg JC, Wu J et al. Novel single nucleotide polymorphisms in the distal IL-10 promoter affect IL-10 production and enhance the risk of systemic lupus erythematosus. J Immunol 2001; 166:3915-22.
33. Eskdale J, Keijsers V, Huizinga T et al. Microsatellite alleles and single nucleotide polymorphisms (SNP) combine to form four major haplotype families at the human interleukin-10 (IL-10) locus. Genes Immun 1999; 1:151-5.
34. D'Alfonso S, Rampi M, Rolando V et al. New polymorphisms in the IL-10 promoter region. Genes Immun 2000; 1:231-3.
35. Eskdale J, Gallagher G, Verweij CL et al. Interleukin 10 secretion in relation to human IL-10 locus haplotypes. Proc Natl Acad Sci USA 1998; 95:9465-70.
36. de Jong BA, Westendorp RG, Eskdale J et al. Frequency of functional interleukin-10 promoter polymorphism is different between relapse-onset and primary progressive multiple sclerosis. Hum Immunol 2002; 63:281-5.
37. Edwards-Smith CJ, Jonsson JR, Purdie DM et al. Interleukin-10 promoter polymorphism predicts initial response of chronic hepatitis C to interferon alfa. Hepatology 1999; 30:526-30.
38. Eskdale J, Wordsworth P, Bowman S et al. Association between polymorphisms at the human IL-10 locus and systemic lupus erythematosus. Tissue Antigens 1997; 49:635-9.
39. Eskdale J, McNicholl J, Wordsworth P et al. Interleukin-10 microsatellite polymorphisms and IL-10 locus alleles in rheumatoid arthritis susceptibility. Lancet 1998; 352:1282-3.
40. Mok CC, Lanchbury JS, Chan DW et al. Interleukin-10 promoter polymorphisms in Southern Chinese patients with systemic lupus erythematosus. Arthritis Rheum 1998; 41:1090-5.
41. Meenagh A, Williams F, Ross OA et al. Frequency of cytokine polymorphisms in populations from western Europe, Africa, Asia, the Middle East and South America. Hum Immunol 2002; 63:1055-61.
42. Padyukov L, Hahn-Zoric M, Lau YL et al. Different allelic frequencies of several cytokine genes in Hong Kong Chinese and Swedish Caucasians. Genes Immun 2001; 2:280-3.

43. Reynard MP, Turner D, Navarrete CV. Allele frequencies of polymorphisms of the tumour necrosis factor-alpha, interleukin-10, interferon-gamma and interleukin-2 genes in a North European Caucasoid group from the UK. Eur J Immunogenet 2000; 27:241-9.
44. MacKay K, Milicic A, Lee D et al. Rheumatoid arthritis susceptibility and interleukin 10: A study of two ethnically diverse populations. Rheumatology (Oxford) 2003; 42:149-53.
45. Rieckmann P, Albrecht M, Kitze B et al. Tumor necrosis factor-alpha messenger RNA expression in patients with relapsing-remitting multiple sclerosis is associated with disease activity. Ann Neurol 1995; 37:82-8.
46. van Boxel-Dezaire AH, Hoff SC, van Oosten BW et al. Decreased interleukin-10 and increased interleukin-12p40 mRNA are associated with disease activity and characterize different disease stages in multiple sclerosis. Ann Neurol 1999; 45:695-703.
47. Katsikis PD, Chu CQ, Brennan FM et al. Immunoregulatory role of interleukin 10 in rheumatoid arthritis. J Exp Med 1994; 179:1517-27.
48. Cohen SB, Katsikis PD, Chu CQ et al. High level of interleukin-10 production by the activated T cell population within the rheumatoid synovial membrane. Arthritis Rheum 1995; 38:946-52.
49. Cush JJ, Splawski JB, Thomas R et al. Elevated interleukin-10 levels in patients with rheumatoid arthritis. Arthritis Rheum 1995; 38:96-104.
50. Llorente L, Richaud-Patin Y, Fior R et al. In vivo production of interleukin-10 by nonT cells in rheumatoid arthritis, Sjogren's syndrome, and systemic lupus erythematosus. A potential mechanism of B lymphocyte hyperactivity and autoimmunity. Arthritis Rheum 1994; 37:1647-55.
51. Pickard C, Mann C, Sinnott P et al. Interleukin-10 (IL10) promoter polymorphisms and multiple sclerosis. J Neuroimmunol 1999; 101:207-10.
52. Maurer M, Kruse N, Giess R et al. Genetic variation at position -1082 of the interleukin 10 (IL10) promotor and the outcome of multiple sclerosis. J Neuroimmunol 2000; 104:98-100.
53. Myhr KM, Vagnes KS, Maroy TH et al. Interleukin-10 promoter polymorphisms in patients with multiple sclerosis. J Neurol Sci 2002; 202:93-7.
54. Martinez Doncel A, Rubio A, Arroyo R et al. Interleukin-10 polymorphisms in Spanish multiple sclerosis patients. J Neuroimmunol 2002; 131:168-72.
55. Reboul J, Mertens C, Levillayer F et al. Cytokines in genetic susceptibility to multiple sclerosis: A candidate gene approach. French Multiple Sclerosis Genetics Group. J Neuroimmunol 2000; 102:107-12.
56. He B, Xu C, Yang B et al. Linkage and association analysis of genes encoding cytokines and myelin proteins in multiple sclerosis. J Neuroimmunol 1998; 86:13-9.
57. McDonnell GV, Kirk CW, Hawkins SA et al. An evaluation of interleukin genes fails to identify clear susceptibility loci for multiple sclerosis. J Neurol Sci 2000; 176:4-12.
58. Almeras L, Meresse B, Seze J et al. Interleukin-10 promoter polymorphism in multiple sclerosis: Association with disease progression. Eur Cytokine Netw 2002; 13:200-6.
59. Ebers GC, Sadovnick AD, Risch NJ. A genetic basis for familial aggregation in multiple sclerosis. Canadian Collaborative Study Group. Nature 1995; 377:150-1.
60. Sawcer S, Jones HB, Feakes R et al. A genome screen in multiple sclerosis reveals susceptibility loci on chromosome 6p21 and 17q22. Nat Genet 1996; 13:464-8.
61. Haines JL, Ter-Minassian M, Bazyk A et al. A complete genomic screen for multiple sclerosis underscores a role for the major histocompatability complex. The Multiple Sclerosis Genetics Group. Nat Genet 1996; 13:469-71.
62. Ebers GC, Kukay K, Bulman DE et al. A full genome search in multiple sclerosis. Nat Genet 1996; 13:472-6.
63. Steinberg AD, Klinman DM. Pathogenesis of systemic lupus erythematosus. Rheum Dis Clin North Am 1988; 14:25-41.
64. Mills JA. Systemic lupus erythematosus. N Engl J Med 1994; 330:1871-9.
65. Shai R, Quismorio Jr FP, Li L et al. Genome-wide screen for systemic lupus erythematosus susceptibility genes in multiplex families. Hum Mol Genet 1999; 8:639-44.
66. Deapen D, Escalante A, Weinrib L et al. A revised estimate of twin concordance in systemic lupus erythematosus. Arthritis Rheum 1992; 35:311-8.
67. Johanneson B, Lima G, von Salome J et al. A major susceptibility locus for systemic lupus erythemathosus maps to chromosome 1q31. Am J Hum Genet 2002; 71:1060-71.
68. Rousset F, Garcia E, Defrance T et al. Interleukin 10 is a potent growth and differentiation factor for activated human B lymphocytes. Proc Natl Acad Sci USA 1992; 89:1890-3.
69. Llorente L, Richaud-Patin Y, Wijdenes J et al. Spontaneous production of interleukin-10 by B lymphocytes and monocytes in systemic lupus erythematosus. Eur Cytokine Netw 1993; 4:421-7.
70. Llorente L, Richaud-Patin Y, Garcia-Padilla C et al. Clinical and biologic effects of anti-interleukin-10 monoclonal antibody administration in systemic lupus erythematosus. Arthritis Rheum 2000; 43:1790-800.

71. Lazarus M, Hajeer AH, Turner D et al. Genetic variation in the interleukin 10 gene promoter and systemic lupus erythematosus. J Rheumatol 1997; 24:2314-7.
72. van der Linden MW, Westendorp RG, Sturk A et al. High interleukin-10 production in first-degree relatives of patients with generalized but not cutaneous lupus erythematosus. J Investig Med 2000; 48:327-34.
73. Crawley E, Woo P, Isenberg DA. Single nucleotide polymorphic haplotypes of the interleukin-10 5' flanking region are not associated with renal disease or serology in Caucasian patients with systemic lupus erythematosus. Arthritis Rheum 1999; 42:2017-8.
74. Mehrian R, Quismorio Jr FP, Strassmann G et al. Synergistic effect between IL-10 and bcl-2 genotypes in determining susceptibility to systemic lupus erythematosus. Arthritis Rheum 1998; 41:596-602.
75. D'Alfonso S, Rampi M, Bocchio D et al. Systemic lupus erythematosus candidate genes in the Italian population: Evidence for a significant association with interleukin-10. Arthritis Rheum 2000; 43:120-8.
76. Johansson C, Castillejo-Lopez C, Johanneson B et al. Association analysis with microsatellite and SNP markers does not support the involvement of BCL-2 in systemic lupus erythematosus in Mexican and Swedish patients and their families. Genes Immun 2000; 1:380-5.
77. Ou TT, Tsai WC, Chen CJ et al. Genetic analysis of interleukin-10 promoter region in patients with systemic lupus erythematosus in Taiwan. Kaohsiung J Med Sci 1998; 14:599-606.
78. Alarcon-Riquelme ME, Lindqvist AK, Jonasson I et al. Genetic analysis of the contribution of IL10 to systemic lupus erythematosus. J Rheumatol 1999; 26:2148-52.
79. Davidson A, Diamond B. Autoimmune diseases. N Engl J Med 2001; 345:340-50.
80. Coakley G, Mok CC, Hajeer AH et al. Interleukin-10 promoter polymorphisms in rheumatoid arthritis and Felty's syndrome. Br J Rheumatol 1998; 37:988-91.
81. Martinez A, Pascual M, Pascual-Salcedo D et al. Genetic polymorphisms in Spanish rheumatoid arthritis patients: An association and linkage study. Genes Immun 2003; 4:117-21.
82. Cantagrel A, Navaux F, Loubet-Lescoulie P et al. Interleukin-1beta, interleukin-1 receptor antagonist, interleukin-4, and interleukin-10 gene polymorphisms: Relationship to occurrence and severity of rheumatoid arthritis. Arthritis Rheum 1999; 42:1093-100.
83. Hajeer AH, Lazarus M, Turner D et al. IL-10 gene promoter polymorphisms in rheumatoid arthritis. Scand J Rheumatol 1998; 27:142-5.
84. Craven NM, Jackson CW, Kirby B et al. Cytokine gene polymorphisms in psoriasis. Br J Dermatol 2001; 144:849-53.
85. Asadullah K, Eskdale J, Wiese A et al. Interleukin-10 promoter polymorphism in psoriasis. J Invest Dermatol 2001; 116:975-8.
86. Hulkkonen J, Pertovaara M, Antonen J et al. Genetic association between interleukin-10 promoter region polymorphisms and primary Sjogren's syndrome. Arthritis Rheum 2001; 44:176-9.
87. Font J, Garcia-Carrasco M, Ramos-Casals M et al. The role of interleukin-10 promoter polymorphisms in the clinical expression of primary Sjogren's syndrome. Rheumatology (Oxford) 2002; 41:1025-30.
88. Anaya JM, Correa PA, Herrera M et al. Interleukin 10 (IL-10) influences autoimmune response in primary Sjogren's syndrome and is linked to IL-10 gene polymorphism. J Rheumatol 2002; 29:1874-6.
89. Tegoshi H, Hasegawa G, Obayashi H et al. Polymorphisms of interferon-gamma gene CA-repeat and interleukin-10 promoter region (-592A/C) in Japanese type I diabetes. Hum Immunol 2002; 63:121-8.
90. Ide A, Kawasaki E, Abiru N et al. Genetic association between interleukin-10 gene promoter region polymorphisms and type 1 diabetes age-at-onset. Hum Immunol 2002; 63:690-5.
91. Mocellin S, Panelli MC, Wang E et al. The dual role of IL-10. Trends Immunol 2003; 24:36-43.
92. Cavaillon JM. Pro- versus anti-inflammatory cytokines: Myth or reality. Cell Mol Biol (Noisy-le-grand) 2001; 47:695-702.
93. Morahan G, Morel L. Genetics of autoimmune diseases in humans and in animal models. Curr Opin Immunol 2002; 14:803-11.
94. Jawaheer D, Seldin MF, Amos CI et al. Screening the genome for rheumatoid arthritis susceptibility genes: A replication study and combined analysis of 512 multicase families. Arthritis Rheum 2003; 48:906-16.
95. Pociot F, McDermott MF. Genetics of type 1 diabetes mellitus. Genes Immun 2002; 3:235-49.
96. Tsao BP. An update on genetic studies of systemic lupus erythematosus. Curr Rheumatol Rep 2002; 4:359-67.
97. A meta-analysis of genomic screens in multiple sclerosis. The Transatlantic Multiple Sclerosis Genetics Cooperative. Mult Scler 2001; 7:3-11.
98. Kurtzke JF. Rating neurologic impairment in multiple sclerosis: An expanded disability status scale (EDSS). Neurology 1983; 33:1444-52.

CHAPTER 3B

Interleukin-10 Gene Polymorphisms and Cancer

W. Martin Howell

Abstract

Interleukin-10 (IL-10) is a multifunctional cytokine with both immunosuppressive and anti-angiogenic functions. In consequence, IL-10 can have both tumor-promoting and tumor-inhibiting properties. Raised levels of serum and peri-tumoral IL-10 production have been reported in many malignancies, which have been interpreted in support of a role for IL-10 in tumor escape from the immune response. However, gene transfection studies in a number of malignancies argue more convincingly for an anti-tumor function of IL-10, possibly via inhibition of pathways of angiogenesis.

A large number of polymorphisms (primarily single nucleotide polymorphisms (SNPs)) have been identified in the IL-10 gene promoter. Convincing evidence that certain of these polymorphisms are associated with differential expression of IL-10 in vitro and in some cases in vivo have been obtained. While a large number of investigations of possible associations between IL-10 genotypes and immune mediated disease have been performed, the literature with regard to IL-10 polymorphisms and cancer is as yet small, but growing. These published studies include both solid tumors and hematological malignancies and common and less common diseases. In this chapter, the results from 15 studies in 10 different malignancies are reviewed. In 12 of these studies, positive associations between IL-10 genotype or haplotype and disease susceptibility and/or progression were reported. In some of these cancers (for example, cutaneous malignant melanoma, prostate cancer, breast cancer, non cardia gastric cancer and nonHodgkin's lymphoma) genotypes associated with low IL-10 expression were a risk factor for disease or disease progression, while in others (for example, cervical cancer, cardia gastric cancer, post-transplant squamous cell carcinoma of the skin and multiple myeloma), genotypes associated with high IL-10 expression were a risk factor).

All results reviewed should be regarded as preliminary, due to the small sample sizes of almost all of the studies and the limited numbers of IL-10 polymorphisms examined. In addition, few of the studies have examined levels of IL-10 production in vivo in the subjects genotyped. However, the preliminary data obtained thus far indicate that much larger studies are required in a number of cancers, in order to confirm initial results, extend studies to include more detailed genotype/haplotype analysis and to combine genotype and gene expression studies in the same subjects. Such studies will contribute significantly to our understanding of the biological role of IL-10 in tumor development, with implications for cytokine therapy in cancer.

Introduction

As considered in more detail elsewhere in this volume, Interleukin-10 (IL-10) is a key regulator of immune responses and was originally described as cytokine synthesis inhibitory factor,[1] being produced by Th2 cells and inhibiting cytokine production by Th1 cells. Later studies showed that the actions of IL-10 on inhibition of pro-inflammatory cytokine production by both T and NK cells were indirect, acting via inhibition of accessory cell function.[2-5] In addition, it was soon shown that IL-10 inhibits a broad range of activated monocyte/macrophage functions, including monokine synthesis, nitric oxide production, major histocompatibility complex (MHC) class II and CD80/CD86 costimulatory molecule expression.[6-11] In vitro and in vivo studies revealed pleiotropic activities of IL-10 on B and T cells and, taken together, that a critical function of IL-10 is to suppress multiple immune responses through individual actions on T cells, B cells, antigen presenting cells and other cell types, and to skew the immune response from Th1 to Th2 (reviewed in detail in ref. 12). In malignancy, this might suggest a priori, that IL-10 might promote tumor development, by acting to suppress anti-tumor immune responses, where these occur. However, a number of other findings suggest that the biological properties of IL-10 are more complex than this and IL-10 may have immunostimulatory or immunosuppressive effects, depending upon the assay used, cell types involved and other concomitant immune events,[13] therefore the actions of IL-10 on tumor development may be more complex. In particular, animal models suggest that IL-10 can induce NK cell activation and so facilitate anti-tumor responses, leading to tumor cell destruction.[14,15] Evidence for both tumor promoting and anti-tumor functions of IL-10 is briefly reviewed below.

IL-10 and Cancer

There are a number of reports describing elevated levels of IL-10 expression in patients with particular cancers, including malignant melanoma,[16-19] ovarian cancer[20] and other carcinomas,[21-24] lymphoma and myeloma.[25,26] These elevated levels have been reported both in the serum and/or tumor lesions. Furthermore, a negative correlation between circulating levels of IL-10 and prognosis has been reported in patients with solid tumors, including lung cancer,[22] renal carcinoma[24] and gastrointestinal tumors[27] and hematological malignancies.[28-29] A simplistic interpretation of these data would be that elevated IL-10 levels are associated with suppression of anti-tumor immune responses. However, elevated IL-10 expression can occur for a number of reasons. Production by tumor and other cells may indeed result in suppression of anti-tumor immune responses, but IL-10 may also act as a tumor growth factor, as evidenced by the action of exogenous IL-10 on human melanoma cells lines.[30] In addition, IL-10 can also be produced by activated cells involved in anti-tumor immune responses and so may be indicative of a potent anti-tumor immune response. Indeed, it should be noted that elevated IL-10 levels do not correlate with prognosis in all studies[31] and in some cases favorable prognosis has been associated with elevated IL-10 expression.[32]

More convincing data come from studies of IL-10 gene therapy in animal models of tumor growth and establishment, which consistently demonstrate an anti-tumor role for IL-10. In a colon carcinoma mouse model, transfection of tumor cells with IL-10 reduces the malignant potential of the tumor cells and induces a predominant Th2-mediated tumor rejection response.[33] Similarly, IL-10 transfected cell lines derived from mouse mammary adenocarcinoma,[34] ovarian carcinoma,[35] malignant melanoma,[36] Burkitt's lymphoma,[37,38] prostate[39] and colon cancers[33] show significant inhibition of tumor growth. In support of this, systemic administration of IL-10 has inhibited tumor metastasis in various murine models, including melanomas,[14,40] sarcomas and colorectal carcinomas.[40]

The mechanisms behind these antitumor effects are still incompletely understood. Many researchers attribute the antitumor effects of IL-10 to its effects on NK cell activation,[14,15] although actions on T cells,[40] macrophages[41] and nitric oxide[42] have also been implicated. In addition, IL-10 enhances the susceptibility of target cells to NK cell lysis by reducing cell surface MHC expression.[43,44] However, an increasing body of evidence suggests that IL-10 exerts an antitumor effect by inhibition of angiogenesis. For example, in vitro studies of prostate tumor cells show that IL-10 stimulates tissue inhibitors of metalloproteinases (TIMPs) and inhibits matrix metalloproteinase (MMP) expression, so affecting induction of angiogenesis.[45,46] Similarly, IL-10 gene transfection studies in malignant melanoma have shown that inhibition of tumor growth by inhibition of angiogenesis is accompanied by downregulation of synthesis of vascular endothelial growth factor (VEGF)—one of the most potent angiogenic factors—along with IL-1β, tumor necrosis factor-α (TNFα), IL-6 and MMP-9 (all known to have angiogenic properties) in tumor-associated macrophages.[17] In addition, in Burkitt's lymphoma, a lymphoid malignancy, introduction of human or viral IL-10 genes into tumors in SCID mice revealed an inhibition of VEGF-induced neovascularisation of the tumors.[37]

Accordingly, while the mechanisms remain unclear, there is a considerable and growing body of evidence for the antitumor properties of IL-10 and this may result at least in part from inhibition of angiogenesis, possibly by inhibition of production of angiogenic cytokines, growth factors and MMPs and stimulation of production of inhibitors of angiogenesis. Based on this, several investigators have suggested therapeutic use of IL-10 in cancer patients,[14,15,17,40,47] but at present no clinical trials have been performed.

An alternative strategy to determine the role of IL-10 in the development of particular malignancies is via genetic approaches. In recent years a considerable number of genetic polymorphisms have been identified within the IL-10 gene, particularly within the promoter region of the gene. Certain of these polymorphisms have been shown to be associated with differential levels of IL-10 expression. A considerable number of studies have been performed to determine whether IL-10 polymorphisms are associated with susceptibility to a large number of immune-mediated diseases (reviewed in refs. 48 and 49) and a small number of investigations have been performed in solid tumors and hematological malignancies and this literature is briefly reviewed below.

IL-10 Gene Polymorphisms

The IL-10 gene is comprised of 5 exons, spans approximately 5.2 kB and is located on chromosome 1, at 1q31-1q32.[50] Due to the critical role of IL-10 in regulating immune responses in health and immune-mediated diseases, a number of groups have pursued intensive studies to identify naturally occurring gene polymorphisms in the IL-10 gene and flanking regions. To date, at least 49 IL-10-associated polymorphisms have been reported in the literature and these are summarised in Table 1. An even larger number of polymorphisms are recorded in single nucleotide polymorphism (SNP) databases (e.g., the Wellcome Trust Sanger Institute/European Bioinformatics Institute SNP database: Ensembl Genome Browser).

From Table 1 it can be seen that of the 49 polymorphisms listed, 46 are SNPs, 2 are microsatellite polymorphisms and 1 is a small (3 bp) deletion. Twenty-eight polymorphisms occur in the promoter region of the gene, 20 polymorphisms are noncoding intronic or synonymous substitutions and only 1 polymorphism results in a change in amino acid sequence. The promoter polymorphisms have been subject to the most scrutiny, particularly with regard to possible influences upon gene transcription and expression. For example, the IL-10 -1082 SNP and -1082, -819, -592 haplotype have been reported to be associated with differential IL-10 expression in vitro, with the -1082 A, -819 T, -592 A haplotype associated with decreased IL-10 expression, compared with the -1082 G, -819 C, -592 C haplotype.[55] This is thought to reflect—at least in part—differential transcription factor binding associated with the -1082 SNP.[60] In addition, IL-10 R and G microsatellite haplotypes have also been shown to be associated with differential levels of IL-10 expression in vitro.[61] Some workers have suggested that as much as 75% of inter-individual variation in IL-10 expression may be due to genetic variation,[62] although others believe that the contribution of individual SNPs—such as the best-described -1082 SNP—may be much less than this.[60] Accordingly, the role of IL-10 polymorphism in determining susceptibility to and prognostic outcome in nonmalignant immune-mediated diseases has been the subject of intense interest and a plethora of case-control studies have demonstrated a number of positive associations in a diverse range of diseases, including asthma,[63,64] systemic lupus erythematosus,[52] reactive arthritis[65] and outcome of clinical renal,[66,67] heart[68] and bone marrow transplantation.[69] Similarly, other investigations have failed

Table 1. IL-10 gene polymorphisms

Region of Gene	Polymorphism	Nucleotide Change	Original Reference
Promoter	IL-10 - 8571	C/T	Kube et al[51]
	-8531	G/A	Kube et al[51]
	-7400	3 bp deletion	Kube et al[51]
	-6752	A/T	Kube et al[51]
	-6208	G/C	Kube et al[51]
	-5402	C/G	Kube et al[51]
	-3715	A/T	Gibson et al[52]
	-3575	T/A	Gibson et al[52]
	-3533	A/T	D'Alfonso et al[53]
	-2849	G/A	Gibson et al[52]
	-2776	A/G	Gibson et al[52]
	-2769	A/G	D' Alfonso et al[53]
	-2763	C/A	Gibson et al[52]
	-2739	A/G	D' Alfonso et al[53]
	-2100	C/A	Gibson et al[52]
	-2050	G/A	Gibson et al[52]
	-2013	A/G	D' Alfonso et al[53]
	-1466	C/T	Gibson et al[52]
	-1349	A/G	D' Alfonso et al[53]
	-1255	C/T	D' Alfonso et al[53]
	-1082	G/A	Tountas & Cominelli,[54] Turner et al[55]
	-851	G/A	D' Alfonso et al[53]
	-819	C/T	Eskdale et al,[56] Turner et al[55]
	-657	A/G	D' Alfonso et al[53]
	-592	C/A	Eskdale et al,[50] Turner et al[55]
	-429	G/T	Gibson et al[52]
	IL – 10 R	CA repeat	Eskdale et al[56]
	IL – 10 G	CA repeat	Eskdale & Gallagher[57]
Coding sequence	+78	G/A (Gly/Arg)	Donger et al[58]
Non-coding Polymorphisms	+919	G/T	Lazarus et al[59]
(introns and 3'	+1135	G/A	Lazarus et al[59]
flanking region	+1547 (intron 3, +19)	C/T	Donger et al[58]
	+1668	A/T	Lazarus et al[59]
	+1703	G/A	Lazarus et al[59]
	+1812	G/T	Lazarus et al[59]
	+2483 (intron 3, +953)	T/C	D' Alfonso et al[53]
	+2664	A/G	Lazarus et al[59]
	+3835	T/C	Lazarus et al[59]
	+3916 (+117 after stop codon)	T/C	D' Alfonso et al,[53] Lazarus et al[59]
	+3990	A/G	Lazarus et al[59]
	+4144	A/T	Lazarus et al[59]
	+4251	A/G	Lazarus et al[59]
	+4571	A/G	Lazarus et al[59]
	+4949	T/C	Lazarus et al[59]
	+5333	G/A	Lazarus et al[59]
	+5466	C/T	Lazarus et al[59]
	+5470	G/A	Lazarus et al[59]
	+5876	C/T	Lazarus et al[59]
	+6474	C/G	Lazarus et al[59]

Location of polymorphisms given according to original reference (position in brackets if alternative numbering system is used in one of the references).

to implicate IL-10 promoter polymorphisms in susceptibility to various diseases including multiple sclerosis,[70,71] while in other diseases results are conflicting e.g., rheumatoid arthritis.[72,73] Summaries of studies of IL-10 (and other) cytokine polymorphisms and disease have been published by Bidwell et al[48] and Haukim et al,[49] both of which contain extensive bibliographies.

IL-10 Gene Polymorphisms and Cancer

The literature concerning IL-10 polymorphism in cancer is very recent and is therefore still relatively small, but growing rapidly, with all publications dating from 2001. Results from these studies are summarised in Table 2 and each disease is considered in more detail below. From a casual inspection of Table 2, it will be noted that while all except one study is of case-control design, several studies have also investigated associations between particular IL-10 polymorphisms and markers of disease prognosis. Of the 15 studies listed, 6 have studied the IL-10 -1082 SNP alone and 8 have studied the IL-10 -1082, -819, -592 SNPs and haplotypes in case-control studies (in one study, cases only) of the malignancy in question. The IL-10G and IL-10R microsatellites were examined in the remaining study. Therefore all studies published thus far have focussed upon those polymorphisms for which there is direct evidence for a causal association with differential IL-10 expression (IL-10 -1082), or polymorphisms and haplotypes which act as markers for differential IL-10 expression (IL-10G and IL-10R microsatellites and IL-10 -1082, -819, -592 haplotypes). As yet, no published studies have performed detailed IL-10 SNP analysis or haplotyping across the complete IL-10 promoter and/or gene sequence in any malignancy.

In the following consideration of the studies summarised in Table 2, cutaneous malignant melanoma (CMM), prostate (PC) and breast cancer (BC) are considered first, since angiogenesis is crucial for the development of these tumors[89] and indeed the extent of angiogenesis correlates with the probability of metastasis and/or prognosis in these malignancies.[90-96]

Cutaneous Malignant Melanoma

CMM is the most serious cutaneous malignancy and is increasing in frequency among most Caucasian populations, where the most important risk factor is exposure to ultraviolet light.[97] Relatively little is known of the genetic factors that mediate susceptibility to and prognosis in sporadic CMM, although polymorphisms associated with the melanocortin-1 receptor (*MCR1*),[98,99] *CDKN2A*,[100] *XRCC3* DNA repair gene[101] and glutathione S-transferase Mu phenotype (*GSTM1*)[102] may be associated with susceptibility to CMM. Polymorphisms associated with the Vitamin D receptor and the cytochrome P450 *CYP2D6* genes have been implicated in modulating prognosis in this tumor.[103,104] In addition, several lines of evidence suggest that CMM patients develop an immune response to their tumors[105] (supported by variable HLA-DQB1 allelic associations with CMM susceptibility and prognosis[106,107]), although in individuals with CMM, this anti-tumor immune response is insufficient to abrogate tumor development.

Based on the above, and to distinguish whether high constitutive levels of IL-10 have a tumor-promoting or anti-tumor influence in CMM, we have shown that the IL-10 -1082 AA genotype, associated with low IL-10 expression in vitro is associated with both susceptibility to CMM (OR = 1.78) and is a risk factor for more advanced (OR = 2.24) , and poorer prognosis disease, as evidenced by tumor Breslow thickness (OR = 3.67).[74] IL-10 -1082, -819, -592 haplotypes associated with low IL-10 expression (ACC/ACC, ACC/ATA and ATA/ATA) were also associated with greater tumor Breslow thickness (OR = 3.63), which is the single most important prognostic indicator in CMM.[108] In addition, the IL-10 -1082 GG genotype and IL-10 -1082, -819, -592 GCC/GCC 'high expression' haplotype were associated with noninvasive tumor growth (ORs = 2.42 and 2.31 for noninvasive growth respectively). This study was performed in British Caucasian CMM cases and controls. Some support for these findings is provided by the small, independent study of Martinez-Escribano et al,[75] who showed

Table 2. IL-10 polymorphisms and cancer

Disease	IL-10 Polymorphism	Cases	Controls	Association	Genotype, Allele or Haplotype	Refs.
Cutaneous malignant melonoma	-1082, -819, -592	153	158	Susceptibility, advanced stage of disease, greater tumor thickness	1082 AA	Howell et al[74]
				Greater tumor thickness	ACC/ACC, ACC/ ATA, ATA/ATA	"
				Noninvasive growth phase	1082 GG	"
				Noninvasive growth phase	GCC/GCC	"
Cutaneous malignant melonoma	- 1082, -819, -592	42	48	Survival (shorter)	ACC/ATA	Martinez-Escribano et al[75]
Prostate cancer	-1082	247	263	Susceptibility	-1082 AA	McCarron et al[76]
Breast cancer	-1082	144	263	No	-	Howell et al[77]
Breast cancer	-1082	125	100	Susceptibility	-1082 AA	Giordani et al[78]
Cervical cancer	-1082	77	69	Susceptibility	-1082 AG	Stanczuk et al[79]
Cervical cancer	-1082, -819, -592	144	179	No	-	Roh et al[80]
Gastric carcinoma	-1082, -819, -592	220	230	Susceptibility, advanced stage	GCC (1 or 2 copies)	Wu et al[81]
Gastric carcinoma	-1082	150	220	Association with EBV—negative gastric carcinoma	-1082 G allele	Wu et al[82]
Gastric carcinoma	-1082, -819, -592	188	212	Susceptibility (non cardia gastric cancer)	ATA haplotype	El-Omar et al[83]
Squamous cell carcinoma of skin (post renal transplant)	-1082, -819, -592	70	70	Susceptibility Protection	GCC haplotype ATA haplotype	Alamartine et al[84]
Multiple myeloma	IL-10G, IL-10R microsatellites	73	109	Susceptiblity Protection	IL-10 G 136/136, IL-10 R 112/114 IL-10 R 114/116	Zheng et al[85]
Myelodysplasia Acute myeloid leukemia	-1082, -819, -592	150	up to 1000	No	-	Gowans et al[86]
NonHodgkins lymphoma	-1082, -819, -592	126	302	Susceptibility to aggressive disease	-1082 AA, ATA, ACC haplotypes	Cunningham et al[87]
Acute lymphoblastic leukemia	-1082	135	-	Protection from poor response to Prednisone	-1082 GG	Lauten et al[88]

that in Italian CMM patients, the IL-10 -1082, -819, -592 ACC/ATA 'low expression' haplotype was also associated with greater tumor Breslow thickness and was a risk factor for poorer survival. Finally, it should be noted that results from these genetic studies are in accordance with the effects of IL-10 gene transfection in animal models of malignant melanoma, which suggests that intratumor expression of IL-10 abrogates tumor development.[36]

Although the influence of IL-10 on CMM development is likely to be complex, these results support recent findings that IL-10 has an anti-tumor effect in CMM, possibly via inhibition of VEGF expression and angiogenesis.[17] In agreement with this, we have also obtained evidence that gene polymorphisms associated with differential expression of other angiogenic cytokines (in particular, VEGF) may also play a role in predisposition to and tumor growth in CMM.[109]

Prostate Cancer

In Western Europe and the USA, PC is the most common cancer diagnosed in men and the second most common cause of death with a continuing increase in incidence.[110] The evidence that PC has a genetic component is compelling from epidemiological and genetic studies; some high-risk genes have been identified, that when present may predispose a carrier to development of the disease.[111] Examples of PC susceptibility genes include *HPC1* on chromosome 1q24-25,[112] *HPCX* on Xq27-28,[113] *BRCA1* on 17q21 and *BRCA2* on 13q12,[114] CAPB at 1p36,[115] *PCAP* on 1q42.2-43[116] and most recently *ELAC2/HPC2* on chromosome 17p.[117] The association between these high penetrance genes and PC susceptibility highlights the complex and multigenic mode of inheritance of PC, yet more common, lower penetrance susceptibility polymorphisms in genes may be implicated in a higher proportion of the sporadic PC disease burden and so have more relevance to public health.

The prostate was originally thought to be an immunologically privileged site. However, there is now good evidence that the prostate has a lymphatic system, can mount inflammatory immune responses and these responses—as evidenced by density of tumor infiltrating lymphocyes—may be associated with prognosis in PC (reviewed in ref. 118). The immune system may therefore play a role in the pathogenesis of PC, via regulation of tumor growth, while evasion of the immune response may play a role in disease progression. In addition, due to the critical role of angiogenesis in PC development,[93] and based on our findings in CMM, we elected to determine whether polymorphisms in pro- and anti-inflammatory and pro-angiogenic cytokine genes were associated with susceptibility to and/or markers of prognosis in a case-control study of British Caucasian PC patients and population controls. Results indicated that the IL-10 -1082 AA 'low expression' genotype was significantly increased in incidence in the patient group (OR = 1.78), closely paralleling results in CMM.[76] This is again suggestive that genetically determined low levels of IL-10 production may be a risk factor in PC, via down-regulation of VEGF synthesis or enhanced lysis of tumor cells by NK cells. Again, results from this genetic study are in accordance with findings from IL-10 gene transfection studies in this malignancy.[39]

Evidence for a role for polymorphism in pro-angiogenic genes was also provided by this study, which showed significant associations between genotypes associated with low VEGF and low IL-8 expression and protection from PC

Breast Cancer

BC is by far the most common malignancy affecting Western women. A family history of BC is one of the most important and consistent risk factors, highlighting the role of inherited germline susceptibility genes. In the mid 1990s, two BC susceptibility genes, BRCA 1 (chromosome 17) and BRCA 2 (chromosome 13) were identified.[119,120] Mutations that render these genes nonfunctional or absent are inherited in an autosomal dominant manner and confer a high disease risk. However, recent epidemiological studies suggest that BRCA 1 and BRCA 2 mutations only account for a few percent of BC cases.[121] It is highly likely that a number of more prevalent, low penetrance genes contribute to BC susceptibility in a larger population of women and are therefore responsible for a greater proportion of the disease burden.[121-123] Recent modelling of breast cancer inheritance in a population where BRCA1 and 2 gene carriers had been excluded from the cohort revealed a model of inheritance that is polygenic and provides an estimate that nearly 90% of all breast cancer cases will occur in an identifiable subset of perhaps half the general population.[124] As yet, little is known about low penetrance susceptibility genes which contribute to BC susceptibility and only a few have been identified, including genes involved in carcinogen detoxification and oestrogen metabolism.[125-127]

There is accumulating evidence indicating the presence of peritumoral inflammatory infiltrate in BC, which may reflect—at least in part—an antitumor immune response, while angiogenesis is necessary for the development of BC and the extent of angiogenesis correlates with tumor development and patient survival.[94-96] In addition, high levels of IL-10 mRNA are detectable in tumor lesions.[128] Accordingly, we have performed a small study of 144 British Caucasian BC patients and 263 controls, for the same panel of SNPs in pro- and anti-inflammatory and pro-angiogenic cytokine genes as studied in PC, but have failed to demonstrate any associations with susceptibility to BC, for any of these SNPs, including IL-10 -1082—save for the TNFα -308 GG, which was increased in frequency in the BC group, at a marginal level of significance.[77] Conversely, in an independent study of 125 Italian BC patients and 100 controls, Giordani et al[78] have reported a significant association between the IL-10 -1082 AA 'low expression' genotype and BC, analogous to our findings in CMM and PC, but have failed to demonstrate any association between the TNFα -308 SNP and BC.

Therefore the limited data obtained to date with regard to IL-10 polymorphism and development of BC are equivocal, but suggest that a larger study of IL-10 -1082 and additional polymorphisms is merited in this very common cancer.

Cervical Cancer

Most high-grade cervical neoplasms have been shown to contain oncogenic human papilloma virus (HPV) DNA, although only a small proportion of such cases progress to cervical cancer. Factors—including genetic factors—leading to impaired immune responses to HPV may play a role in determining susceptibility to the development of cervical cancer. In support of this, a number of studies have implicated particular HLA polymorphisms (In particular, HLA-DQB1*03 alleles) in conferring susceptibility to squamous cell cervical carcinoma.[129-132] In addition, variation in the secretion of several cytokines, including IL-10, have been reported in the blood and tissues of patients with cervical cancer,[133-135] while angiogenesis is also necessary for the development of cervical neoplasia.[136]

Based on the above, two studies have sought to address whether IL-10 polymorphisms are associated with susceptibility to cervical cancer. In the first published study of 77 Zimbabwean women with histologically proven cancer of the uterine cervix and 69 age- and parity matched controls, the IL-10 -1082 GA genotype was found at a significantly increased frequency of 40.2% in the cases as compared with a frequency of 16% in the controls ($P = 0.001$). Since only one case and no controls were of IL-10 -1082 GG genotype, and the GA genotype is associated with higher IL-10 expression in vitro than the AA genotype, the authors of this study infer that a genetic predisposition to produce higher levels of IL-10 may play an important role in the pathogenesis of cervical cancer.[79] This is consistent with IL-10 contributing to tumor escape from the immune response in the development of this malignancy, but the molecular function of IL-10 in the pathogenesis of HPV infection and cervical cancer may be multifactorial and the authors stress that results should not be interpreted in isolation from studies of other cytokine polymorphisms in this malignancy.[79]

The second published study examined the frequency of the IL-10 -1082, -819 and -592 polymorphisms in 144 Korean women with invasive cervical cancer and 179 ethnically matched noncancer controls. In this study, all individuals were homozygous for the -1082 AA genotype and only two haplotypes (-1082, -819, -592 ATA and ACC) were observed, neither of which were associated with invasive cervical cancer, nor with serum IL-10 concentration.[80] Therefore, in Korean women the IL-10 genotypes studied do not appear to influence susceptibility to *invasive* cervical carcinoma, but due to the lack of IL-10 -1082 polymorphisms observed in this study group and hence lack of comparability with the Zimbabwean study outlined above, a role for IL-10 polymorphism in the development of cervical cancer cannot be ruled out by this study alone.

Gastric Cancer

Gastric carcinoma remains a common disease worldwide[137] and cancers of the upper gastrointestinal tract comprise four distinct entities, namely squamous cell and adenocarcinoma of the esophagus, adenocarcinoma of the gastric cardia and adenocarcinoma of the distal (noncardia) stomach. Environmental and host-related factors interact in disease development, among which *Helicobacter pylori* infection and cigarette smoking are important environmental risks. Host immunogenetic factors have been shown to be associated with increased risk of gastric cancer and its precursors. In particular, El-Omar and colleagues have shown that functional polymorphisms in the pro-inflammatory IL-1β gene and its receptor antagonist (IL-1RN) are associated with increased risk of noncardia gastric tumors.[138,139] These data indicate that genetic control of inflammation and response to *Helicobacter pylori* infection may be important in the development of upper gastrointestinal cancers in general. Accordingly, in a US-based study group, El-Omar and colleagues have also investigated the role of SNPs in the TNFα, IL-1β, IL-4, IL-6 and IL-10 genes and risk of development of upper gastrointestinal tract cancers, including esophageal and gastric cancers. Proinflammatory genotypes of TNFα (carriers of -308 A 'high expression' allele) and IL-10 (carriers of the -1082, -819, -592 ATA 'low expression' haplotype) were associated with a more than doubling risk of developing noncardia gastric cancers (OR for possession of IL-10 ATA haplotype = 2.5). Carriage of multiple proinflammatory polymorphisms of TNFα, IL-1β, IL-1RN and IL-10 conferred greater risk, with ORs of 2.8 for one, 5.4 for two and 27.3 for three or four high risk genotypes. In contrast, these polymorphisms were not consistently related to the risk of esophageal or gastric cardia cancers.[83] The authors interpret these findings to suggest that a proinflammatory host genotype favors the development of a hypochlorydric, atrophic response to gastric infection with *Helicobacter pylori*, which in turn predisposes to noncardia gastric adenocarcinoma, but not to cardia or esophageal tumors. However, an association of IL-10 'low expression' genotypes with noncardia gastric adenocarcinoma is also consistent with anti-angiogenic functions of IL-10 and angiogenesis is known to be important in the development of gastric cancer.[140]

Two similar studies of IL-10 polymorphism in the development of gastric cancer in Taiwanese Chinese have been performed by Wu and colleagues.[81,82] In opposition to the study of El-Omar and colleagues,[83] the studies of Wu report that the IL-10 -1082, -819, -592 GCC 'high expression' haplotype is associated with the development of gastric cancer taken as a single entity (OR = 2.67), and in particular with cardia cancer (OR = 3.21) and advanced stage of disease (OR = 2.29).[82] These results are more consistent with genetically programmed high expression levels of IL-10 contributing to an anti-inflammatory immune response, contributing to tumor escape. However, due to the very different frequencies of IL-10 SNP alleles in the two study groups (US and Chinese), differing environmental factors and associations with different forms of gastric cancer (noncardia v cardia), it is difficult to compare the results from these two principal studies and—as yet—to extract unifying findings with regard to IL-10 polymorphisms and gastric cancer development.

Post-Transplant Squamous Cell Skin Cancer

Malignancies arise in 20 to 40% of transplant recipients within 20 years of receipt of graft.[141] Skin carcinomas account for up to 50% of these cancers. Viruses, including HPVs, Epstein Barr virus and human herpes virus-8 are involved in the pathogenesis of post-transplant tumors, especially skin carcinomas, B-cell lymphomas and Kaposi's sarcoma. Impaired host immune defences, resulting from heavy immunosuppression are also associated with an increased risk of malignancy. Genetic risk factors are also suspected. Accordingly, genetic polymorphisms associated with differential levels of cytokine production may have an important effect on tumorigenesis after organ transplantation. In this context, IL-10 plymorphism is of particular interest, since, in addition to the biological functions of IL-10 discussed in some detail above, ultraviolet-induced DNA damage, a risk factor for skin cancer, also increases production of IL-10.[142]

For the above reasons, Alamartine et al,[84] in a French-based study, investigated possible associations between the IL-10 -1082, -819 and -592 SNPs and the occurrence of post-transplant skin cancers in a series of 70 renal transplant recipients who developed post-transplant squamous (SCC) or basal cell carcinoma (BCC), 70 healthy controls and 70 age, sex and immunosuppression-matched renal recipients without cancer. Taken together, IL-10 -1082, -819, -592 haplotypes were differently distributed when comparing cancer patients with unaffected patients and with controls. Results showed that genotypes associated with low production of IL-10 (IL-10 -1082, -819, -592 GCC negative) were less frequent in cancer patients (23% v 47% in unaffected patients), but only in patients with SCC (12%) and not BCC (37%). In addition, genotypes associated with high IL-10 production (GCC homozygous) were more frequent in cancer patients (24% v 11%), but this difference was significant only when comparing all cancer patients, or patients with SCC, with controls. The frequency of GCC homozygosity was not increased when considering patients with BCC with unaffected patients. The predicted correlation between IL-10 genotype and in vitro secretion of IL-10 by mononuclear cells was also confirmed by the same study.

Therefore results from this study suggest that genetically programmed elevated IL-10 production may favor the development of carcinoma—especially SCC—in renal transplant patients, although, as for most of the studies reviewed in this chapter, this result requires independent confirmation or replication.

Hematological Malignancies

Multiple myeloma (*MM*) is a monoclonal B cell neoplasm, in which immunoglobulin producing malignant plasma cells accumulate in the bone marrow. It is unclear whether genetic factors influence susceptibility to MM or clinical course of the disease. However, IL-10 has been implicated in the growth and differentiation of normal B cells,[143] has been shown to be a growth factor for MM cells[144] and elevated levels of IL-10 have been reported in patients with MM.[26] In addition, IL-10 has been implicated in the pathogenesis of other human B-cell malignancies.[25] Accordingly, Zheng et al[85] analysed the frequency of IL-10 G and R microsatellite alleles in a series of 73 Swedish Caucasian MM patients and 109 ethnically matched controls. Significantly increased frequencies of the IL-10 G 136/136 and IL-10 R 112/114 genotypes were seen in the MM patients, along with a decreased frequency of the IL-10 R 114/116 genotype. In addition, increased production of IL-10 was detected in the supernatants of lipopolysaccharide-stimulated peripheral blood mononuclear cells from MM patients carrying one or two IL-10 G 136 alleles, compared with other IL-10 genotypes. These results suggest that genetically determined elevated levels of IL-10 production may may a role in the development of MM.

The *Myelodysplastic syndromes* (*MDS*) are clonal hematopoietic stem cell disorders, with some subtypes rarely transforming to acute leukemia.[145] There is increasing evidence for an inflammatory component in the pathogenesis of MDS and an autoimmune mechanism is suggested by response to immunosuppression in some MDS patients.[146] Despite this, Gowans et al[86] failed to find any association between HLA class I and II, TNFα, LTα and IL-10 polymorphisms with either susceptibility to or disease progression in MDS or secondary acute myeloid leukemia (AML) in a series of 150 UK MDS/AML patients and up to 1000 controls, depending upon polymorphism investigated. Accordingly, this single study provides no evidence for a role for the IL-10 -1082, -819 and -592 SNPs in determining susceptibility to or disease progression in MDS and AML. Conversely, a study by Lauten et al[88] showed an association between the IL-10 -1082 GG genotype and a protective effect from poor predisone response in a series of 135 German *childhood acute lymphoblastic leukemia (ALL)* patients, where prednisone response has high predictive power of survival in childhood ALL.[147] These preliminary data suggest that genetically programmed high levels of IL-10 production may be associated with treatment outcome in childhood ALL, although large prospective studies will be needed to confirm this.

Finally, Cunningham et al,[87] in an Australian-based study, have reported significant associations between the IL-10 -1082 AA genotype and aggressive *nonHodgkin's lymphoma (NHL)* (OR = 1.97) and between the -1082, -819, -592 ATA and ACC haplotypes and aggressive lymphoma (OR = 1.65). No association was found between IL-10 genotypes and *Hodgkin's disease* or less aggressive forms of lymphoma. Results in NHL are in apparent conflict with reports that serum IL-10 level is an important prognostic indicator in this disease, with high levels of viral IL-10 correlating with poor prognosis.[28,148] However, the genetic results may be interpreted in several ways. The authors consider that a genetically determined 'relative lack' of IL-10 may allow lymphoma to arise or progress under the influence of other 'pro-lymphoma' cytokines, or that in patients with aggressive disease, IL-10 genotypes alone may not be an accurate predictor of of IL-10 production in vivo.

Summary

As a multifunctional cytokine with both immunosuppressive and anti-angiogenic functions, IL-10 has both tumor-promoting and tumor-inhibiting properties. In addition, when considering both serum and peritumoral levels of IL-10 production in individual malignancies, interpretation of apparently raised levels of IL-10 requires caution and should not be considered in isolation from source of production and levels of other biologically relevant cytokines. However, gene transfection studies in a number of malignancies argue more convincingly for an anti-tumor function of IL-10, possibly via inhibition of pathways of angiogenesis.

Much endeavor has been directed towards identification of polymorphisms in the IL-10 gene and a large number of polymorphisms—primarily SNPs—have been identified in the IL-10 gene promoter. Convincing evidence that certain of these polymorphisms—in particular the IL-10 -1082 SNP and associated IL-10 -1082, -819 and -592 haplotype—are associated with differential expression of IL-10 in vitro and in some cases in vivo have been obtained. While a large number of investigations of possible associations between IL-10 genotypes and immune mediated disease have been performed, the literature with regard to IL-10 polymorphisms and cancer is as yet small, but growing. These published studies include both solid tumors and hematological malignancies and common and less common diseases. As yet, most of these studies comprise small, single center case-control investigations, which may be prone to sampling bias and type 1 errors. In addition, in most of the 10 malignancies reviewed in this chapter, only from one to a maximum of three studies have been performed in each disease, in a range of human populations and ethnic groups. Despite this, it is perhaps striking that in 12 of the 15 studies reviewed in this chapter, positive associations between IL-10 genotype or haplotype and disease susceptibility and/or progression were detected—albeit with relatively modest Odds Ratios and probabilities. In some of these cancers (for example, cutaneous malignant melanoma, prostate cancer, breast cancer, non cardia gastric cancer and nonHodgkin's lymphoma) genotypes associated with low IL-10 expression were a risk factor for disease or disease progression, while in others (for example, cervical cancer and cardia gastric cancer, post-transplant squamous cell carcinoma of the skin and multiple myeloma), genotypes associated with high IL-10 expression were a risk factor.

At this stage, all of the above findings should be regarded as highly preliminary, due to the small sample sizes of almost all of the studies and the limited numbers of IL-10 polymorphisms examined. In addition, few of the studies have examined levels of IL-10 production in vivo in the subjects genotyped. However, the preliminary data obtained thus far indicate that much larger studies are required in a number of common and relatively common cancers, in order to confirm initial results, extend studies to include more detailed genotype/haplotype analysis and to combine genotype and gene expression studies in the same subjects. In this way, our understanding of the biological role of IL-10 in tumor development will be greatly aided, with implications for cytokine therapy in cancer.

References

1. Fiorentino DF, Bond MW, Mosmann TR. Two types of mouse helper T cell. IV. Th2 clones secrete a factor that inhibits cytokine production by Th1 clones. J Exp Med 1989; 170:2081-2095.
2. de Waal Malefyt R, Haanen J, Spits H et al. IL-10 and viral IL-10 strongly reduce antigen-specific human T cell proliferation by diminishing the antigen-presenting capacity of monocytes by downregulation of class II MHC expression. J Exp Med 1991; 174:915-924.
3. Fiorentino DF, Zlotnik A, Viera P et al. IL-10 acts on the antigen-presenting cell to inhibit cytokine production by Th1 cells. J Immunol 1991; 146:3444-3451.
4. Ding L, Shevach EM. IL-10 inhibits mitogen-induced T cell proliferation by selectively inhibiting macrophage costimulatory function. J Immunol 1992; 148:3133-3139.
5. Hsu D-H, Moore KW, Spits H. Differential effects of Interleukin-4 and -10 on Interleukin-2-induced interferon-γ synthesis and lymphokine-activated killer activity. Int Immunol 1992; 4:563-569.
6. de Waal Malefyt R, Abrams J, Bennett B et al. IL-10 inhibits cytokine synthesis by human monocytes: an autoregulatory role of IL-10 produced by monocytes. J Exp Med 1991; 174:1209-1220.
7. Fiorentino DF, Zlotnik A, Mossman TR et al. IL-10 inhibits cytokine production by activated macrophages. J Immunol 1991; 147:3815-3822.
8. Ding L, Linsley PS, Huang L-Y et al. IL-10 inhibits macrophage costimulatory activity by selectively inhibiting the up-regulation of B7 expression. J Immunol 1993; 151:1224-1234.
9. Gazzinelli RT, Oswald IP, James SL et al. IL-10 inhibits parasite killing and nitric oxide production by IFN-γ-activated macrophages. J Immunol 1992; 148:1792-1796.
10. Oswald IP, Gazzinelli RT, Sher A et al. IL-10 synergises with IL-4 and TGF-beta to inhibit macrophage cytotoxic activity. J Immunol 1992; 148:3578-3582.
11. Ralph P, Nakoinz I, Sampson-Johannes A et al. IL-10, T lymphocyte inhibitor of human blood cell production of IL-1 and tumor necrosis factor. J Immunol 1992; 148:808-814.
12. Moore KW, de Waal Malefyt R, Coffman RL et al. Interleukin-10 and the Interleukin-10 receptor. Ann Rev Immunol 2001; 19:683-765.
13. Ding YZ, Fu S, Zamarin D et al. Interleukin-10. In: Thomson AW, Lotze MT, eds. The Cytokine Handbook. 4th ed. London: Academic Press, 2003:603-625.
14. Zheng LM, Ojcius DM, Garaud F et al. Interleukin-10 inhibits tumor metastasis through an NK cell-dependent mechanism. J Exp Med 1996; 184:579-584.
15. Kundu N, Beaty TL, Jackson MJ et al. Antimetastatic and antitumor activities of IL-10 in a murine model of breast cancer. J Natl Cancer Inst 1996; 88:536-541.
16. Dummer W, Becker JC, Schwaaf A et al. Elevated serum levels of Interleukin-10 in patients with metastatic malignant melanoma. Melanoma Res 1995; 5:67-68.
17. Huang S, Ullrich SE, Bar-Eli M. Regulation of tumor growth and metastasis by Interleukin-10: The melanoma experience. J Interferon Cytokine Res 1999; 19:697-703.
18. Kruger-Krasagakes S, Krasagakis K, Garbe C et al. Expression of Interleukin-10 in human melanoma. Br J Cancer 1994; 70:1182-1185.
19. Sato T, McCue P, Masuoka K et al. Interleukin-10 production by human melanoma. Clin Cancer Res 1996; 2:1383-1390.
20. Gotlieb WH, Abrams JS, Watson JM et al. Presence of Interleukin-10 (IL-10) in the ascites of patients with ovarian and other intra-abdominal cancers. Cytokine 1992; 4:385-390.
21. Fortis C, Foppoli M, Gianotti L et al. Increased Interleukin-10 serum levels in patients with solid tumors. Cancer Lett 1996; 104:1-5.
22. De Vita F, Orditura M, Galizia G et al. Serum Interleukin-10 levels as a prognostic factor in advanced nonsmall cell lung cancer patients. Chest 2000; 117:365-373.
23. Fujieda S, Sunaga H, Tsuzuki H et al. IL-10 expression is associated with the expression of platelet-derived endothelial cell growth factor and prognosis in oral and oropharyngeal carcinoma. Cancer Lett 1999; 136:1-9.
24. Wittke F, Hoffmann R, Buer J et al. Interleukin-10 (IL-10): An immunosuppressive factor and independent predictor in patients with metastatic renal cell carcinoma. Br J Cancer 1999; 79:1182-1184.

25. Khatri VP, Caligiuri MA. A review of the association between Interleukin-10 and human B-cell malignancies. Cancer Immunol Immunother 1998; 46:239-244.
26. Klein B, Lu ZY, Gu ZJ et al. Interleukin-10 and Gp130 cytokines in human multiple myeloma. Leuk Lymphoma 1999; 34:63-70.
27. De Vita F, Orditura M, Galizia G et al. Serum Interleukin-10 levels in patients with advanced gastrointestinal malignancies. Cancer 1999; 86:1936-1943.
28. Blay JY, Burdin N, Rousset F et al. Serum Interleukin-10 in non-Hodgkin's lymphoma: A prognostic factor. Blood 1993; 82:2169-2174.
29. Bohlen H, Kessler M, Sextro M et al. Poor clinical outcome of patients with Hodgkin's disease and elevated Interleukin-10 serum levels. Clinical significance of Interleukin-10 serum levels for Hodgkin's disease. Ann Hematol 2000; 79:110-113.
30. Yue FY, Dummer R, Geertsen R et al. Interleukin-10 is a growth factor for human melanoma cells and down-regulates HLA class-I, HLA class-II and ICAM-1 molecules. Int J Cancer 1997; 71:630-637.
31. Cortes JE, Talpaz M, Cabanillas F et al. Serum levels of Interleukin-10 in patients with diffuse large cell lymphoma: Lack of correlation with prognosis. Blood 1995; 85:2516-2520.
32. Sjoberg J, Aguilar-Santelises M, Sjogren AM et al. Interleukin-10 mRNA expression in B-cell chronic lymphocytic leukaemia inversely correlates with progression of disease. Br J Haematol 1996; 92:393-400.
33. Adris S, Klein S, Jasnis M et al. IL-10 expression by CT26 colon carcinoma cells inhibits their malignant phenotype and induces a T cell-mediated tumor rejection in the context of a systemic Th2 response. Gene Ther 1999; 6:1705-1712.
34. Giovarelli M, Musiani P, Modesti A et al. Local release of IL-10 by transfected mouse mammary adenocarcinoma cells does not suppress but enhances antitumor reaction and elicits a strong cytotoxic lymphocyte and antibody-dependent immune memory. J Immunol 1995; 155:3112-3123.
35. Richter G, Kruger-Krasagakes S, Hein G et al. Interleukin-10 transfected into Chinese hamster ovary cells prevents tumor growth and macrophage infiltration. Cancer Res 1993; 53:4134-4137.
36. Gerard CM, Bruyns C, Delvaux A et al. Loss of tumorigenicity and increased immunogenicity induced by interleukin-10 gene transfer in B16 melanoma cells. Hum Gene Ther 1996; 7:23-31.
37. Cervenak L, Morbidelli L, Donati D et al. Abolished angiogenicity and tumorigenicity of Burkitt lymphoma by Interleukin-10. Blood 2000; 96:2568-1273.
38. Mucke S, Draube A, Polack A et al. Suppression of the tumorigenic growth of Burkitt's lymphoma cells in immunodeficient mice by cytokine gene transfer using EBV-derived episomal expression vectors. Int J Cancer 2000; 86:s301-306.
39. Stearns ME, Wang M. Antimetastatic and antitumor activities of Interleukin 10 in transfected human prostate PC-3 ML clones: Orthotopic growth in severe combined immunodeficiency mice. Clin Cancer Res 1998; 4:2257-2263.
40. Berman RM, Suzuki T, Tahara H et al. Systemic administration of cellular IL-10 induces an effective, specific, and long-lived immune response against established tumors in mice. J Immunol 1996; 157:231-238.
41. Di Carlo E, Coletti A, Modesti A et al. Local release on Interleukin-10 by transfected mouse adenocarcinoma cells exhibit pro- and anti-inflammatory activity and results in a delayed tumor rejection. Eur Cytokine Netw 1998; 9:61-68.
42. Kundu D, Dorsey R, Jackson MJ et al. Interleukin-10 gene transfer inhibits murine mammary tumors and elevates nitric oxide. Int J Cancer 1998; 76:713-719.
43. Petersson M, Charo J, Salazar-Onfray F et al. Constitutive IL-10 production accounts for the high NK sensitivity, low MHC class I expression and poor transporter associated with antigen processing (TAP)-1/2 function in the prototype NK target YAK-1. J Immunol 1998; 161:2099-2105.
44. Salazar-Onfray F, Petersson M, Franksson L et al. IL-10 converts mouse lymphoma cells to a CTL-resistant, NK-sensitive phenotype with low but peptide-inducible MHC class I expression. J Immunol 1995; 154:6291-6298.
45. Stearns ME, Fudge K, Garcia F et al. IL-10 inhibition of human prostate PC-3 ML cell metastases in SCID mice: IL-10 stimulation of TIMP-1 and inhibition of MMP-2/MMP-9 expression. Invasion Metastasis 1997; 17:62-74.
46. Stearns ME, Rhim J, Wang M. Interleukin 10 (IL-10) inhibition of primary human prostate cell-induced angiogenesis: IL-10 stimulation of tissue inhibitor of metalloproteinase-1 and inhibition of matrix metalloproteinase (MMP)2/MMP-9 secretion. Clin Cancer Res 1999; 5:189-196.
47. Kaufman HL, Rao JB, Irvine KR. Interleukin-10 enhances the therapeutic effectiveness of a recombinant poxvirus-based vaccine in an experimental murine tumor model. J Immunother 1999; 22:489-496.

48. Bidwell J, Keen L, Gallagher G et al. Cytokine gene polymorphism in human disease: On-line databases. Genes and Immun 1999; 1:3-19.
49. Haukim N, Bidwell JL, Smith AJP et al. Cytokine gene polymorphism in human disease: On-line databases, Supplement 2. Genes and Immun 2002; 3:313-330.
50. Eskdale J, Kube D, Tesch H et al. Mapping of the human IL-10 gene and further characterisation of the 5' flanking sequence. Immunogenetics 1997; 46:120-128.
51. Kube D, Rieth H, Eskdale J et al. Structural characterisation of the distal 5' flanking region of the human Interleukin-10 gene. Genes and Immun 2001; 2:181-190.
52. Gibson AW, Edberg JC, Wu J et al. Novel single nucleotide polymorphisms in the distal IL-10 promoter affect IL-10 production and enhance the risk of systemic lupus erythematosus. J Immunol 2001; 166:3915-3922.
53. D'Alfonso S, Rampi M, Rolando V et al. New polymorphisms in the IL-10 promoter region. Genes and Immun 2000; 1:231-233.
54. Tountas NA, Cominelli F. Identification and initial characterisation of two polymorphisms in the human Interleukin-10 promoter. Eur Cytokine Netw 1996; 7:578.
55. Turner DM, Williams DM, Sankaran D et al. An investigation of polymorphism in the Interleukin-10 gene promoter. Eur J Immunogenetics 1997; 24:18.
56. Eskdale J, Kube D, Gallagher G. A second dinucleotide polymorphic repeat in the 5' flanking region of the human IL10 gene. Immunogenetics 1996; 45:82-83.
57. Eskdale J, Gallagher G. A polymorphic dinucleotide repeat in the human IL-10 promoter. Immunogenetics 1995; 52:444-445.
58. Donger C, Georges J-L, Nicaud V et al. New polymorphisms in the Interleukin-10 gene - relationships to myocardial infarction. Eur J Clin Invest 2001; 31:9-14.
59. Lazarus R, Klimecki WT, Palmer LJ et al. Single nucleotide polymorphisms in the Interleukin-10 gene: Differences in frequencies, linkage disequilibrium patterns and haplotypes in three United States ethnic groups. Genomics 2002; 80:223-228.
60. Reuss E, Fimmers R, Kruger A et al. Differential regulation of Interleukin-10 production by genetic and environmental factors: A twin study Genes and Immun. 2002; 3:407-413.
61. Eskdale J, Gallagher G, Verweij CL et al. Interleukin 10 secretion in relation to human IL-10 locus haplotypes. Proc Natl Acad Sci USA 1998; 95:9465-9470.
62. Westendorp RG, Langermans JA, Huizinga TW et al. Genetic influence on cytokone production and fatal meningococcal disease. Lancet 1997; 349:170-173.
63. Borish L, Aarons A, Rumbyrt J et al. Interleukin-10 regulation in normal subjects and patients with asthma. J Allergy Clin Immunol 1996; 97:1288-1296.
64. Lim S, Crawley E, Woo P et al. Haplotype associated with low Interleukin 10 production in patients with severe asthma. Lancet 1998; 352:113.
65. Kaluza W, Leirisalo-Repo M, Marker-Hermann E et al. IL-10.G microsatellites mark promoter Haplotypes associated with protection against the development of reactive arthritis in Finnish patients. Arthritis Rheum 2001; 44:1209-1214.
66. Poole KL, Gibbs PJ, Evans PR et al. Influence of patient and donor cytokine genotypes on renal allograft rejection: Evidence from a single centre study. Transplant Immunol 2001; 8:259-265.
67. Pelletier R, Pravica V, Perrey C et al. Evidence for a genetic predisposition towards acute rejection after kidney and simultaneous kidney-pancreas transplantation. Transplantation 2000; 70:674-680.
68. Awad MR, Webber S, Boyle G et al. The effect of cytokine gene polymorphisms on pediatric heart allograft outcome. J Heart Lung Transplant 2001; 20:625-630.
69. Middleton PG, Taylor PR, Jackson G et al. Cytokine gene polymorphisms associating with severe acute graft-versus-host disease in HLA-identical siblings. Blood 1998; 92:3943-3948.
70. He B, Xu C, Yang B et al. Linkage and association analysis of genes encoding cytokines and myelin proteins in multiple sclerosis. J Neuroimmunol 1998; 86:13-19.
71. Maurer M, Kruse N, Giess R et al. Genetic variation at position -1082 of the Interleukin 10 (IL10) promotor and the outcome of multiple sclerosis. J Neuroimmunol 2000; 104:98-100.
72. Eskdale J, McNicholl J, Wordsworth P et al. Interleukin-10 microsatellite polymorphisms and IL-10 locus alleles in rheumatoid arthritis susceptibility. Lancet 1998; 352:1282-1283.
73. Coakley G, Mok CC, Hajeer AH et al. Interleukin-10 promoter polymorphisms in rheumatoid arthritis and Felty's syndrome. Br J Rheumatol 1998; 37:988-991.
74. Howell WM, Turner SJ, Bateman AC et al. IL-10 promoter polymorphisms influence tumor development in cutaneous malignant melanoma. Genes and Immun 2001; 2:25-31.
75. Martinez-Escribano JA, Moya-Quiles MR, Muro M et al. Interleukin-10, Interleukin-6 and interferon-γ gene polymorphisms in melanoma patients. Melanoma Res 2002; 12:465-469.
76. McCarron SL, Edwards S, Evans PR et al. Influence of cytokine gene polymorphisms on the development of prostate cancer. Cancer Res 2002; 62:3369-3372.

77. Howell WM, Smith KC, Fussell HM et al. Influence of cytokine gene polymorphisms on susceptibility to and prognosis in breast cancer. Genes and Immun 2003; 4(Suppl. 1):S35.
78. Giordani L, Bruzzi P, Lasalandra C et al. Association of breast cancer and polymorphisms of interleukin-10 and tumor necrosis factor-alpha genes. Clin chem 2003; 49:1664-1667.
79. Stanczuk GA, Sibanda EN, Perrey C et al. Cancer of the uterine cervix may be significantly associated with a gene polymorphism coding for increased IL-10 production. Int J Cancer 2001; 94:792-794.
80. Roh JW, Kim MH, Seo SS et al. Interleukin-10 promoter polymorphisms and cervical cancer risk in Korean women. Cancer Lett 2002; 184:57-63.
81. Wu M-S, Wu C-Y, Chen C-J et al. Interleukin-10 genotypes associate with the risk of gastric carcinoma in Taiwanese Chinese. Int J Cancer 2003; 104:617-623.
82. Wu M-S, Huang S-P, Chang Y-T et al. Tumor necrosis factor-α and Interleukin-10 promoter polymorphisms in Epstein-Barr virus-associated gastric carcinoma. J Infect Dis 2002; 185:106-109.
83. El-Omar EM, Rabkin CS, Gammon MD et al. Increased risk of noncardia gastric cancer associated with proinflammatory cytokine gene polymorphisms. Gastroenterology 2003; 124:1193-1201.
84. Alamartine E, Berthoux E, Mariat C et al. Interleukin-10 promoter polymorphisms and susceptibility to skin squamous cell carcinoma after renal transplantation. J Invest Dermatol 2003; 120:99-103.
85. Zheng C, Huang D, Liu L et al. Interleukin-10 gene promoter polymorphisms in multiple myeloma. Int J Cancer 2001; 95:184-188.
86. Gowans D, O'Sullivan A, Rollison S et al. Allele and Haplotype frequency at human leucocyte antigen class I/II and immunomodulatory cytokine loci in patients with myelodysplasia and acute myeloid leukaemia: In search of an autoimmune aetiology. Br J Haematol 2002; 117:541-545.
87. Cunningham LM, Chapman C, Dunstan R et al. Polymorphisms in the Interleukin 10 gene promoter are associated with susceptibility to aggressive non-Hodgkin's lymphoma. Leukemia & Lymphoma 2003; 44:251-255.
88. Lauten M, Matthias T, Stanulla M et al. Association of initial response to prednisone treatment in childhood acute leukaemia and polymorphism within the tumor necrosis factor and the Interleukin-10 gene. Leukemia 2002; 16:1437-1442.
89. Folkman J. Angiogenesis in cancer, vascular, rheumatoid and other diseases. Nat Med 1995; 1:27-31.
90. Srivastava A, Laidler P, Hughes LE et al. Neovascularisation in human cutaneous melanoma: A quantitative morphological and Doppler ultrasound study. Eur J Cancer Clin Oncol 1986; 22:1205-1209.
91. Srivastava A, Laidler P, Davies R et al. The prognostic significance of tumor vascularity in intermediate-thickness (0.76-4.0 mm thick) skin melanoma. Am J Pathol 1988; 133:419-423.
92. Herlyn M, Clark WH, Rodeck U et al. Biology of tumor progression in human melanocytes. Lab Invest 1987; 56:461-474.
93. Weidner N, Carroll PR, Flax J et al. Tumor angiogenesis correlates with metastasis in invasive prostate carcinoma. Am J Pathol 1993; 143:401-409.
94. Weidner N, Semple JP, Welch WR et al. Tumor angiogenesis and metastasis - correlation in invasive breast carcinoma. N Eng J Med 1991; 324:1-8.
95. Bosari S, Lee AKC, DeLellis RA et al. Microvessel quantitation and prognosis in invasive breast cancer. Hum Pathol 1992; 23:755-761.
96. Horak E, Leek R, Klenk N et al. Angiogenesis, assessed by platelet/endothelial cell adhesion molecule antibodies, as indicator of node metastases and survival in breast cancer. Lancet 1992; 340:1120-1124.
97. Cress RD, Holly EA. Incidence of cutaneous melanoma among nonHispanic whites, Hispanics, Asians and blacks: An analysis of California cancer registry data, 1988-93. Cancer Causes Control 1997; 8:246-252.
98. Valverde P, Healy E, Sikkink S et al. The Asp84glu variant of the melanocortin-1 receptor is associated with melanoma. Hum Mol Genet 1996; 5:1663-1666.
99. Kennedy C, ter Huurne J, Berkhout J et al. Melanocortin 1 receptor (MCR1) gene variants are associated with an increased risk for cutaneous melanoma which is largely independent of skin type and hair color. J Invest Dermatol 2001; 117:294-300.
100. Kumar R, Smeds J, Berggren P et al. A single nucleotide polymorphism in the 3' untranslated region of the CDKN2A gene is common in sporadic primary melanomas, but mutations in the CDK2NB, CDK2NC, CDK4 and P53 genes are rare. Int J Cancer 2001; 95:388-393.
101. Winsey SL, Haldar NA, Marsh HP et al. A variant within the DNA repair gene XRCC3 is associated with the development of melanoma skin cancer. Cancer Res 2000; 60:5612-5616.
102. Lafuente A, Molina R, Palou J et al. Phenotype of glutathione S-transferase Mu (GSTM1) and susceptibility to malignant melanoma. Br J Cancer 1995; 72:324-326.

103. Hutchinson PE, Osborne JE, Lear JT et al. Vitamin D receptor polymorphisms are associated with altered prognosis in patients with malignant melanoma. Clin Cancer Res 2000; 6:498-504.
104. Strange RC, Ellison T, Ichii-Jones F et al. Cytochrome P450 CYP2D6 genotypes: Association with hair colour, Breslow thickness and melanocyte stimulating hormone receptor alleles in patients with malignant melanoma. Pharmacogenetics 1999; 9:269-276.
105. Wolfel T, Hauer M, Klehmann E et al. Analysis of antigens recognised on human melanoma cells by A2-restricted cytolytic T lymphocytes (CTL). Int J Cancer 1993; 55:237-244.
106. Lee JE, Reveille JD, Ross MI et al. HLA-DQB1*0301 association with increased cutaneous melanoma risk. Int J Cancer 1994; 59:510-513.
107. Bateman AC, Turner SJ, Theaker JM et al. HLA-DQB1*0303 and *0301 alleles influence susceptibility to and prognosis in cutaneous malignant melanoma in the British caucasian population. Tissue Antigens 1998; 52:67-73.
108. Mackie R, Hunter JA, Aitchison TC et al. Cutaneous malignant melanoma, Scotland, 1979-1983. The Scottish Melanoma Group. Lancet 1992; 339:971-975.
109. Howell WM, Bateman AC, Turner SJ et al. Influence of vascular endothelial growth factor single nucleotide polymorphisms on tumor development in cutaneous malignant melanoma. Genes and Immun 2002; 3:229-32.
110. Hegarty NJ, Fitzpatrick JM, Richie JP et al. Future prospects in prostate cancer. The Prostate 1999; 40:261-268.
111. Singh R, Eeles RA, Durocher F et al. High risk genes predisposing to prostate cancer development - do they exist? Prostate Cancer Prostatic Dis 2000; 3:241-247.
112. Smith JR, Freije D, Carpten JD et al. Major susceptibility locus for prostate cancer on chromosome 1 suggested by a genome-wide search. Science 1996; 274:1371-1374.
113. Xu J, Meyers D, Freije D et al. Evidence for a prostate cancer susceptibility locus on the X chromosome. Nat Genet 1998; 20:175-179.
114. Ford D, Easton DF, Bishop DT et al. Risk of cancer in BRCA1 mutation carriers. Lancet1994; 343:692-695.
115. Gibbs M, Stanford JL, McIndoe RA et al. Evidence for a rare prostate cancer-susceptibility locus at chromosome 1p36. Am J Hum Genet 1999; 64:776-787.
116. Berthon P, Valeri A, Cohen-Akenine A et al. Predisposing gene for early-onset prostate cancer, localized on chromosome 1q42.2-43. Am J Hum Genet 1998; 62:1416-1424.
117. Tavtigian SV, Simard J, Teng DH et al. A strong candidate prostate cancer susceptibility gene at chromosome 17p. Nat Genet 2001; 27:172-180.
118. Hrouda D, Perry M, Dalgleish AG. Gene therapy for prostate cancer. Semin Oncol 1999; 26:455-471.
119. Miki Y, Swensen J, Shattuck-Eidens D et al. A strong candidate for the breast and ovarian cancer susceptibility gene BRCA1. Science 1994; 266:66-71.
120. Wooster R, Bignell G, Lancaster J et al. Identification of the breast cancer susceptibility gene BRCA2. Nature 1995; 378:789-792.
121. Blackwood MA, Weber BL. BRCA1 and BRCA2: From molecular genetics to clinical medicine. Journal of Clinical Oncology 1998; 16:1969-1977.
122. Nathanson K, Weber BL. "Other" breast cancer susceptibility genes: Searching for more holy grail. Hum Mol Genet 2001; 10:715-720.
123. Wooster R, Weber BL. Breast and ovarian cancer. New Eng J Med 2003; 348:2339-2347.
124. Pharoah PD, Antoniou A, Bobrow M et al. Polygenic susceptibility to breast cancer and implications for prevention. Nat Genet 2002; 31:33-36.
125. Charrier J, Maugard CM, LeMevel B et al. Allelotype influence at glutathione S-transferase M1 locus on breast cancer susceptibility. Br J Cancer 1999; 79:346-353.
126. Bergman-Jungestrom M, Gentile M, Lundin A et al. The South East Breast Cancer Group, Wingren S. Association between CYP17 gene polymorphism and risk of breast cancer in young women. Int J Cancer 1999; 84:350-353.
127. Dunning AM, Healey CS, Pharaoh PDP et al. A systematic review of genetic polymorphisms and breast cancer risk. Cancer Epidemiol Biomarkers Prev 1999; 8:843-854.
128. Venetsanakos E, Beckman I, Bradley J et al. High incidence of Interleukin-10 mRNA but not Interleukin-2 mRNA detected in human tumors. Br J Cancer 1997; 75:1826-1830.
129. Wank R, Thomssen C. High risk of squamous cell carcinoma of the cervix for women with HLA-DQw3. Nature 1991; 352:723-725.
130. Mehal WZ, Lo Y-MD, Herrington CS et al. Role of human papillomavirus in determining the HLA associated risk of cervical carcinogenesis. J Clin Pathol 1994; 47:1077-1081.
131. Apple RJ, Erlich HA, Klitz W et al. HLA DR-DQ associations with cervical carcinoma show papillomavirus-type specificity. Nat Genet 1994; 6:157-163.

132. Nawa A, Nishiyama Y, Kobayashi T et al. Association of human leukocyte antigen-B1*03 with cervical cancer in Japanese women aged 35 years and younger. Cancer 1995; 75:518-521.
133. Jacobs N, Giannini SL, Doyen A et al. Inverse modulation of IL-10 and IL-12 in the blood of women with preneoplastic lesions of the uterine cervix. Clin Exp Immunol 1998; 111:219-224.
134. Giannini SL, Al-Saleh H, Piron N et al. Cytokine expression in squamous intraepithelial lesions of the uterine cervix: Implications for the generation of local immunosuppression. Clin Exp Immunol 1998; 113:183-189.
135. De Gruijl TD, Bontkes HJ, van den Muysenberg AJC et al. Differences in cytokine mRNA profiles between premalignant and malignant lesions of the uterine cervix. Eur J Cancer 1999; 35:490-497.
136. Van Trappen PO, Ryan A, Carroll M et al. A model for coexpression pattern analysis of genes implicated in angiogenesis and tumor cell invasion in cervical cancer. Br J Cancer 2002; 87:537-544.
137. Parkin DM, Pisani P, Ferlay J. Estimates of the worldwide incidence of 25 major cancers in 1990. Int J Cancer 1999; 80:827-841.
138. El-Omar EM, Carrington M, Chow WH et al. Interleukin-1 polymorphisms associated with increased risk of gastric cancer. Nature 2000; 404:398-402.
139. Furuta T, El-Omar EM, Xiao F et al. Interleukin 1β polymorphisms increase risk of hypochlorhydria and atrophic gastritis and reduce risk of duodenal ulcer recurrence in Japan. Gastroenterology 2002; 123:92-105.
140. Yu HG, Li JY, Yang YN. Increased abundance of cyclooxygenase-2 correlates with vascular endothelial growth factor-A abundance and tumor angiogenesis in gastric cancer. Cancer Lett 2003; 195:43-51.
141. London NJ, Farmery SM, Will EJ et al. Risk of neoplasia in renal transplant patients. Lancet 1995; 346:403-406.
142. Nishigori C, Yarosh DB, Ullrich SE et al. Evidence that DNA damage triggers Interleukin-10 cytokine production in UV-irradiated murine keratinocytes. Proc Natl Acad Sci USA 1996; 93:10354-10359.
143. Emilie D. Production and roles of IL-6, IL-10, and IL-13 in B-lymphocyte, malignancies and in B-lymphocyte hyperactivity of HIV infection and autoimmunity. Methods 1997; 11:133-142.
144. Lu ZY, Gu ZJ, Xhang XG et al. Interleukin-10 induces Interleukin-11 responsiveness in human myeloma cell lines. FEBS Lett 1995; 377:515-518.
145. Germing U, Gattermann N, Aivado M et al. Two types of acquired idiopathic sideroblastic anaemia (AISA): A time tested distinction. Br J Haematol 2000; 108:724-728.
146. Jonasova A, Neuwirtova R, Cermak J et al. Cyclosporin A therapy in hypoplastic MDS patients and certain refractory anaemias without hypoplastic bone marrow. Br J Haematol 1998; 100:304-309.
147. Dordelmann M, Reiter A, Borkhardt A et al. Prednisone response is the strongest predictor of treatment outcome in infant acute lymphoblastic leukaemia. Blood 1999; 94:1209-1217.
148. Cortes J, Kurzrock R. Interleukin-10 in non-Hodgkin's lymphoma. Leukemia & Lymphoma 1997; 26:251-259.

CHAPTER 3C

IL-10 Gene Polymorphisms in Transplantation

Andrey Morgun, Natalia Shulzhenko and Maria Gerbase-DeLima

Abstract

Transplantation is currently the treatment of choice for end-stage insufficiency of various organs and severe hematologic disorders. Cytokine gene polymorphisms have been extensively explored in transplantation because they are thought to possibly explain the heterogeneity of courses and outcomes and to help in immunosuppression individualization. Although many polymorphisms have been described in the IL-10 gene, the -1082 single nucleotide polymorphism is the most investigated in transplantation. Furthermore, acute rejection is the endpoint most frequently used in cytokine gene polymorphisms studies in organ transplantation. It appears that factors such as effect of HLA and minor histocompatibility mismatch, type of donor, and type of immunosuppression are important to determine the influence of IL-10 gene polymorphisms on transplant outcome. Although there are several discrepancies between the results of studies of IL-10 gene polymorphisms and transplant outcome, it seems that recipient's IL-10 polymorphisms may indicate the severity of acute graft versus host disease in bone marrow transplantation, frequency of acute rejection in organ transplantation and even could be associated with long-term renal graft outcome. In conclusion, data collected up to now demonstrate substantial evidence of IL-10 gene polymorphism involvement in post-transplant outcome. Multi-centric prospective studies using multi-locus approaches might ascertain when and how IL-10 genotype information should be considered in clinical transplant practice.

Introduction

Transplantation and Cytokine Polymorphisms

Transplantation is currently the treatment of choice for end-stage insufficiency of various organs and severe hematologic disorders. Despite high short-term survival rates for most of the transplants, long-term results still need to be improved. Among different possibilities for improvement of transplant outcome, the individualization of immunosuppression therapy and tolerance induction seem to be the most promising ones.

Regulated by different cytokines, the post-transplant immune response is a very important variable influencing graft outcome. Thus, any variations in cytokine levels may directly influence the course of transplant. Therefore, cytokine gene polymorphisms have been extensively explored in transplantation because they are thought to possibly explain the heterogeneity of courses and outcomes and to help in immunosuppression adjustment.

Interleukin-10, edited by Francesco M. Marincola. ©2006 Eurekah.com.

Why Study Gene Polymorphisms of IL-10?

In studies with twins, it has been suggested that 50-75% of the variation in IL-10 production is genetically determined.[1,2] Although the basis for heritable differences has not been completely defined, the polymorphisms located within and around the IL-10 gene have been implicated. Indeed, variations in the distal and proximal IL-10 gene promoter regions were shown to be associated with transcriptional activity and/or production of IL-10.[3-8]

Which IL-10 Polymorphisms Are Usually Analyzed in Transplantation?

Although many single nucleotide polymorphisms (SNPs) and microsatellites have been described in the IL-10 gene, the SNP at -1082 position is the polymorphism most frequently investigated in transplantation. SNPs at the -819, -592 positions and microsatellites at the -1064 position are also popular, while SNPs in the distal 5' promoter region and other microsatellites are rarely analyzed.

Which Follow-Up Parameters Are Usually Analyzed?

Several parameters are used to evaluate the transplant follow-up. They include incidence and severity of acute and chronic rejection, graft and patient survival, occurrence of infection episodes, and cancer. There are also particular parameters for a given type of transplantation, like Graft Versus Host Disease (GVHD) in hematopoietic stem cell transplantation, or functional characteristics specific for each organ. Acute rejection is the endpoint most frequently used in cytokine gene polymorphisms studies in organ transplantation.

IL-10 Polymorphisms in Different Transplant Models

Allogeneic Stem Cell Transplantation

Allogeneic hematopoietic stem cell transplantation, which includes the use of bone marrow, cord blood, or peripheral blood stem cells, can cure many patients with a variety of hematologic malignancies and nonmalignant bone marrow failure syndromes. The major complication of this type of transplantation is acute GVHD (aGVHD) which is caused by the immune response of the injected allograft against the recipient's tissues and is characterized by the presence of dermatitis, hepatitis, and gastroenteritis developing within the first 100 days after transplantation. Lesions in the same organs, occurring more than 100 days after transplantation, are called chronic GVHD.[9] The level of IL-10 was shown to be associated with the development of aGVHD and even demonstrated a predictive value. Indeed, two groups independently observed an association between increased pretransplant production of IL-10 and low incidence of aGVHD.[10,11] Furthermore, the therapeutic use of this cytokine has been suggested as an additional or alternative prophylaxis for GVHD.[12]

Promising results were obtained in the first study on IL-10 polymorphism in bone marrow transplantation with HLA-identical donors. The authors analyzed the -1064 dinucleotide CA repeat of the IL-10 gene. An increased frequency of alleles with 12 to 16 repeats was found in the group of recipients with more severe aGVHD.[13] This finding was confirmed in two other independent groups of patients by the same investigators (Table 1).[14,15] In contrast, no similar association was observed in Japanese patients who received transplants from related or unrelated donors.[16]

In a more recent study of transplants involving sibling and unrelated donors, Nordlander and collaborators[17] reported that patients homozygous for the IL-10 -1064 allele 13 showed an increased risk of developing grade II-IV aGVHD compared to all other patients. Even though these investigators did not find exactly the same phenomenon as demonstrated previously, their results agreed with the initial studies. In contrast to HLA-identical sibling bone marrow transplantation, in mismatched unrelated cord blood transplantation IL-10 -1064 alleles were not associated with aGVHD (Table 1).[18]

Another widely explored IL-10 polymorphism is -1082 G/A. The data regarding this SNP are controversial. Although three studies did not show any relation of this SNP to aGVHD,[14,17,19]

Table 1. *IL-10 gene polymorphisms in relation to acute graft versus host disease (aGVHD)*

References	Type of Transplantation	No. of Patients	Alleles/Genotypes Analyzed	Allelic/Genotypic Frequencies		p
				Mild aGVHD*	Severe aGVHD*	
Middleton et al, 1998[13]	BMT, HLA-identical siblings	49	-1064 (12-16 repeats)	47% (18/38)	82% (9/11)	0.02
Cavet et al, 1999[14]	BMT, HLA-identical siblings	127	-1064 (12-16 repeats)	55% (59/108)	89% (17/19)	0.005
Cavet et al, 2001[15]	BMT, HLA-identical siblings	80	-1064 (12-16 repeats)	48% (31/64)	81% (13/16)	0.017
Takahashi et al, 2000[16]	BMT, HLA identical & non-identical	62	-1064 (13-17 repeats)	27% (14/51)	17% (1/6)	0.99
Nordlander et al, 2002[17]	BMT & PBT, HLA-identical siblings & non-siblings	196	-1064 (13 repeats, homozygous)	1.4% (2/140)	7% (4/56)	0.02
Kogler et al, 2002[18]	CBT, HLA mismachted	115	-1064 (11-16 repeats)	63% (49/78)	64% (23/36)	0.83
Cavet et al, 1999[14]	BMT, HLA-identical siblings	127	-1082 G/G	27% (28/105)	42% (8/19)	0.18
Nordlander et al, 2002[17]	BMT & PBT, HLA-identical siblings	85	-1082 G/G	No data reported	No data reported	NS
	BMT & PBT, HLA-identical non-siblings	111	-1082 G/G	18% (13/74)	22% (8/37)	0.6
Tambur et al, 2001[19]	Pediatric BMT, HLA-identical siblings	24	-1082 G/G	42% (5/12)	8% (1/12)	0.16
Socie et al, 2001[20]	BMT & PBT, HLA-identical siblings	100	-1082 G/G	absence of aGVHD 15% (9/59)	aGVHD of any grade 30% (12/40)	0.04
Rocha et al, 2002[21]	BMT, HLA-identical	107	-1082 G/G	16% (10/62)	29% (13/45)	0.08

*Mild aGVHD, grades 0-II; severe aGVHD, grades III-IV; Abbreviations: BMT: bone marrow transplantation; PBT: peripheral blood transplantation; CBT: cord blood transplantation a group from France reported association between G/G genotype and severity of aGVHD (Table 1).[20,21] There is also no agreement between the results of three studies that analyzed chronic GVHD in relation to IL-10 polymorphisms.[14,16,21]

Briefly, studies conducted on bone marrow transplantation have shown that factors such as effect of HLA mismatch, type of donor (siblings or unrelated subjects), type of GVHD prophylaxis, and minor histocompatibility are important to determine the influence of cytokine gene polymorphisms on transplant outcome. It seems that recipient's IL-10 -1064 alleles may indicate the severity of aGVHD in HLA-identical sibling bone marrow transplantation, while there is no consensus about other IL-10 polymorphisms.

Kidney

Renal transplantation is a widely used therapy for patients with end-stage renal insufficiency. An important cause of morbidity early after renal transplantation is acute allograft rejection, which is induced by the immune response of the recipient against the transplanted organ and leads to allograft deterioration and functional insufficiency.

Increased IL-10 intragraft expression was shown during acute rejection.[22,23] It is interesting to note that, although the strong candidates to produce IL-10 in the graft are cells from the mononuclear infiltrate, it is possible that graft cells could also contribute to the increased IL-10 levels during rejection. This raises the idea of the analysis not only of recipient's but also of donor's genotypes, as well as a combination of both. In contrast to bone marrow transplantation, the most frequently studied IL-10 polymorphism in renal transplantation is -1082 SNP. Table 2 summarizes the results of IL-10 polymorphisms genetic association studies in renal transplantation.

The IL-10 -1082 G/G genotype (corresponding to the high producer phenotype in vitro) was associated with an increase in the incidence of multiple acute rejection episodes in recipients with HLA-DR-mismatched transplants.[24] In the cited study, the same genotype was associated with the severity of acute rejection in combination with TNF -308 high production genotype. Similar results were obtained by a Chinese group that observed a higher incidence of rejection in recipients with high/intermediate IL-10 producer genotypes.[25] In agreement, Pelletier and colleagues[26] reported that a combination of TNF low and IL-10 high producer genotypes was an indicator of increased frequency of acute rejection. On the other hand, in an African American population the -1082 A/A genotype (low producer) alone was associated with the absence of acute rejection.[27]

In contrast, in another study, the -1082 A/A genotype was more frequent in the group of rejectors than in the group free of rejection.[28] A similar finding was also reported by Poole and colleagues,[29] who observed a tendency to a higher frequency of rejection in recipients with -1082 A/A and A/G genotypes. Other studies did not show any association between -1082 SNP and acute rejection.[30-33]

The only study of -1064 microsatellite in renal transplantation with living related haplo-identical donors demonstrated a marked difference in the frequencies of the allele with 12 repeats between rejectors and nonrejectors, with an increase in this difference observed in a subgroup of steroid-resistant rejections.[34]

Some studies have explored the effect of donor (graft) polymorphisms on transplant outcome. Neither -1082 SNP nor -1064 repeat polymorphism was shown to be associated with acute renal rejection in three studies.[24,34,35] However, in the study of Poole and colleagues,[29] -1082 A donor allele frequency was decreased in multiple rejectors with HLA-DR mismatch. In addition, the combination of recipient -1082A negative/donor -1082A positive (recipient high producer/donor low producer) was also significantly decreased in multiple rejectors but independently of HLA match.[29]

Other parameters whose evaluation could be even more important than incidence of acute rejection in renal transplants are long-term graft function and patient and graft survival. Unfortunately, up to now only a few studies have explored these endpoints. Three studies that analyzed 3 and/or 5 years graft survival did not observe any association with IL-10

Table 2. IL-10 gene polymorphisms in renal transplantation

References	No. of Patients	Endpoints Evaluated	Findings	Notes
Sankaran et al, 1999[24]	88	Multiple acute rejection episodes; graft survival	Recipient -1082 G/G associated with multiple rejections ($p=0.02$); Recipient -1082 G/G and TNF -308A associated to steroid-resistant rejections	Only in HLA-DR mismatched transplants
Tian et al, 2002[25]	115	Acute rejection	Recipient -1082 G/G and G/A associated with higher incidence of acute rejection ($p<0.01$)	
Pelletier et al, 2000[26]	100	Acute rejection; serum creatinine	Association of recipient -1082 G/G with increased incidence of acute rejection ($p=0.02$); No association with serum creatinine	Only in recipients with TNF -308 G/G
Hutchings et al, 2002[27]	105	Acute rejection	Association of recipient -1082 A/A with decreased incidence of acute rejection ($p=0.006$)	Only in African-American patients
George et al, 2001[28]	105	Acute rejection	Association of recipient -1082 A/A with increased incidence acute rejection ($p=0.01$)	
Poole et al, 2001[29]	120	Acute rejection incidence; multiple rejection episodes	Recipient -1082 A/A and A/G associated (trend) with multiple rejection episodes ($p=0.07$) Donor -1082 A frequency decreased in multiple rejectors group ($p=0.06$) Recipient -1082 A negative/donor -1082 A positive decreased in multiple rejectors group ($p=0.04$)	Only in HLA-DR mismatched transplants
Marshall et al, 2000[30]	209	Acute rejection	No association with recipient genotype, irrespective of HLA-DR mismatch	Two independent sets of patients studied

Table continued on next page

Table 2. Continued

References	No. of Patients	Endpoints Evaluated	Findings	Notes
Hahn et al, 2001[31]	120	Acute rejection	No association with recipient genotype, irrespective of HLA-DR mismatch	
Asderakis et al, 2001[32]	88	Acute rejection; graft survival	No association of recipient genotype either with incidence or severity of acute rejection or with graft survival	
Poli et al, 2001[33]	169	Acute rejection, serum creatinine, long-term survival	No association of recipient genotype with acute rejection or serum creatinine levels Recipient -1082 A/A genotype associated with survival of more than 15 years (p=0.03)	
Kobayashi et al, 1999[34]	120	Incidence and severity of acute rejection	Association of recipient - 1064 allele 12 repeats with incidence and severity of acute rejection (p=0.05 and 0.0009, respectively)	HLA-haploidentical living-related transplants
Marshall et al, 2001[35]	145	Acute rejection	No association with donor genotype	
Mytilineos & Opelz, 2002[36]	4700	Graft survival	No association with recipient genotype	2100 second transplants included
Alamartine et al, 2003[37]	140	Skin carcinoma after renal transplantation	Association of recipient -1082, -819,-592 GCC genotype with squamous skin carcinoma (p=0.003)	Correlation of IL-10 in vitro production with predicted phenotype

polymorphisms at -1082.[24,32,36] However, an increased frequency of the -1082 A/A genotype was demonstrated in the group of patients with graft survival of more than 15 years compared to patients transplanted more recently.[33]

Finally, it is worth citing an interesting single study that explored the relationship between skin cancer after renal transplantation and IL-10 SNPs.[37] In fact, the investigators demonstrated the association of skin squamous cell carcinoma with the presence of -1082, -819, -592 GCC haplotype, as well as with higher IL-10 production in vitro by the individuals that possessed this marker.

Liver

The role of IL-10 in liver transplantation is controversial. Despite promising results in animal models, such as prolongation of graft survival by IL-10 gene transfer,[38] there is no clear picture in clinical transplants. Indeed, while some investigations[39] have shown increased IL-10 mRNA levels just before rejection, others[40] have failed to detect any changes in IL-10 level related to rejection or even observed an association of increased levels with a rejection-free transplant course.[41]

Controversies are also observed in studies of IL-10 polymorphisms and liver transplant outcome. Among five studies published on this issue, three did not show any relationship between IL-10 -1082 polymorphism and acute rejection.[42-44] One study detected an increased frequency of the -1082 G/G genotype in rejectors compared to nonrejectors,[45] while another observed a trend to a higher incidence of IL-10 A/A genotype in patients with rejection compared to control individuals.[46]

Although about 50% of allograft recipients suffer from acute rejection in liver transplantation, in this type of transplantation patients more frequently develop a state of tolerance than in other organ transplants. A tolerance induction protocol and/or assay that could identify a subset of patients in whom immunosuppression could be successfully withdrawn is the Holy Grail of transplant immunology. Authors from Pittsburgh University demonstrated that none of the clinically tolerant patients, in whom immunosuppression was successfully withdrawn, had the low producer IL-10 -1082 genotype, whereas about 27% of the patients requiring immunosuppression presented this genotype.[47] Despite the retrospective character of this investigation and the small number of tolerant patients under study, these results are promising and warrant future investigation.

Heart and Lung

The first study on cytokine gene polymorphism in transplantation was done in 1995 by Hutchinson's group in heart transplant recipients.[48] Shortly after, they showed that a combination of high TNF -308 and low IL-10 -1082 genotypes was associated with acute cardiac rejection.[49] Similar data were also reported in pediatric transplantation by investigators from Pittsburgh University. In fact, a higher frequency of the low producer genotype IL-10 was observed in rejectors (41%) than in nonrejectors (18%).[50] Recently our group also evaluated IL-10 SNPs in heart transplant recipients, observing a trend to a lower frequency of intermediate IL-10 producer GCC/ATA genotype in rejectors (8%) versus nonrejectors (29%).[51] However, no association with acute rejection was detected by other investigators examining either SNPs -1082, -819, -592 or microsatellite -1064 of the IL-10 gene.[52] Furthermore, the analysis of IL-10 polymorphisms in relation to coronary vasculopathy, which is considered to represent chronic allograft cardiac rejection, also failed to reveal any difference.[53] No association was seen in relation to IL-10 -1082, -819 and -592 polymorphisms in the sole investigation conducted on lung allograft recipients.[54]

Discussion

Reasons for the Discrepant Results

Although a lot of data supporting the involvement of IL-10 polymorphisms in transplantation have been reported, some results are not obvious and are even discrepant. It is important to note that this situation is not a particular characteristic of the transplantation field. Moreover, the problem of replicates in genetic association studies is contemporary and has been the subject of various recent reviews.[55-57] Herein we discuss the reasons for the discrepant results that are specific for genetic associations in transplantation.

Evidently, only case-control and not family-based studies can be done in transplantation, a fact that implies the problem of population stratification. In transplantation, HLA matching, type of donor, and immunosuppression should be taken into account. This, however, sometimes is not done but, when considered, it leads to small numbers of subjects under study, decreasing the power of the statistical analysis. Also the failure to see consistent associations might be due in part to misinterpretation of the endpoints analyzed, such as different criteria for the diagnosis of acute rejection or different patient management approaches at transplant centers. Accordingly, some apparently discrepant results are actually results of studies with different designs that cannot be compared to each other. Finally, it seems that even stratification by sex could be important in IL-10 polymorphism studies in transplantation. Indeed, a gender-specific association between -1082 IL-10 promoter polymorphism and longevity has been recently demonstrated.[58]

Perspectives

It seems that investigations of IL-10 polymorphism in transplantation are approaching the end of the exploratory stage. Indeed, hypotheses are already being generated based on the results of numerous preliminary studies.

Usually two solutions are offered to confirm a hypothesis: meta-analysis and multi-centric prospective studies. However, we believe that meta-analysis may be difficult to conduct. First, there is a traditional bias of nonpublishing of negative results. Second, many details specific for the transplantation field such as HLA match, immunosuppression, ethnicity and several others are not reported in some studies but are essential for correct implementation of meta-analysis.

In contrast, multi-centric prospective studies seem to be the most promising solution. Large numbers of patients should permit careful stratification based on well known risk factors as well as the uniform analysis of independent sets of patients that should greatly increase the statistical power of the study. Contrary to the current state-of-the-art, the long-term evaluation and composite endpoints should be prioritized.[59] Regarding the IL-10 gene, it is important to note that the polymorphisms already explored in transplantation are only partially responsible for the heredity-related differences in IL-10 production. Thus, other polymorphisms of IL-10 should be included in future investigations, permitting the analysis of extended haplotypes of the IL-10 gene that may also reduce the amount of contradictory results. Considering the pleiotropic character of IL-10 function and the fact that the outcome of transplantation itself is a complex phenotype, multi-locus approaches should be used.[60] The analyses combining IL-10 polymorphisms with TNF SNPs are very common. There are some other strong candidates for multi-locus analysis like the recently described IL-10 receptor polymorphisms that have functional significance for TNF production.[61]

Conclusion

In conclusion, data collected up to now demonstrate substantial evidence of IL-10 gene polymorphism involvement in post-transplant outcome. Multi-centric prospective studies using multi-locus approaches might ascertain when and how IL-10 genotype information should be considered in clinical transplant practice.

References

1. Westendorp RG, Langermans JA, Huizinga TW et al. Genetic influence on cytokine production and fatal meningococcal disease. Lancet 1997;349:170-173. Erratum in: Lancet 1997; 349:656.
2. Reuss E, Fimmers R, Kruger A et al. Differential regulation of interleukin-10 production by genetic and environmental factors—a twin study. Genes Immun 2002; 3:407-413.
3. Turner DM, Williams DM, Sankaran D et al. An investigation of polymorphism in the interleukin-10 gene promoter. Eur J Immunogenet 1997; 24:1-8.
4. Eskdale J, Gallagher G, Verweij CL et al. Interleukin 10 secretion in relation to human IL-10 locus haplotypes. Proc Natl Acad Sci USA 1998; 95:9465-9470.
5. Crawley E, Kay R, Sillibourne J et al. Polymorphic haplotypes of the interleukin-10 5' flanking region determine variable interleukin-10 transcription and are associated with particular phenotypes of juvenile rheumatoid arthritis. Arthritis Rheum 1999; 42:1101-1108.
6. Lim S, Crawley E, Woo P et al. Haplotype associated with low interleukin-10 production in patients with severe asthma. Lancet 1998; 352:113.
7. Gibson AW, Edberg JC, Wu J et al. Novel single nucleotide polymorphisms in the distal IL-10 promoter affect IL-10 production and enhance the risk of systemic lupus erythematosus. J Immunol 2001; 166:3915-3922.
8. Suarez A, Castro P, Alonso R et al. Interindividual variations in constitutive interleukin-10 messenger RNA and protein levels and their association with genetic polymorphisms. Transplantation 2003; 75:711-717.
9. Goker H, Haznedaroglu IC, Chao NJ. Acute graft-vs-host disease: Pathobiology and management. Exp Hematol 2001; 29:259-277.
10. Holler E, Roncarolo MG, Hintermeier-Knabe R et al. Prognostic significance of increased IL-10 production in patients prior to allogeneic bone marrow transplantation. Bone Marrow Transplant 2000; 25:237-241.
11. Ju X, Wang J, Xu B et al. Roles of interleukin-10 in acute graft-versus-host disease and graft rejection. Chin Med J 2003; 116:534-537.
12. Wang XN, Lange C, Schulz U et al. Interleukin-10 modulation of alloreactivity and graft-versus-host reactions. Transplantation 2002; 74:772-778.
13. Middleton PG, Taylor PR, Jackson G et al. Dickinson Cytokine gene polymorphisms associating with severe acute graft-versus-host disease in HLA-identical sibling transplants. Blood 1998; 92:3943-3948.
14. Cavet J, Middleton PG, Segall M et al. Recipient tumor necrosis factor-alpha and interleukin-10 gene polymorphisms associate with early mortality and acute graft-versus-host disease severity in HLA-matched sibling bone marrow transplants. Blood 1999; 94:3941-3946.
15. Cavet J, Dickinson AM, Norden J et al. Interferon-gamma and interleukin-6 gene polymorphisms associate with graft-versus-host disease in HLA-matched sibling bone marrow transplantation. Blood 2001; 98:1594-1600.
16. Takahashi H, Furukawa T, Hashimoto S et al. Contribution of TNF-alpha and IL-10 gene polymorphisms to graft-versus-host disease following allo-hematopoietic stem cell transplantation. Bone Marrow Transplant 2000; 26:1317-1323.
17. Nordlander A, Uzunel M, Mattsson J et al. The TNFd4 allele is correlated to moderate-to-severe acute graft-versus-host disease after allogeneic stem cell transplantation. Br J Haematol 2002; 119:1133-1136.
18. Kogler G, Middleton PG, Wilke M et al. Recipient cytokine genotypes for TNF-alpha and IL-10 and the minor histocompatibility antigens HY and CD31 codon 125 are not associated with occurrence or severity of acute GVHD in unrelated cord blood transplantation: A retrospective analysis. Transplantation 2002; 74:1167-1175.
19. Tambur AR, Yaniv I, Stein J et al. Cytokine gene polymorphism in patients with graft-versus-host disease. Transplant Proc 2001; 33:502-503.
20. Socie G, Loiseau P, Tamouza R et al. Both genetic and clinical factors predict the development of graft-versus-host disease after allogeneic hematopoietic stem cell transplantation.Transplantation 2001; 72:699-706.
21. Rocha V, Franco RF, Porcher R et al. Host defense and inflammatory gene polymorphisms are associated with outcomes after HLA-identical sibling bone marrow transplantation. Blood 2002; 100:3908-3918.
22. Xu GP, Sharma VK, Li B et al. Intragraft expression of IL-10 messenger RNA: A novel correlate of renal allograft rejection. Kidney Int 1995; 48:1504-1507.
23. Strehlau J, Pavlakis M, Mark Lipman et al. Quantitative detection of immune activation transcripts as a diagnostic tool in kidney transplantation. Proc Natl Acad Sci USA 1997; 94:695-700.

24. Sankaran D, Asderakis A, Ashraf S et al. Cytokine gene polymorphisms predict acute graft rejection following renal transplantation. Kidney Int 1999; 56:281-288.
25. Tian Y, Ma W, Zhang Y. Influence of cytokine gene polymorphism on renal transplantation. Zhonghua Wai Ke Za Zhi 2002; 40:256-258.
26. Pelletier R, Pravica V, Perrey C et al. Evidence for a genetic predisposition towards acute rejection after kidney and simultaneous kidney-pancreas transplantation Transplantation 2000; 70:674-680.
27. Hutchings A, Guay-Woodford L, Thomas JM et al. Association of cytokine single nucleotide polymorphisms with B7 costimulatory molecules in kidney allograft recipients. Pediatr Transplant 2002; 6:69-77.
28. George S, Turner D, Reynard M et al. Significance of cytokine gene polymorphism in renal transplantation. Transplant Proc 2001; 33:483-484.
29. Poole KL, Gibbs PJ, Evans PR et al. Influence of patient and donor cytokine genotypes on renal allograft rejection: Evidence from a single centre study. Transpl Immunol 2001; 8:259-265.
30. Marshall SE, McLaren AJ, Haldar NA et al. The impact of recipient cytokine genotype on acute rejection after renal transplantation. Transplantation 2000; 70:1485-1491.
31. Hahn AB, Kasten-Jolly JC, Constantino DM et al. TNF-alpha, IL-6, IFN-gamma, and IL-10 gene expression polymorphisms and the IL-4 receptor alpha-chain variant Q576R: Effects on renal allograft outcome. Transplantation 2001; 72:660-665.
32. Asderakis A, Sankaran D, Dyer P et al. Association of polymorphisms in the human interferon-gamma and interleukin-10 gene with acute and chronic kidney transplant outcome: The cytokine effect on transplantation. Transplantation 2001; 71:674-677.
33. Poli F, Boschiero L, Giannoni F et al. TNF-alpha IFN-gamma IL-6, IL-10, and TGF-beta1 gene polymorphisms in renal allografts. Transplant Proc 2001; 33:348-349.
34. Kobayashi T, Yokoyama I, Hayashi S et al. Genetic polymorphism in the IL-10 promoter region in renal transplantation. Transplant Proc 1999; 31:755-756.
35. Marshall SE, McLaren AJ, McKinney EF et al. Donor cytokine genotype influences the development of acute rejection after renal transplantation. Transplantation 2001; 71:469-476.
36. Mytilineos J, Opelz G. Impact of cytokine gene single nucleotide polymorphisms (SNP) in kidney transplant recipients. Transplantation 2002; 74:43.
37. Alamartine E, Berthoux P, Mariat C et al. Interleukin-10 promoter polymorphisms and susceptibility to skin squamous cell carcinoma after renal transplantation. J Invest Dermatol 2003; 120:99-103.
38. Shinozaki K, Yahata H, Tanji H et al. Allograft transduction of IL-10 prolongs survival following orthotopic liver transplantation. Gene Ther 1999; 6:816-822.
39. Platz KP, Mueller AR, Rossaint R et al. Cytokine pattern during rejection and infection after liver transplantation—improvements in postoperative monitoring? Transplantation 1996; 62:1441-1450.
40. Conti F, Boulland ML, Leroy-Viard K et al. Low level of interleukin 10 synthesis during liver allograft rejection. Lab Invest 1998; 78:1281-1289.
41. Minguela A, Torio A, Marin L et al. Implication of soluble and membrane HLA class I and serum IL-10 in liver graft acceptance. Hum Immunol 1999; 60:500-509.
42. Bathgate AJ, Pravica V, Perrey C et al. Polymorphisms in tumour necrosis factor alpha, interleukin-10 and transforming growth factor beta1 genes and end-stage liver disease. Eur J Gastroenterol Hepatol 2000; 12:1329-1333.
43. Jonsson JR, Hong C, Purdie DM et al. Role of cytokine gene polymorphisms in acute rejection and renal impairment after liver transplantation. Liver Transpl 2001; 7:255-263.
44. Fernandes H, Koneru B, Fernandes N et al. Investigation of promoter polymorphisms in the tumor necrosis factor-alpha and interleukin-10 genes in liver transplant patients. Transplantation 2002; 73:1886-1891.
45. Warle MC, Farhan A, Metselaar HJ et al. Cytokine gene polymorphisms and acute human liver graft rejection. Liver Transpl 2002; 8:603-611.
46. Tambur AR, Ortegel JW, Ben-Ari Z et al. Role of cytokine gene polymorphism in hepatitis C recurrence and allograft rejection among liver transplant recipients. Transplantation 2001; 71:1475-1480.
47. Mazariegos GV, Reyes J, Webber SA et al. Cytokine gene polymorphisms in children successfully withdrawn from immunosuppression after liver transplantation. Transplantation 2002; 73:1342-1345.
48. Turner DM, Grant SC, Lamb WR et al. A genetic marker of high TNF-alpha production in heart transplant recipients. Transplantation 1995; 60:1113-1117.
49. Turner D, Grant SC, Yonan N et al. Cytokine gene polymorphism and heart transplant rejection. Transplantation 1997; 64:776-779.
50. Awad MR, Webber S, Boyle G et al. The effect of cytokine gene polymorphisms on pediatric heart allograft outcome. J Heart Lung Transplant 2001; 20:625-630.

51. Morgun A, Shulzhenko N, Rampim GF et al. Influence of cytokine genetic polymorphisms on cardiac transplant outcome. In: Hansen JA, Dupont B, eds. HLA 2002, Immunobiology of the human MHC. Scattle, WA: IHWG press, in press
52. Bijlsma FJ, Bruggink AH, Hartman M et al. No association between IL-10 promoter gene polymorphism and heart failure or rejection following cardiac transplantation. Tissue Antigens 2001; 57:151-3.
53. Densem CG, Hutchinson IV, Yonan N et al. Influence of interleukin-10 polymorphism on the development of coronary vasculopathy following cardiac transplantation. Transpl Immunol 2003; 11:223-228.
54. Lu KC, Jaramillo A, Lecha RL et al. Interleukin-6 and interferon-gamma gene polymorphisms in the development of bronchiolitis obliterans syndrome after lung transplantation. Transplantation 2002; 74:1297-12302.
55. Daly AK, Day CP. Candidate gene case-control association studies: Advantages and potential pitfalls. Br J Clin Pharmacol 2001; 52:489-499.
56. Cardon LR, Palmer LJ. Population stratification and spurious allelic association. Lancet 2003; 361:598-604.
57. Colhoun HM, McKeigue PM, Davey Smith G. Problems of reporting genetic associations with complex outcomes. Lancet 2003; 361:865-872.
58. Lio D, Scola L, Crivello A et al. Gender-specific association between -1082 IL-10 promoter polymorphism and longevity. Genes Immun 2002; 3:30-33.
59. Hariharan S, McBride MA, Cohen EP. Evolution of endpoints for renal transplant outcome. Am J Transplant 2003; 3:933-941.
60. Hoh J, Ott J. Mathematical multi-locus approaches to localizing complex human trait genes. Nat Rev Genet 2003; 4:701-709.
61. Gasche C, Grundtner P, Zwirn P et al. Novel variants of the IL-10 receptor 1 affect inhibition of monocyte TNF-alpha production. J Immunol 2003; 170:5578-5582.

Chapter 3D

IL-10, Genetic Polymorphism and Its Relevance to Age-Related Diseases

Domenico Lio and Calogero Caruso

Abstract

The process of life for the individual is the struggle to preserve its integrity. However, the preservation of the integrity of the organism comes with a price, systemic inflammation. Accordingly, ageing is associated with chronic, low-grade inflammatory activity and the major age-related diseases such as Alzheimer's diseases, Parkinson's diseases, atherosclerosis and type 2 diabetes, are initiated or worsened by systemic inflammation, thus suggesting the critical importance of unregulated systemic inflammation in the shortening of survival. In the present review, the influence of the anti-inflammatory cytokine interleukin(IL)-10 on these major age-related diseases development is reviewed in the light of the inflammation mechanisms involved in their pathogenesis. The experimental and clinical observations reported suggest that IL-10 may play a central role in protection against major age-related inflammatory diseases. So, typing of IL-10 functionally relevant polymorphisms might allow to evaluate the individual risk for inflammation mediated disease leading to unsuccessful ageing. In fact, it is an emergent evidence that polymorphic alleles of inflammatory cytokines, involved in high cytokine production, are related to unsuccessful ageing and, reciprocally, controlling inflammatory status may allow to better attain successful ageing. So, the major findings reported in recent papers on cytokine polymorphisms and longevity suggest that those individuals who are genetically predisposed to produce low levels of inflammatory cytokines or high levels of anti-inflammatory cytokines (i.e., IL-10) have an increased capacity to reach the extreme limit of human life-span.

Introduction

The process of life for the individual is the struggle to preserve its integrity. However, the preservation of the integrity of the organism comes with a price, systemic inflammation.[1] Accordingly, ageing is associated with chronic, low-grade inflammatory activity, as documented by inflammatory markers predicting disability and mortality in the elderly cohorts.[2] In fact, increased plasma levels of acute phase proteins have frequently been found in ageing people and age-related diseases such as Alzheimer's diseases, Parkinson's diseases, atherosclerosis, type 2 diabetes, sarcopenia and osteoporosis are initiated or worsened by systemic inflammation, thus suggesting the critical importance of unregulated systemic inflammation in the shortening of survival in human beings.[1-5] So, pro-inflammatory and regulatory cytokines are believed to play a pathogenetic role in age-related diseases and data are accumulating that show that genetic variations in pro- or anti-inflammatory cytokines may influence both unsuccessful ageing (i.e., age-related inflammatory diseases) and successful ageing (i.e., longevity). In other terms it is hypothesising that subjects genetically equipped for a low pro-inflammatory and/or

Interleukin-10, edited by Francesco M. Marincola. ©2006 Eurekah.com.

high anti-inflammatory production may have a successful aging and reach the extreme edge of human life duratio.[3-6]

In this scenario is emerging Interleukin-10 (IL-10), a centrally operating anti-inflammatory cytokine that plays a crucial role in the regulation of the innate immune system. IL-10, a cytokine with anti-inflammatory and B cell stimulating activity, is produced by activated T cells, B cells, monocytes/macrophages and dendritic cells. IL-10 is thought to block the ability of monocytes/macrophages and dendritic cells to act as antigen presenting cells by down regulation of MHC products and costimulatory molecules via suppression of the Mitogen-Activated Protein Kinase cascade. Thus, the principal function of IL-10 appears to be to limit and ultimately terminate the inflammatory signal.[7,8]

The IL-10 gene is located on chromosome 1 at 1q31-32 and is highly polymorphic.[9] Stimulation of human blood samples with bacterial lipopolysaccharide showed large interindividual variations of IL-10 production, suggesting a genetic component of approximately 75%. Interindividual differences in the regulation of IL-10 production may be critical regarding the final outcome of an inflammatory response, i.e., within physiological limits or pathological one.[10] The IL-10 production, independently by interaction with other cytokine gene products,[11] is genetically controlled by polymorphisms in IL-10 promoter sequence.[9] Several polymorphisms located close to or within the IL-10 genes are potentially associated with transcription levels, however, the molecular mechanisms linking these polymorphisms to gene expression and posttranscriptional processing are only partially known. Recently, three polymorphisms in the proximal region of IL-10, that might be probable targets for transcription regulating factors, have been described even if conflicting results have been reported on what of these may influence, alone or in reciprocal linkage, the IL-10 transcription.[12,13] However, in the proximal promoter region, three well documented haplotypes, involving three single nucleotide polymorphisms (SNP)s at -1082 (G→A), -819 (C→T) and -592 (C→A) nucleotides of IL-10 promoter, have been identified.[9,14-20] In the presence of allele -1082A, stimulation of lymphocytes with concanavalin A resulted in lower IL-10 production than in allele -1082A-negative cells. In the original paper of Turner et al[14] it was claimed that this correlation was independent on the polymorphisms at position -819 and -592. However, in further studies GCC haplotype has been associated with higher and the ATA haplotype with lower production of IL-10 by peripheral blood mononuclear cells than ACC haplotype.[15] In any case, IL-10 low producer genotypes seem to play a particular role in the susceptibility to inflammatory and infectious disease, where it might be relevant to control of inflammatory responses (ref. in 21-25). Accordingly, it has been reported an increased frequency of the homozygote -1082GG genotype in Italian centenarian men. The -1082GG genotype was argued to confer an anti-inflammatory status which it was postulated enhanced the possibility to attain the extreme longevity.[21,23]

In the next paragraphs, the influence of IL-10 on major age-related diseases atherosclerosis, type 2 diabetes, Alzheimer's disease and Parkison's disease development is reviewed in the light of the inflammation mechanisms involved in their pathogenesis.[1-5]

Atherosclerosis

For long time, hypercholesterolaemia has been claimed as be the most important risk factor for atherogenesis. However, recently it has been appreciated that inflammatory mechanisms couple dyslipidaemia to atheroma formation and inflammatory processes play an important role for the progression of atherosclerosis.[26] Thus, atherosclerotic patients display a pro-inflammatory phenotype and the control of inflammation might play a protective role in atherosclerosis complications, including acute myocardial infarction. Indeed, atherosclerosis is now better described as an inflammatory disease of the vessel wall, characterized by the accumulation of lipid-laden macrophages and fibrous material in the large arteries and flared plaque inflammation is considered a cause of intimal erosion and rupture and therefore of acute ischaemia.[26-28] Studies in vitro and in experimental animals are supported by the clinical finding of increased inflammatory markers in patients with chronic stable angina, severe unstable angina, and acute myocardial infarction and by the predictive value of such a marker as C reactive protein (CRP) for subsequent coronary events.[29,30] An initiating event is the accumulation of lipids in the vessel wall, which subsequently will become modified and trigger an inflammatory process. Low-density lipoproteins (LDL) are taken up by macrophages through scavenger receptors, leading to foam cells; the lipid deposits become oxidized forming pro-inflammatory lipid peroxides. Monocytes are attracted from the blood and differentiate into macrophages that take up the modified LDL and will form lipid-laden foam cells, which is the first hallmark of atherosclerotic plaque development. Later on, inflammatory mediators will increase, other immune cells will be attracted, and smooth muscle cells will be activated and become involved. More advanced stages of plaque development are characterized by increased deposition of extra cellular lipid cores, fibrous material, and often necrosis. Subsequently, these macrophages are further activated, leading to the production of a wide range of cytokines and growth factors. Myocardial infarction may occur as a result of erosion or uneven thinning and rupture of the fibrous cap, often at the shoulders of the lesion where macrophages enter, accumulate and are activated and where apoptosis may occur.[26-28,31]

Several lines of evidence indicate an involvement of IL-10 in the development of atherosclerosis. In fact, IL-10 can limit the progression of experimental atherosclerosis.[7] IL-10–deficient knockout mice have a thirty-fold increase in susceptibility to atherosclerosis, due either to high cholesterol diet,[32] or to a cardiac allograft vasculopathy setting.[33] The transfection of adenovirus-mediated over expression of IL-10 into the LDL-receptor knockout mouse strongly inhibits the otherwise rapidly growing atherosclerotic lesions. IL-10 has several anti-atherogenic effects including inhibition of adhesion of LDL–activated monocytes to endothelium and down regulation of fibrinogen biosynthesis.[34-36]

IL-10 is detectable in human atherosclerotic plaques[26] and has been claimed to plays a regulation role in progression of atherosclerosis human coronary artery disease. A study compared 50 patients with stable and 45 patients with unstable angina pectoris. Plasma IL-6 levels where significantly higher and plasma IL-10 levels lower in those patients with the unstable form of the disease.[37]

Besides, serum level of IL-10 may be an important prognostic determinant in patients with acute coronary syndromes.[38] In this study, IL-10, CRP, and troponin T were measured at baseline and before discharge in 547 patients. Patients with elevated IL-10 levels were at significantly lower risk compared with patients with low IL-10 levels. The predictive value of IL-10 showed also a significant interaction with well known risk factor CRP levels. CRP-positive patients with the higher IL-10 serum levels were protected from the increased cardiac risk of CRP-positive patients with low IL-10 levels.[38]

Recently it has been reported that protective effect of IL-10 in atherosclerosis may be linked to the effect of NF-κB regulation pathway.[39] The transcription factor NF-κB is one of the key regulators of inflammation, immune responses, and cell survival. Upon activation, NF-κB can mediate the induction of more than 160 genes, many of which have a documented role in atherogenesis.[40] Activated NF-κB has been detected in endothelial cells, smooth muscle cells, and macrophages in atherosclerotic plaques.[41] In resting cells, NF-κB dimers are kept inactive associated with inhibitory proteins, the IκBs. NF-κB activation is mediated by the IκB kinase (IKK) complex containing two catalytic subunits, IKK1 (IKKα) and IKK2 (IKKβ), and a regulatory subunit called NF-κB essential modulator (NEMO or IKKγ). Upon stimulation, the IKK complex phosphorylates IκB, inducing its ubiquination and subsequent degradation. NF-κB is then free to the translocation to the nucleus where it facilitates the transcription of many genes, including cytokines, chemokines, and anti-apoptotic factors.[42]

A lot of circumstantial evidence has indicated a potential role for NF-κB in atherosclerosis. First of all, activated NF-κB has been demonstrated in human atherosclerotic plaques, in macrophages, smooth muscle cells, and endothelial cells.[40] Moreover, several NF-κB regulated genes have been demonstrated to be unregulated in plaques, including pro-inflammatory cytokines, such as tumor necrosis factor(TNF)-α and IL-6.[43] Additionally, receptors that can signal to NF-κB are also present in lesions, including several member of the toll-like receptor family, which also mediate signal-induced activation of macrophages.[44] In an animal experimental model the inhibition or deletion of IKK2 regulatory protein in macrophages enhances atherosclerosis.[39] One of the effects of IKK2 deletion in macrophages results in a reduction in IL-10 and may be regarded as pro-atherogenic phenomenon.[39] However, through which mechanisms and genes NF-κB is involved in the development of atherogenesis and its role in pro-inflammatory/anti-inflammatory balance remain challenging questions for future research.[39]

IL-10 seems also able to attenuate CD40-mediated interleukin-12 synthesis in human endothelial cells through induction of nitric-oxide synthase expression.[45] It has been reported, in fact, that under pro-inflammatory conditions expression of the human IL-10 receptor gene is enhanced in endothelial cells in vitro and in vivo and exposure to IL-10 results in an up-regulation of both endothelial nitric-oxide synthase (e-NOS) expression and activity through activation of the transcription factor STAT-3. CD154-induced IL-12 p40 expression is enhanced after blockade of e-NOS activity but attenuated in the presence of exogenous nitric oxide. Increased e-NOS-3 expression may, thus, be one mechanism by which IL-10 exerts its anti-inflammatory effects in type 1 cell-mediated chronic inflammatory diseases.[45]

Moreover the latter effect might attribute a role for IL-10 secretion in the control of blood pressure. It has been hypothesized that a rise in blood pressure (BP) activates a vicious cycle, causing chronic inflammation of the endothelium, which in turn might be responsible for a further damage of endothelium and worsening of BP control with activation of pro-and anti-inflammatory molecule production.[46] A sustained IL-10 production by endothelial cells might induce in the same cells nitric oxide production by autocrine effect with potentially positive effects on BP control.

Accumulating evidence suggests that inflammation plays an important role not only in the development of cardiovascular but also in cerebrovascular disease.[47-49] Markers of inflammation, such as CRP,[29,30,48,49] and pro-inflammatory cytokines are associated with stroke.[50] Analysis of the association between low IL-10 production levels and natural history of stroke episodes.[48] shows that low IL-10 production levels are associated with both a history of stroke, in a cross-sectional study, and increased mortality due to stroke, as obtained in a prospective follow-up study. These associations persisted after adjustment for known risk factors for stroke. An inverse relationship was found between CRP, one of the most sensitive detectable signals of inflammation, and IL-10 levels in association with stroke.[47,48,51] CRP and IL-10 at least partly represent the effect of an inflammatory response on stroke. Studies of cerebral ischemia emphasize the relevance of an inflammatory response in regard to lesion size, in which IL-10 is a key regulator.[7,52,53] IL-10, as a powerful suppressor of proinflammatory cytokines such as TNF-α and IL-6,[7] may also inhibit CRP, since it has been suggested that IL-6 is implied in regulation of CRP production.[54]

Similar results were obtained evaluating the IL-10 levels in patients with neurological worsening after acute ischemic stroke. Lower plasma concentrations of IL-10 were associated with clinical worsening and in particular early worsening was, independently by other risk and clinical factors, associated with lower IL-10 plasma levels in patients with subcortical infarcts or lacunar stroke.[55]

These reports seem to attribute a protective role to anti-inflammatory effects of IL-10 in stroke, and, considering its effect on the reduction of brain ischemic damage size in experimental model of cerebral arteries occlusion,[53] may indicate a potential therapeutic role for this cytokine. In Figure 1 we show our suggestion on the multiple inhibitory effects of IL-10 on pathogenetic mechanisms of atherosclerosis.

Concerning the role of the genetic control of IL-10 production in atherosclerosis development, case-control studies across Europe do not support any role for IL-10 in either atherosclerotic-related disease or myocardial infarction.[56,57] However, a more recent study suggests that -1082AA genotype, which is associated with low production of the cytokine IL-10 and elevated markers of systemic inflammation such as C reactive protein, was predictive for a

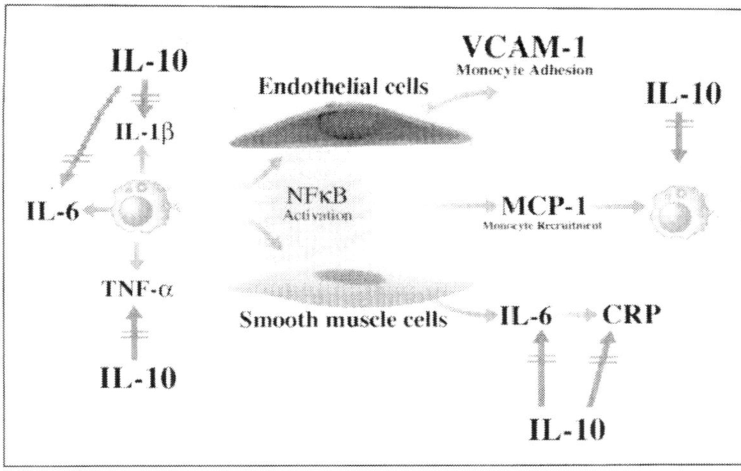

Figure 1. The figure depicts the multiple inhibitory effects of IL-10 on pathogenetic mechanisms of atherosclerosis. The role of NFκB, cytokines, chemokines (MCP-1), adhesion molecules (VCAM-1) is outlined.

higher cardiovascular morbidity in dialysis patients compared to the -1082GG genotype. In fact, in a group of haemodyalised patients sequential measures of plasmatic C-reactive protein levels allowed to find that patients with the IL-10 "high-producer" genotype had significantly less variation and less elevation of CRP levels.[58] Inflammatory mediators are induced in all patients with chronic renal failure due to uraemia and renal replacement therapy. However, while those with the IL-10 "high-producer" genotype may effectively limit inflammation, this is not the case in the "low-producer" group. So, it was observed a significantly higher risk both for events and cardiovascular mortality in patients with the IL-10 "low-producer" genotype. These data impressively confirm the important role of IL-10 for the protection of the individual against the long-term effects of systemic inflammation.[58] A recent report in two Italian samples of patients affected by myocardial infarction seems also to suggest a role for IL-10 polymorphism linked to low cytokine production in the occurrence of the disease.[59] These last data are in agreement with the role in Italian population of IL-10 low-producer genotype in attainment of longevity.[21,23]

Type 2 Diabetes

The aetiology of type 2 diabetes is not yet completely understood. Although low physical activity and hyperalimentation are lifestyle factors associated with an increased risk of the disease, type 2 diabetes has been claimed to be a manifestation of an inflammatory response.[60-62] A recent study examined prospectively the effects of the classical pro-inflammatory cytokines IL-1β, IL-6, and TNF-α on the development of type 2 diabetes. Data suggest that the pattern of circulating inflammatory cytokines modifies the risk for type 2 diabetes. In particular, a combined elevation of IL-1β and IL-6 independently increases the risk of type 2 diabetes. These data strongly support the hypothesis that a subclinical inflammmatory reaction plays a role in the pathogenesis of type 2 diabetes.[63]

Several mechanisms may be implied in the involvement of inflammatory cytokines in the pathogenesis of the disease. For instance, both IL-6 and IL-1β act on the liver to produce the characteristic dyslipidemia of the metabolic syndrome, with increased very low-density lipoproteins high-density lipoproteins.[62] Carriers of the G SNP at -174 at 5'-upstream of IL-6, characterized by high IL-6 plasma levels, appear to be prone to develop lipid abnormalities and have a worse glucose handling capacity, a higher blood glycosylated hemoglobin, a higher fasting insulin levels, and an higher insulin sensitivity.[64,65]

Accordingly, in the Leiden 85-Plus Study[66] it has been demonstrated that low production of IL-10 by stimulated peripheral blood cells, associates with the metabolic syndrome and type 2 diabetes. It was found that individuals with raised blood glucose tended to have a low capacity to produce IL-10. A similar association was shown between high plasma triglyceride concentrations and low production capacity of IL-10. Even if this study does not consider the possibility that the elevation of blood glucose concentration itself could have a direct effect upon cellular IL-10 production, it seems to confirm the role of inflammation in the aetiology of type 2 diabetes.

A partial confirmation of the role that low production of IL-10 may play in metabolic diabetes was recently offered by immunogenetic studies on association of interleukin-10 (IL-10) gene promoter region polymorphisms and adult onset of type 1 diabetes.[67] The frequency of the ATA haplotype at -1082G/A, -819C/T, and -592C/A SNP was increased in adult-onset patients. Even if the pathogeneses of the two disease are very different, considering that up to 75% of IL-10 production capacity in humans derives from genetic factors,[10] the possibility that IL-10 polymorphisms may play a role in genetic predisposition or protection against type 2 diabetes seems to be a conceivable hypothesis.

On the other hand, a confirmation on the role of low IL-10 in type 2 diabetes pathogenesis came from animal models. Experimental studies using gene-targeted mice[68] provide direct evidence that IL-10 offers some protection against endothelial dysfunction during diabetes and that this effect is mediated by inhibition of increases in reactive oxygen species in blood vessels. It was also demonstrated that endogenous IL-10 is an important counterbalance to mechanisms that produce endothelial dysfunction during diabetes[68] and that mucosal administration of IL-10 enhances oral tolerance in diabetes.[68,69]

Alzheimer's Disease

The neurodegenerative inflammatory age-related disease, Alzheimer's disease (AD), is the most common cause of dementia in the elderly. It is a progressive neurological disease that results in the irreversible loss of neurons, particularly in the cortex and hippocampus. The pathological hallmarks are neuronal loss, extracellular senile plaques containing the peptide amyloid-beta (Aβ) and neurofibrillary tangles; the latter are composed of a hyperphosphorylated form of the microtubular protein tau. Amyloid in senile plaques is the product of cleavage of a much larger protein, the β-amyloid precursor protein, by a series of proteases, the α-,β-,γ-secretases. The γ-secretase, in particular, appears to be responsible for generating the $A\beta_{42}$ peptide, that is 42 amino acids in length and has pathogenetic importance, because it can form insoluble toxic fibrils and accumulates in the senile plaques of AD brains.[70-72]

Inflammation clearly occurs in pathologically vulnerable regions of the AD brain, and it does so with the full complexity of local peripheral inflammatory responses. In the periphery, degenerating tissue and the deposition of highly insoluble abnormal materials are classical stimulants of inflammation. Likewise, in the AD brain damaged neurons and neurites and highly insoluble $A\beta_{42}$ peptide deposits and neurofibrillary tangles provide obvious stimuli for inflammation. Senile plaques in AD brains are associated with reactive astrocytes and activated microglial cells and cytokines and acute phase proteins are overexpressed in microglia and astrocytes surrounding neuropathological lesions in AD brains. There is accumulating evidence that Aβ peptide may promote or exacerbate inflammation by inducing glial cells to release immune mediators (Table 1). Moreover, microglial and astroglial cells surrounding mature plaques in AD brains have been found to express activation markers.[73-76] The role of inflammation is further emphasized by a number of clinical studies demonstrating that the long-term use of nonsteroidal anti-inflammatory drugs may protect against AD.[77] Enriched populations of human microglial cells isolated from mixed cell cultures prepared from embryonic human telencephalon tissues are able to express constitutively mRNA transcripts for cytokines and

Table 1. Markers of inflammation in the brain of patients with neuropathological diagnosis of AD

Acute Phase Proteins	Cytokines	Complement Proteins
α-1-antichymotripsin	IL-1	Complement activation products
α-1-antitrypsin	IL-6	Membrane attack complex
α-2-macroglobulin	TNF-α	C1 inhibitor
C-reactive protein	TGF-β1	CD59
Serum amyloid substance P	IL-10	Clusterin
Amyloid precursor protein	Chemokines	Vitronectin

chemokines and treatment with pro-inflammatory stimuli as lipopolysaccharide or Aβ peptide led to increased expression of mRNA levels of these inflammatory molecules.[78]

However, glial cells are also capable of producing anti-inflammatory cytokines including IL-10 and transforming growth factor (TGF)-β and balance among pro- and anti-inflammatory events may be a crucial point for development of an overt AD.[79,80]

IL-10 is synthesized in the central nervous system (CNS) and may act to limit inflammatory responses that occur during microbial infections and immune-inflammatory diseases, including stroke and multiple sclerosis. In fact, expression of IL-10 is elevated during the course of most major diseases in the CNS promoting survival of neurons and glial cells, expressing cell survival signals and limiting inflammation in the brain. In the CNS, the anti-inflammatory mechanism of IL-10 may follow three major pathways, i.e., reducing synthesis of pro-inflammatory cytokines, suppressing cytokine receptor expression, and inhibiting receptor activation.[81]

Data obtained using transgenic mice that develop Aβ plaques in neocortex and hippocampus of the aged brain may represent an useful tool to study such regulatory mechanisms. The Aβ-induced glial responses observed in these mice were similar to those described in AD brains. Activated microglia was localized in and around plaques only, while reactive astrocytes were found in close proximity to both fibrillary and diffuse Aβ deposits. The number of IL-10-positive glial cells around Aβ peptide plaques seems to be increased with age indicating the development of anti-inflammatory mechanisms.[82,83]

Following corticectomy, local application of IL-10 has been found to decrease microglia activation and TNF-α production, while in vitro IL-10 has been shown to inhibit microglial synthesis of pro-inflammatory cytokines IL-1, IL-6 and TNF-α.[81,84-86]

It is more than a decade that genes responsible for familial AD have been identified. Thus, a major challenge is to identify possible genetic factors involved in the most common form of the disease, the sporadic one occurring late in life.[72,75,87,88] The most important genetic factor identified in sporadic AD is the allele ε4 of apolipoprotein E (ApoE). In fact, the APOE ε4 allele increases and the APOE ε2 allele decreases the risk of AD.[70,88] A number of associations of the disease with variant of genes other than ApoE have also been reported but remain to be confirmed and are the subject of ongoing research.[70] However, a large number of studies has shown a significant association between the presence of polymorphisms linked to a high production of pro-inflammatory cytokines IL-1, IL-6 and TNF-α and acute phase proteins and the development of AD. In most studies the association between AD and polymorphisms has been shown evident (or more evident) in APOE ε4 allele noncarrier or carrier patients, suggesting a complex interaction between the genes under study and ApoE.[4,5,72,74,75,87-92]

These findings suggest that a genetic background leading for an excessive inflammation response may play a central role in the induction of degenerative lesions in the brain of AD patients. So, control of pro-inflammatory cytokines production may be a key mechanism in controlling and delaying disease onset. Whole blood samples from AD and controls were stimulated ex vivo with endotoxin under standard conditions and cytokine levels assessed. Patients with AD had seven- to ten-fold higher IL-1β production relative to the amount of IL-10. These data suggest that an absolute (or relative) reduced production of Il-10 may contribute to the development of AD.[93]

Recently, we have demonstrated an association between SNPs located at position –1082, -819 and –592 of the promoter region of IL-10 gene and AD. Our data demonstrate a significant increase of percentage of –1082A carrier subjects among AD patients and that this increase is mainly due to increase of the frequency at IL-10 promoter of ATA haplotype, associated to a low production of IL-10.[22] So, our data strongly suggest that a genetically determined low IL-10 production may be implied in the AD susceptibility. Our findings have been recently confirmed in a study on a large cohort of North-European patients affected by AD.[94] In fact, the authors published only a not-significant trend for the association between -1082A allele and AD, but a statistical analysis of their published data allows to find a strong statistically significant increase of homozygous -1082A genotype frequency in AD patient group.

The overall chance of a subject to develop AD might be profoundly affected by a "susceptibility profile" reflecting the combined influence of inheriting multiple high-risk alleles.[89] IL-1 has been claimed as a key molecule in AD pathogenesis[90] based on findings of an IL-1 overexpression in AD brain that is directly related to plaque progression and tangle formation, and on findings that IL-1 induces excessive synthesis, translation, and processing of neuronal amyloid precursor protein as well as synthesis of most known plaque-associated proteins. In addition, IL-1 activates astrocytes, with the important consequence of overexpression of the neuritogenic cytokine S100β and overgrowth of dystrophic neurites in neuritic plaques. As further evidence of the importance of IL-1 in AD, several genetic studies have demonstrated that gene polymorphisms of IL-1α and IL-1β are associated with increased risk of AD and influence the age at onset of the disease.[4,5,74,90] In Figure 2, we suggest how IL-10 can be able to counterbalance this pro-inflammatory effect of IL-1 (and of the other pro-inflammatory cytokines).

Multiple functional relationships may link these molecules and their genes influencing different but converging mechanisms involved in degeneration of the CNS might suggest novel therapeutic approach for AD.

As above reported, the deposit of Aβ peptide, derived from the proteolysis of the amyloid precursor protein (APP), as deposits in senile plaques is a pathological hallmark of AD.[70] In a seminal report, Schenk et al[95] demonstrated that immunization of transgenic PDAPP mice with Aβ led to the production of Aβ antibodies and a decrease of amyloid plaque pathology. In subsequent studies, cognitive improvement was documented in experimental animals following immunization.[71] High titres of IgG anti-Aβ were found in immunized animals and small amounts of these antibodies crossed the blood/brain barrier. This observation, along with additional ex vivo evidence, led to one proposed mechanism for $A\beta_{42}$ immunization-induced reduction of Aβ plaques whereby anti-Aβ IgG enter the brain and opsonize Aβ, thereby promoting microglia-mediated phagocytosis and clearance of Aβ.[95,96] A better understanding of the nature of the T helper response after Aβ immunization might allow for additional therapeutic targets and/or optimisation of immunization to provide greatest therapeutic benefit. In an experimental model[98] after stimulation with mitogen, primary splenocyte cultures from mice repeatedly immunized with Aβ secreted less interferon(IFN)-γ and IL-12, and more IL-4 and IL-10 and immunized mice showed reduced type 1 (IFN-γ) and increased type 2 (IL-10) levels in blood.

These findings have implications for the use of Aβ immunization as a therapy for AD. It has recently been shown that Aβ immunization of mice with appreciable preexisting Aβ deposits did not result in significant plaque clearance as opposed to a prophylactic regimen where mice received Aβ vaccination prior to amyloid deposition.[96]

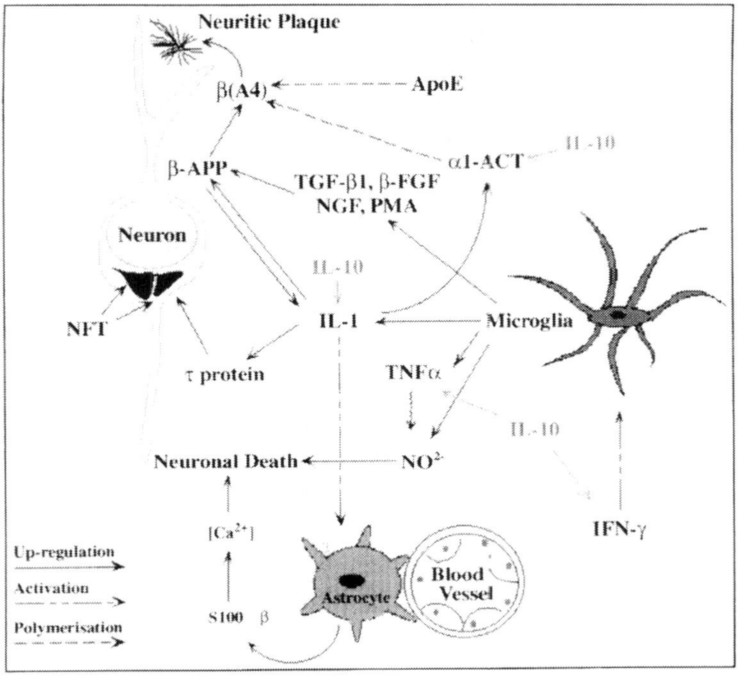

Figure 2. Schematic representation of the role of pro-inflammatory cytokines (esemplified by the key cytokine IL-1) and anti-inflammatory cytokine IL-10 on pathogenetic mechanisms of Alzheimer's disease. Astrocytes or microglial cells become activated upon contact with Aβ, which induces the production of inflammatory molecules. IL-1 overexpression in AD brain is directly related to excessive synthesis, translation, and processing of APP as well as synthesis of most known plaque-associated proteins. In addition, IL-1 activates astrocytes, with the important consequence of overexpression of the neuritogenic cytokine S100β and overgrowth of dystrophic neurites in neuritic plaques. The multiple inhibitory effects of IL-10 are shown.

If Aβ deposition in transgenic mouse models of AD is representative of the clinical syndrome, the suggestion arises that patients with AD may be immunologically tolerant to Aβ vaccination, and that an Aβ "immune barrier" has to be overcome in order for this therapy to be effective.[97] Perhaps supplementing the Aβ vaccine with type 2 cytokines or adjuvants known to suppress type 1 and/or evoke type 2 responses might enhance its therapeutic efficacy for AD, provided that other forms of the vaccine could be demonstrated to be safe in humans.

On the other hand, the blood/brain barrier might condition the effectiveness of this kind of approach. In a recent paper, it has been reported that IL-10 does not cross the intact barrier after intravenous peripheral delivery.[98] Thus, peripheral IL-10 probably can serve as a CNS therapeutic only when the barrier is disrupted as in case of stroke but not in AD if not administered in a way or pharmacological presentation able to cross the formidable obstacle of blood/brain barrier.

However, clinical trials on AD patients were started in 2001. Immunization with $A\beta_{42}$ in its fibrillary form was less well tolerated by aged humans than by experimental animals. Side effects were noted during phase II of the trials, including an inflammatory reaction of the CNS in 15 of 360 vaccinated patients. These side effects were presumably due to excessive activation of the specific immune response against Aβ.[71] So, these results have to taken into account in designing new trials.

Table 2. Potential point of attack for anti-inflammatory effect of IL-10 in atherosclerosis and AD

	Atherosclerosis	Alzheimer's Disease
Cells		
Astrocytes	0	++
Myofibroblast	++	0
Macrophage (microglia)	+++	++
T-cell	++	?
Cytokines and Acute Phases Mediators		
CRP	+++	+
Complement proteins	+	+
Clotting factors	++	0
Cytokines: IL-1, IL-6 TNF-α	+++	+++

Cells or immune mediators: 0: absent; +: weak; ++: moderate; +++: extensive.

The inflammatory reaction may also play a role in the development of Parkinson's Disease (PD). By now there are only few and contrasting data on relationship among IL-10 production and PD development. In an animal model, two peaks of IL10 mRNA, immediately (6 h) and at the 3-day post N-methyl-4-phenyl-1,2,3,6-tetrahydropyridine induction of nigro-striatal dopamine neurodegeneration that might be linked to a protective effect of the cytokine were found.[99] On the other hand, treatment with amantidine (a well known anti-PD therapy) of newly diagnosed patients seems do not affect IL-10 production by peripheral blood cells.[100]

Conclusions

The experimental and clinical observations above reported suggest that IL-10 may play a central role in protection against major age-related inflammatory diseases. So, typing of IL-10 functionally relevant polymorphisms might allow to evaluate the individual risk for inflammation mediated disease leading to unsuccessful ageing. It is an emergent evidence that polymorphic alleles of inflammatory cytokines, involved in high cytokine production, are related to unsuccessful ageing and, reciprocally, controlling inflammatory status may allow to better attain successful ageing. In fact, the major findings reported in recent papers on cytokine polymorphisms and longevity suggest that those individuals who are genetically predisposed to produce low levels of inflammatory cytokines or high levels of anti-inflammatory cytokines (i.e., IL-10) have an increased capacity to reach the extreme limit of human life-span.[4,5,21,23,101,102]

After reaching maturity, humans experience progressive declines in physiological processes that universally limit survival. Evidence from longitudinal studies reveals, however, that there are large variations in the preservation of physical and mental health and in the maintenance of functional abilities in ageing individuals that appear related to the influence both of life-style and genetics. The bulk of the research indicates that health, functional maintenance and survival depend on the ability of host defences to sustain homeostasis via their ability to adapt to or resist stresses.[3,70]

Under a therapeutical point of view this evidence for the pervasiveness of inflammatory processes in "normal" ageing and in AD and vascular disease makes plausible an otherwise unexpected emergent trend: some drugs and perhaps cytokines (Table 2) that protect against heart disease may also used to reduce the risk of Alzheimer's disease.

On the other hand, a new perspective is emerging by studies of the recently identified T CD4+CD25+ subset of regulator cells. Actually, IL-10 represents an important growth factor for these cells, so induction of the differentiation of these cells depends by an efficient IL-10 production and this may allow to overcome the immune-inflammation response leading to the major age related diseases.[103]

Acknowledgments

This work was supported by a grnt from the Italian Ministry of Education, University and Research (FIRB project on age-related diseases).

References

1. Brod SA. Unregulated inflammation shortens human functional longevity. Inflamm Res 2000; 49:561-570.
2. Bruunsgaard H, Pedersen M, Pedersen BK. Aging and proinflammatory cytokines. Curr Opin Hematol 2001; 8:131-136.
3. Franceschi C, Bonafè M, Valensin S et al. Inflamm-aging. An evolutionary perspective on immunosenescence. Ann N Y Acad Sci 2000; 908:244-254.
4. Pawelec G, Barnett Y, Forsey R et al. T cells and aging, January 2002 update. Front Biosci 2002; 7:d1056-1183.
5. Candore G, Colonna-Romano C, Lio D et al. Immunological and immunogenetic markers of successful and unsuccessful ageing. Adv Cell Age Gerontol 2003; 13:29-45.
6. Pawelec G, Ouyang Q, Colonna-Romano G et al. Is human immunosenescence clinically relevant? Looking for 'immunological risk phenotypes'. Trends Immunol 2002; 23:330-332.
7. Moore KW, de Waal Malefyt R, Coffman RL et al. Interleukin-10 and the interleukin-10 receptor. Annu Rev Immunol 2001; 19:683-765.
8. Sato K, Nagayama H, Tadokoro K et al. Extracellular signal-regulated kinase stress-activated protein kinase/c-jun N-terminal kinase and p38 Mapk are involved in IL-10 mediated selective repression of TNF-alpha induced activation and maturation of human peripheral blood monocyte-derived dendritic cells. J Immunol 1999; 162:3865-3872.
9. Kube D, Rieth H, Eskdale J et al. Structural characterisation of the distal 5' flanking region of the human interleukin-10 gene. Genes Immunity 2001; 2:181-190.
10. Westendorp RG, Langermans JA, Huizinga TW et al. Genetic influence on cytokine production and fatal meningococcal disease Lancet. 1997; 349:170–173.
11. Lio D, Candore G, Colombo A et al. A genetically determined high setting of TNF-α influences immunological parameters of HLA-B8,DR3 positive subjects: Implications for autoimmunity. Human Immunol 2001; 62:705-713.
12. Westendorp RG, van Dunne FM, Kirkwood TB et al. Optimizing human fertility and survival. Nat Med 2001; 7:873.
13. Gibson AW, Edberg JC, Wu J et al. Novel single nucleotide polymorphisms in the distal IL-10 promoter affect IL-10 production and enhance the risk of systemic lupus erythematosus J Immunol 2001; 166:3915-3922.
14. Turner DM, Williams DM, Sankaran D et al. An investigation of polymorphism in the interleukin-10 gene promoter. Eur J Immunogenet 1997; 24:1-8.
15. Crawley E, Kay R, Sillibourne J et al. Polymorphic haplotypes of the interleukin-10 5' flanking region determine variable interleukin-10 transcription and are associated with particular phenotypes of juvenile rheumatoid arthritis. Arthritis Rheum 1999; 42:1101-1108.
16. Eskdale J, Gallagher G, Verweij CL et al. Interleukin 10 secretion in relation to human IL-10 locus haplotypes. Proc Natl Acad Sci USA 1998; 95:9465-9470.
17. Eskdale J, Keijsers V, Huizinga T et al. Microsatellite alleles and single nucleotide polymorphisms (SNP) combine to form four major haplotype families at the human interleukin-10 (IL-10) locus. Genes Immunity 1999; 1:151-155.
18. Eskdale J, Kube D, Tesch H et al. Mapping of the human IL-10 gene and further characterization of the 5' flanking sequence. Immunogenetics 1997; 46:120-128.
19. Donger C, Georges JL, Nicaud V et al. New polymorphisms in the interleukin-10 gene—relationships to myocardial infarction. Eur J Clin Invest 2001; 31:9-14.
20. D'Alfonso S, Rampi M, Rolando V et al. New polymorphisms in the IL-10 promoter region. Genes Immun 2000; 1:231–233.
21. Lio D, Scola L, Crivello A et al. Gender-specific association between -1082 IL-10 promoter polymorphism and longevity. Genes Immun 2002; 3:30-33.
22. Lio D, Licastro F, Scola L et al. Interleukin-10 promoter polymorphism in sporadic Alzheimer's disease. Genes Immun 2003; 4:234-238.

23. Lio D, Scola L, Crivello A et al. Inflammation, genetics, and longevity: Further studies on the protective effects in men of IL-10 -1082 promoter SNP and its interaction with TNF-alpha -308 promoter SNP. J Med Genet 2003; 40:296-299.
24. Scola L, Crivello A, Marino V et al. IL-10 and TNF-alpha polymorphisms in a sample of sicilian patients affected by tuberculosis: Implication for ageing and life span expectancy. Mech Ageing Dev 2003; 124:569-572.
25. Lio D, Caruso C, Di Stefano R et al. IL-10 and TNF-alpha polymorphisms and the recovery from HCV infection. Hum Immunol 2003; 64:674-680
26. Ross R. Atherosclerosis: An inflammatory disease. N Engl J Med 1999; 340:115-126.
27. Lusis AJ. Atherosclerosis. Nature 2000; 407:233-241.
28. Libby P. Inflammation in atherosclerosis. Nature 2002; 420:868–874.
29. Ridker PM, Rifai N, Rose L et al. Comparison of C-reactive protein and low-density lipoprotein cholesterol levels in the prediction of first cardiovascular events. N Engl J Med 2002; 347:1557-1565.
30. Ridker PM, Morrow DA. C-reactive protein, inflammation, and coronary risk. Cardiol Clin. 2003; 21:315-325.
31. Lee RT, Libby P. The unstable atheroma. Arterioscler Thromb Vasc Biol 1997; 17:1859-1867.
32. Mallat Z, Besnard S, Duriez M et al. Protective role of interleukin-10 in atherosclerosis. Circ Res 1999; 85:e17–e24.
33. Raisanen-Sokolowski A, Glysing-Jensen T, Russell ME. Leukocyte-suppressing influences of interleukin (IL)-10 in cardiac allografts: Insights from IL-10 knockout mice. Am J Pathol 1998; 153:1491–1500.
34. Pinderski LJ, Fischbein MP, Subbanagounder G et al. Overexpression of interleukin-10 by activated T lymphocytes inhibits atherosclerosis in LDL receptor-deficient mice by altering lymphocyte and macrophage phenotypes. Circ Res 2002; 90:1064-1071.
35. Tedgui A, Mallat Z. Interleukin-10: An anti-atherogenic cytokine Pathol Biol (Paris) 2001; 49:107-108.
36. Vasse M, Paysant I, Soria J et al. Down-regulation of fibrinogen biosynthesis by IL-4, IL-10 and IL-13. Br J Haematol 1996; 93:955-961.
37. Smith DA, Irving SD, Sheldon J et al. Serum levels of the anti-inflammatory cytokine interleukin-10 are decreased in patients with unstable angina. Circulation 2001; 104:746–749.
38. Heeschen C, Dimmeler S, Hamm CW et al. for the CAPTURE Study Investigators. Serum level of the antiinflammatory cytokine interleukin-10 is an important prognostic determinant in patients with acute coronary syndromes. Circulation 2003, 107:2109-2114.
39. Kanters E, Pasparakis M, Gijbels MJ et al. Inhibition of NF-kappaB activation in macrophages increases atherosclerosis in LDL receptor-deficient mice. J Clin Invest 2003; 112:1176-1185.
40. Collins T, Cybulsky MI. NF-kappaB: Pivotal mediator or innocent bystander in atherogenesis? J Clin Invest 2001; 107:255-264.
41. Brand K, Page S, Rogler G et al. Activated transcription factor nuclear factor-kappa B is present in the atherosclerotic lesion. J Clin Invest 1996; 97:1715-1722.
42. Karin M, Ben-Neriah Y. Phosphorylation meets ubiquitination: The control of NF-[kappa]B activity. Annu Rev Immunol 2000; 18:621-663.
43. Zhou X, Hansson GK. Detection of B cells and proinflammatory cytokines in atherosclerotic plaques of hypercholesterolaemic apolipoprotein E knockout mice. Scand J Immunol 1999; 50:25-30.
44. Edfeldt K, Swedenborg J, Hansson GK et al. Expression of toll-like receptors in human atherosclerotic lesions: A possible pathway for plaque activation. Circulation. 2002; 105:1158-1161.
45. Cattaruzza M, Slodowski W, Stojakovic M et al. Interleukin-10 induction of nitric-oxide synthase expression attenuates CD40-mediated interleukin-12 synthesis in human endothelial cells. J Biol Chem 2003; 278:37874-37880.
46. Barbieri M, Ferrucci L, Corsi AM et al. Is chronic inflammation a determinant of blood pressure in the elderly? Am J Hypertens 2003; 16:537-543.
47. Ford ES, Giles WH. Serum-reactive protein and self-reported stroke: Findings from the Third National Health and Nutrition Examination Survey. Arterioscler Thromb Vasc Biol 2000; 20:1052–1056.
48. Gussekloo J, Schaap MC, Frolich M et al. C-reactive protein is a strong but nonspecific risk factor of fatal stroke in elderly persons. Arterioscler Thromb Vasc Biol 2000; 20:1047–1051.
49. Ridker PM, Cushman M, Stampfer MJ et al. Plasma concentration of C-reactive protein and risk of developing peripheral vascular disease. Circulation 1998; 97:425–428.
50. Vila N, Castillo J, Davalos A et al. Pro-inflammatory cytokines and early neurological worsening in ischemic stroke. Stroke 2000; 31:2325–2329.
51. van Exel E, Gussekloo J, de Craen AJ et al. Inflammation and stroke: The Leiden 85-Plus Study. Stroke. 2002; 33:1135-1138.

52. Grilli M, Barbieri I, Basudev H et al. Interleukin-10 modulates neuronal threshold of vulnerability to ischaemic damage. Eur J Neurosci 2000; 12:2265–2272.
53. Spera PA, Ellison JA, Feuerstein GZ et al. IL-10 reduces rat brain injury following focal stroke. Neurosci Lett 1998; 251:189–192.
54. Yudkin JS, Kumari M, Humphries SE et al. Inflammation, obesity, stress and coronary heart disease: Is interleukin-6 the link? Atherosclerosis 2000; 148:209–214.
55. Vila N, Castillo J, Davalos A et al. Levels of anti-inflammatory cytokines and neurological worsening in acute ischemic stroke. Stroke 2003; 34:671-675.
56. Donger C, Georges JL, Nicaud V et al. New polymorphisms in the interleukin-10 gene-relationships to myocardial infarction. Eur J Clin Invest 2001; 31:9-14.
57. Koch W, Kastrati A, Bottiger C et al. Interleukin-10 and tumor necrosis factor gene polymorphisms and risk of coronary artery disease and myocardial infarction. Atherosclerosis 2001; 159:137-144.
58. Girndt M, Ulrich C, Kaul H et al. Uremia-associated immune defect: The IL-10-CRP axis. Kidney Int Suppl 2003; 84:S76-79.
59. Lio D, Candore G, Crivello A et al. Opposite effects of IL-10 common gene polymorphisms in cardiovascular disease and in successful ageing: Genetic background of male centenarians is protective against coronary heart disease. J Med.Genet 2004, in press.
60. Pickup JC, Mattock MB, Chusney GD et al. NIDDM as a disease of the innate immune system: Association of acute-phase reactants and interleukin-6 with metabolic syndrome X. Diabetologia 1997; 40:1286–1292.
61. Pickup JC, Crook MA. Is type II diabetes mellitus a disease of the innate immune system? Diabetologia 1998; 41:1241–1248.
62. Schmidt MI, Duncan BB, Sharrett AR et al. Markers of inflammation and prediction of diabetes mellitus in adults (Atherosclerosis Risk in Communities study): A cohort study. Lancet 1999; 353:1649–1652.
63. Spranger J, Kroke A, Mohlig M et al. Inflammatory Cytokines and the Risk to Develop Type 2 Diabetes Results of the Prospective Population-Based European Prospective Investigation into Cancer and Nutrition (EPIC)-Potsdam Study. Diabetes 2003; 52:812–817.
64. Fishman D, Faulds G, Jeffery R et al. The effect of novel polymorphisms in the interleukin-6 (IL-6) gene on IL-6 transcription and plasma IL-6 levels, and an association with systemic-onset juvenile chronic arthritis. J Clin Invest 1998; 102:1369-1376.
65. Fernandez-Real JM, Broch M, Vendrell J et al. Interleukin-6 gene polymorphism and insulin sensitivity. Diabetes 2000; 49:517-520.
66. van Exel E, Gussekloo J, de Craen AJM et al. Low production capacity of interleukin-10 associates with the metabolic syndrome and type 2 diabetes. The Leiden 85-Plus Study. Diabetes 2002; 51:1088–1092.
67. Ide A, Kawasaki E, Abiru N et al. Genetic association between interleukin-10 gene promoter region polymorphisms and type 1 diabetes age-at-onset. Hum Immunol 2002; 63:690-695.
68. Gunnett CA, Heistad DD, Faraci FM. Interleukin-10 protects nitric oxide-dependent relaxation during diabetes: Role of superoxide. Diabetes 2002; 51:1931-1937.
69. Slavin AJ, Maron R, Weiner HL. Mucosal administration of IL-10 enhances oral tolerance in autoimmune encephalomyelitis and diabetes. Int Immunol 2001; 13:825-833.
70. Nussbaum RL, Ellis CE. Alzheimer's disease and Parkinson's disease. N Engl J Med 2003; 348:1356-1364.
71. Wick G, Berger P, Jansen-Durr P et al. A Darwinian-evolutionary concept of age-related diseases. Exp Gerontol 2003; 38:13-25.
72. Candore G, Balistreri CR, Colonna Romano G et al. Major histocompatibilty complex and sporadic Alzheimer's disease: A critical reappraisal. Exp Gerontol 2004; 35:645-652.
73. Akiyama H, Barger S, Barnum S et al. Inflammation and Alzheimer's disease. Neurobiol Ageing 2000; 21:383-421.
74. Franceschi C, Valensin S, Lescai F et al. Neuroinflammation and the genetics of Alzheimer's disease: The search for a pro-inflammatory phenotype. Aging Clin Exp Res 2001; 13:163-170.
75. Licastro F, Chiappelli M. Brain immune responses cognitive decline and dementia: Relationship with phenotype expression and genetic background. Mech Ageing Dev 2003; 124:539-548.
76. Teunissen CE, de Vente J, Steinbusch HW et al. Biochemical markers related to Alzheimer's dementia in serum and cerebrospinal fluid. Neurobiol Aging 2002; 23:485-508.
77. in t' Veld BA, Ruitenberg A, Hofman A et al. Nonsteroidal antiinflammatory drugs and the risk of Alzheimer's disease N Engl J Med 2001; 345:1515-1521.
78. Lee YB, Nagai A, Kim SU. Cytokines, chemokines, and cytokine receptors in human microglia. J Neurosci Res 2002; 69:94-103.

79. Mattson MP, Barger SW, Furukawa K et al. Cellular signaling roles of TGF-beta, TNF-alpha and beta-APP in brain injury responses and Alzheimer's disease. Brain Res Rev 1997; 23:47–61.
80. Mizuno T, Sawada M, Marunouchi T et al. Production of IL-10 by mouse glial cells in culture. Biochem Biophys Res Commun 1994; 205:1907–1915.
81. Strle K, Zhou JH, Shen WH et al. Interleukin-10 in the brain. Crit Rev Immunol 2001; 21:427-449.
82. Hsiao K, Chapman P, Nilsen S et al. Correlative memory deficits, Aβ elevation and amyloid plaques in transgenic mice. Science 1996; 274:99–102.
83. Apelt J, Schliebs R. β-Amyloid-induced glial expression of both pro- and anti-inflammatory cytokines in cerebral cortex of aged transgenicTg2576 mice with Alzheimer plaque pathology. Brain Research 2001; 894:21–30.
84. Balasingam V, Yong VW. Attenuation of astroglial reactivity by interleukin-10. J Neurosci 1996; 16:2945–2955.
85. Chao CC, Hu S, Sheng WS et al. Tumor necrosis factor-alpha production by human fetal microglial cells: Regulation by other cytokines. Dev Neurosci 1995; 17:97–105.
86. Frei K, Lins H, Schwerdel C et al. Antigen presentation in the central nervous system. The inhibitory effect of IL-10 on MHC class II expression and cytokine production depends on the inducing signals and the type of cell analysed. J Immunol 1994; 152:2720–2728.
87. Myers AJ, Goate AM. The genetics of late-onset Alzheimer's disease. Curr Opin Neurol 2001; 14:433-440.
88. Rosenberg RN. The molecular and genetic basis of AD: The end of the beginning. Neurology 2000; 54:2045-2054.
89. McGeer PL, McGeer EG. Polymorphisms in inflammatory genes and the risk of Alzheimer disease. Arch Neurol 2001; 58:1790-1792.
90. Griffin WS, Nicoll JA, Grimaldi LM et al. The pervasiveness of interleukin-1 in Alzheimer pathogenesis: A role for specific polymorphisms in disease risk. Exp Gerontol 2000; 35:481-487.
91. Licastro F, Grimaldi LM, Bonafe M et al. Interleukin-6 gene alleles affect the risk of Alzheimer's disease and levels of the cytokine in blood and brain. Neurobiol Aging 2003; 24:921-926.
92. Caruso C, Franceschi C, Licastro F. Genetics of neurodegenerative disorders. N Engl J Med 2003; 349:193-194 (Letter).
93. Remarque EJ, Bollen EL, Weverling-Rijnsburger AW et al. Patients with Alzheimer's disease display a pro-inflammatory phenotype. Exp Gerontol 2001; 36:171-176.
94. Depboylu C, Du Y, Muller U et al. Lack of association of interleukin-10 promoter region polymorphisms with Alzheimer's disease. Neurosci Lett 2003; 342:132-134.
95. Schenk D, Barbour R, Dunn W et al. Immunization with amyloid-beta attenuates Alzheimer-disease-like pathology in the PDAPP mouse. Nature 1999; 400:173–177.
96. Bard F, Cannon C, Barbour R et al. Peripherally administered antibodies against amyloid beta-peptide enter the central nervous system and reduce pathology in a mouse model of Alzheimer disease. Nat Med 2000; 6:916–919.
97. Town T, Vendrame M, Patel A et al. Reduced Th1 and enhanced Th2 immunity after immunization with Alzheimer's β-amyloid1–42. J Neuroimmunol 2002; 132:49–59.
98. Kastin AJ, Akerstrom V, Pan W. Interleukin-10 as a CNS therapeutic: The obstacle of the blood-brain/blood-spinal cord barrier. Brain Res Mol Brain Res 2003; 114:168-171.
99. Mandel S, Grunblatt E, Youdim M. cDNA microarray to study gene expression of dopaminergic neurodegeneration and neuroprotection in MPTP and 6-hydroxydopamine models: Implications for idiopathic Parkinson's disease. J Neural Transm Suppl 2000; 60:117-124.
100. Wandinger KP, Hagenah JM, Kluter H et al. Effects of amantadine treatment on in vitro production of interleukin-2 in de-novo patients with idiopathic Parkinson's disease J Neuroimmunol 1999; 98:214–220.
101. Bonafe M, Olivieri F, Cavallone L et al. A gender—dependent genetic predisposition to produce high levels of IL-6 is detrimental for longevity. Eur J Immunol 2001; 31:2357-2361.
102. Lio D, Scola L, Crivello A et al. Allele frequencies of +874T—>A single nucleotide polymorphism at the first intron of interferon-gamma gene in a group of Italian centenarians. Exp Gerontol 2002; 37:315-319.
103. Wakkach A, Fournier N, Brun V et al. Characterization of dendritic cells that induce tolerance and T regulatory 1 cell differentiation in vivo. Immunity 2003; 18:605-617.

CHAPTER 4

IL-10:
Role in Infectious Diseases

M. Kryworuchko, W. David Creery and Ashok Kumar

Abstract

Interleukin-10 (IL-10) is a pleiotropic cytokine produced by monocytes/macrophages, B cells, and various subsets of T cells including T helper type 2 (Th2), Th0 cells, and regulatory T (Tr) cells and to a lesser extent by Th1 cells. IL-10 exerts diverse effects on a number of hemopoeitic cell types that include regulation of proliferation and differentiation of most immune cells. The main biological function of IL-10 appears to be immunosuppressive and anti-inflammatory in nature and therefore, it has been exploited therapeutically for the treatment of various autoimmune disorders such as inflammatory bowel disease, psoriasis and rheumatoid arthritis in Phase II and Phase III clinical trials. Studies of IL-10 manipulation by employing IL-10 knock out and IL-10 transgenic animal models have revealed a central role for IL-10 in the immunopathogenesis of a number of infectious diseases. Overall, the results suggest that in infectious diseases wherein inflammation constitutes a critical component of pathogenesis, IL-10 appears to be playing a beneficial role by limiting the extent of inflammation. The significant role of IL-10 in host immune responses against infectious agents has led to phase I clinical trials in a number of human infectious diseases. To our knowledge, results of IL-10 administration in clinical trials are limited to infections with human immunodeficiency virus (HIV), hepatitis C virus (HCV) and experimental endotoxemia/sepsis. Therefore, in this review, we will discuss the role of IL-10 in the immunopathogenesis of these diseases, and describe its therapeutic potential.

Introduction

The resolution of a disease following infection is dependent upon the production of an appropriate balance of pro- and anti-inflammatory cytokines. Since most cytokines are pleuripotent, it is difficult to assess their protective role per se in a particular disease process. However, there are a small number of cytokines whose role has been demonstrated in animal models. As a result, several such cytokines (e.g., IL-2, IFN-α, IFN-γ and GM-CSF) have been evaluated in randomized controlled trials for the treatment of human diseases.[1-3] Interleukin-10 is a cytokine with potent inhibitory activity on inflammation and cell-mediated immune responses. The biological properties of IL-10 and the molecular mechanism by which it exerts its biological effects on various cell types are reviewed elsewhere.[4-6] Because of its potent anti-inflammatory effects, IL-10 has been used therapeutically in the recent past for the treatment of various autoimmune disorders such as inflammatory bowel disease and rheumatoid arthritis in Phase II and Phase III clinical trials.[7,8] Since IL-10 has been implicated in the pathogenesis of a number of infectious diseases, this review will describe its role in infection, focussing on results from studies in experimental animal models. Preclinical trials of IL-10 have also been undertaken in a few human infectious diseases such as endotoxaemia, and infections with human immunodeficiency-(HIV) and hepatitis C viruses (HCV). Therefore, we

Interleukin-10, edited by Francesco M. Marincola. ©2006 Eurekah.com.

will describe the role of IL-10 in immunopathogenesis and the potential for therapeutic manipulation of IL-10 in these diseases.

Regulatory Role of IL-10 in Infectious Diseases: Results from Experimental Mouse Models

In general, the patho-physiological effects of a particular cytokine can be investigated by modulating its levels through (a) exogenous administration of the cytokine, blocking cytokine activity by anti-cytokine or anti-cytokine receptor antibodies, or (b) through selectively disrupting (knock-out models) or over-expressing (transgenic models) the cytokine gene. The global immunosuppressive effects of IL-10 have been demonstrated by employing both IL-10 knock out (IL-10 KO) and IL-10 transgenic mice (IL-10 Tg). IL-10 KO mice demonstrate a state of chronic inflammation characterized by an overexpression of Th1 cytokines.[9,10] These mice are growth-retarded, anemic, and feature a form of enterocolitis which is similar to human inflammatory bowel disease. This condition can be partially ameliorated by exogenous administration of IL-10.[10] IL-10 Tg mice, on the other hand, exhibit impaired cell-mediated and humoral immune responses to a number of intracellular pathogens including Leishmania, and Listeria.[11] In addition, these mice are resistant to the induction of experimental autoimmune encephalomyelitis.[12] Similarly, IL-10 Tg mice were unable to limit the growth of immunogenic tumor cells, and administration of anti-IL-10 antibodies restored anti-tumor responses indicating that cell mediated immunity (CMI) was globally suppressed in these mice.[13] Furthermore, systemic administration of IL-10 inhibited footpad swelling [indicative of delayed type hypersensitivity (DTH)], which was accompanied by an inhibition of vascular permeability and production of Th1 cytokines in the footpads.[14]

The role of IL-10 in a number of bacterial, fungal, protozoan and viral infections has been extensively studied by employing both IL-10 KO and IL-10 Tg animals. The induction of protective or harmful effects of IL-10 in these infectious disease models may depend upon the nature of the host immunologic response to the infectious agent. The effects of IL-10 in terms of disease resistance or susceptibility following challenge with the infectious organism are summarized in Table 1 and Figure 1.

Bacterial Sepsis

Sepsis is a clinical diagnosis marked by a constellation of clinical and laboratory findings suggestive of detrimental immune activation by pro-inflammatory mediators associated with bacterial infection. Numerous animal studies with widely varying models of sepsis have been performed. In general, animal models featuring inflammation in the absence of active infection (such as endotoxin administration) demonstrate benefit when IL-10 is present and detriment when IL-10 is antagonized.[31,58-61] Systemic administration of IL-10 reproducibly protected mice from a lethal intraperitoneal injection of endotoxin, and this protection could be reversed by prior administration of anti-IL-10 antibodies.[31] Furthermore, IL-10 KO mice demonstrate increased mortality upon administration of endotoxin[62] or staphylococcal enterotoxin B,[32] and show exaggerated production of proinflammatory Th1-type cytokines. In contrast, models using live pathogens, such as Klebsiella, Listeria and Pneumococcus[15,19-21] demonstrate a primarily harmful effect of IL-10, which is reversed upon its neutralization. Cecal ligation and puncture is a model that mimics the human septic process of peritonitis, as necrotic tissue is present and endogenous gut organisms are released into the circulation. Several studies have demonstrated a beneficial role of IL-10 in this model.[22,33,63]

Viral Infections

The role of IL-10 in animal models of viral infection has not been investigated extensively. Nevertheless, a study of Herpes Simplex Virus-1 (HSV-1) induced stromal keratitis in mice, characterized by ocular inflammation leading to blindness, demonstrated that intraperitoneal administration of IL-10 reduced the incidence of blindness, cellular infiltration and levels of IL-2 and IL-6 in the cornea. However, IL-10 did not influence virus clearance from the eye.[46]

Table 1. Summary of the role of IL-10 in disease outcome for various pathogens in experimental mouse models

Organism / Disease Model	Method of IL-10 Modulation	Disease Status Resistance (R)/Susceptibility (S)	References
Bacterial Infections			
Listeria monocytogenes	rIL-10	S	15
	IL-10 Tg	S	11
	IL-10 KO	R	16
	Anti-IL-10	R	17
Chlamydia trachomatis	IL-10 KO	R	9
	Anti-IL-10	R	18
Klebsiella pneumoniae	Anti-IL-10	R	19,20,21
Streptococcus pneumoniae	rIL-10	S	21,22
	Anti-IL-10	R	22
Mycobacterium bovis	IL-10 Tg	S	23
	IL-10 KO	R	24
M. tuberculosis	IL-10 KO	No difference	25
M. avium	Anti-IL-10	R	26,27
Salmonella choleraesuis	Anti-IL-10	R	28
Staphylococcus aureus	Anti-IL-10	R	29
Borrelia bugdorferi	IL-10 KO	R	30
Endotoxin	rIL-10	S	31
	Anti-IL-10	R	31
Staphylococcus enterotoxin B	IL-10 KO	S	32
Caecal ligation & puncture	IL-10	R	22,33
Protozoal Infections			
Leishmania major	rIL-10	S	34
	IL-10 Tg	S	11,13
	Anti-IL-10	R	11,35,36
	IL-10 KO	R	36,37
Trypanosoma cruzi	rIL-10	S	38
	Anti-IL-10	R	39
	IL-10 KO	R	38,40,41
Trypanosoma congolense	Anti-IL-10	R	42
Toxolplasma gondii	IL-10 KO	S	43
Plasmodium yoelii	IL-10 KO	No difference	44
Plasmodium chabaudi	IL-10 KO	Inconclusive	44,45
Viral Infections			
Herpes Simplex Virus	rIL-10 / plasmid	R	46,47
Vaccinia Virus	IL-10 KO	R	48
JHMV	IL-10 KO	S	49
EMCV	rIL-10	R	50
Fungi			
Candida albicans	rIL-10	S	51
	Anti-IL-10	R	52
	IL-10 KO	R	53,54
Cryptococcus neoformans	IL-10 KO	R	55
Coccidiodes immitis	IL-10 KO	R	56
Aspergillus fumigatus	Anti-IL-10	R	57
	IL-10 KO	R	54

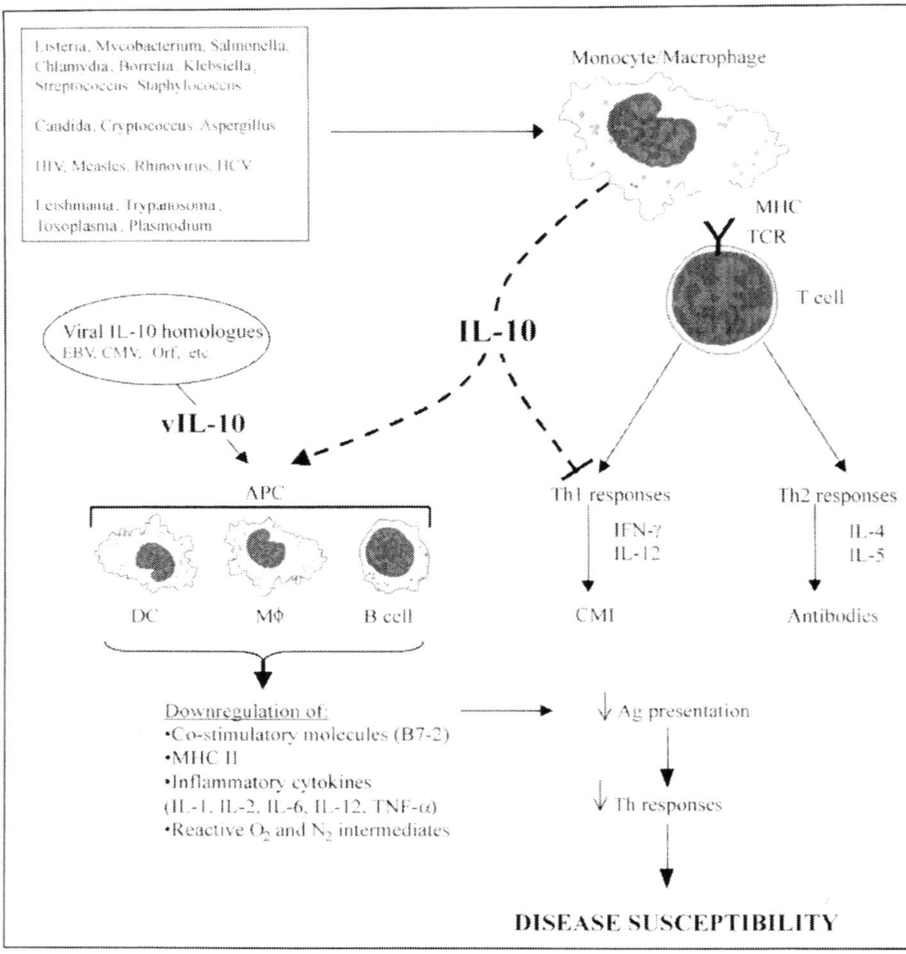

Figure 1. Schematic depicting the impact of IL-10 on the immune response to infectious agents.

In another study of herpetic stromal keratitis, established ocular lesions could be controlled in the majority of animals by the administration of a plasmid DNA encoding IL-10.[47] In a recent study on acute vaccinia virus infection, Van der Broek et al demonstrated that virus replication was impaired in IL-10 KO mice, which was associated with its impact on IL-12 expression and the induction of cytotoxic T lymphocyte (CTL) responses.[48] Neurotropic mouse hepatitis virus (JHMV)-induced demyelinating encaphalomyelitis has also been investigated in IL-10 KO mice wherein increased morbidity and mortality were observed.[49] This was associated with increased central nervous system (CNS) inflammation, and production of inflammatory mediators (TNF-α, IFN-γ and iNOS), however, IL-10 was not required for elimination of infectious virus from the brain.[49] In a murine model of viral myocarditis caused by encephalomyocarditis virus (EMCV), IL-10 administration reduced myocardial lesions resulting in enhanced survival compared to control animals but did not affect virus replication.[50] Taken together, in the viral infections exhibiting inflammatory pathology, these observations suggest that although IL-10 is not required for viral clearance, it appears to be one of the mechanisms responsible for inhibiting the extent of inflammation.

Intracellular Bacterial and Protozoan Pathogens

In certain infections with intracellular pathogens characterized by protective Th1 responses, such as *Listeria monocytogenes,*[16] *Chlamydia trachomatis*[9] or *Candida albicans,*[53] IL-10 KO animals demonstrate improved survival, suggesting a detrimental role for endogenous IL-10. Concordantly, overproduction of IL-10 by macrophages has been suggested to explain the increased susceptibility of neonatal mice to infection with Listeria.[64] Furthermore, IL-10 Tg mice are highly susceptible to *L. monocytogenes* or *Leishmania major* infection,[13] and administration of IL-10 to resistant strains of mice abolished protection from *L. monocytogenes.*[15]

There are conflicting studies with respect to the role of IL-10 in infections with Mycobacterium and Plasmodium, perhaps relating to differences in the strains tested. *Mycobacterium bovis*-infected IL-10 KO mice showed lower bacterial burdens compared to control mice,[24] whereas IL-10 Tg mice were highly susceptible to infection with this organism.[23] However, in another study, IL-10 KO mice infected with *M. tuberculosis* did not show any differences in the course of pulmonary infection compared to controls.[25] Murine infection with malarial parasites have been characterized by their requirement for either antibody- or cell-mediated immunity for suppression of acute parasitemia, with *Plasmodium yoelii* reportedly requiring antibodies for suppression and *P. chabaudi* requiring CMI. IL-10 KO mice infected with *P. chabaudi* demonstrated exacerbated pathology and increased mortality.[45] However, in another study, IL-10 KO mice showed a similar time course of infection with either parasite compared to that seen in controls.[44]

Involvement of IL-10 in the Persistence of Infection with Intracellular Pathogens

As mentioned above, resistance to *L. major, M. tuberculosis,* or *L. monocytogenes,* is generally enhanced in IL-10 KO mice. However, severe inflammatory responses against some of these intracellular pathogens persisted in IL-10-deficient mice resulting in enhanced mortality at later stages, suggesting that IL-10 can also have a beneficial impact on the outcome of such infections. A critical role for IL-10 in the persistence of these organisms in the host and disease progression has been demonstrated in infection with Leishmania as summarized in Figure 2.[35,65-68] Genetically resistant strains of mice mount protective Th1 responses; however, infection persists in these mice for several weeks, and complete eradication of the parasite is never observed. These characteristics are reminiscent of the clinical recurrences in human leishmaniasis. Elimination of the organism, resulting in a sterile cure was observed only in IL-10 KO mice.[66] In contrast, IL-10 Tg mice are highly susceptible to infection with Leishmania, and self-healing resistant B6 mice rendered transgenic for IL-10 are converted from a resistant to a noncuring phenotype.[66] Furthermore, administration of anti-IL-10 receptor blocking antibodies induced parasite killing within liver macrophages.[35,69] Along similar lines, signal transducer and activator of transcription (STAT)-1 KO mice in which IFN-γ-mediated biological effects and Th1 development are compromised, failed to control *L. major* infection.[70,71] Similar observations were made in mice infected experimentally with *P. chabaudi,*[45] *T. cruzi, T. gondii,*[41,72] *Aspergillus fumigatus,*[73] and *Schistosoma mansoni.*[74] Taken together, these studies suggest that endogenous IL-10 may exert beneficial effects through the control of inflammation.

Role of IL-10-Producing Regulatory T Cells in the Persistence of Infection

In the anti-leishmanial response of genetically resistant strains of mice, IL-10 has been suggested to be produced primarily by the infiltrating CD4+ IFN-γ-expressing T cells in the lesions.[65,66] IL-10 and IFN-γ-producing cells constitute approximately 25% of the total IFN-γ-expressing population of CD4+ cells in the lesions.[66] These cells may correspond to the CD4+ CD25+ regulatory T cells (Tr) observed in cancer, autoimmune diseases, and transplantation.[75,76] IL-10 and IFN-γ producing cells may differ from Tr cells in that Tr cells produce IL-10 but may exert immunosuppressive effects via IL-10-independent, cell-contact-dependent

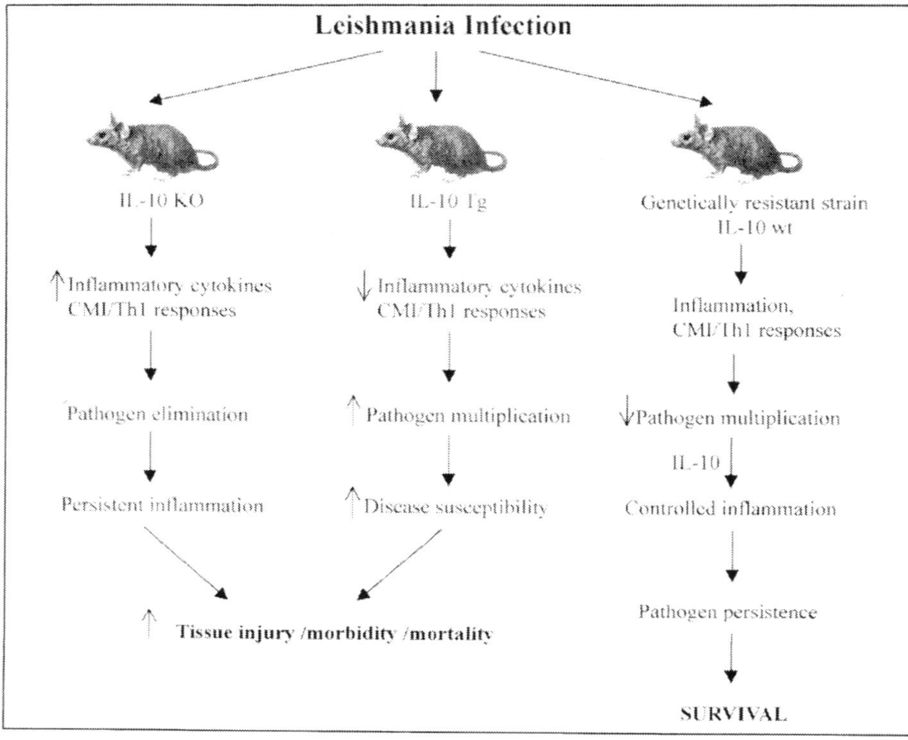

Figure 2. Influence of IL-10 on disease outcome following infection with leishmania, as determined in experimental mouse models.

mechanisms. IL-10 and IFN-γ producing cells have also been shown to produce TGF-β in certain cases, however, their immunosuppressive properties have been attributed primarily to IL-10.[77,78]

Recently, antigen-specific IL-10-producing Tr cells have been detected in a number of infectious diseases. Pathogen-specific CD4+ T cell clones producing both IL-10 and IFN-γ have been detected in broncho-alveolar lavages from patients infected with *M. tuberculosis.*[79] These T cell clones produced IL-10 and IFN-γ in an antigen-dependent as well as an antigen-independent manner. High frequencies of antigen-specific T cells producing both IL-10 and IFN-γ have also been detected in T cell lines derived from patients infected with *Borrelia burgdorferi,* (Lyme disease),[80,81] *P. falciparum,*[82] *Pneumocystis carinii,*[83] HCV,[84] Leishmania,[85,86] HIV,[87] and other retroviral infections.[86,88] In addition, both of these cytokines have been detected in the lesions and plasma of patients with Leishmaniasis, suggesting the involvement of IL-10 in clinical persistence and disease reactivation.[68] These observations suggest that IL-10 and IFN-γ producing cells enriched at the site of infection or lesions may inhibit elimination of the pathogen from the host but at the same time may serve to curtail the harmful inflammatory effects as observed in the IL-10 KO mouse model of *L. major* infection.

The Role of IL-10 in Human Infectious Diseases

In general, human pathogens induce IL-10 production. However, the clinical significance of alterations in IL-10 production is only becoming clear for certain pathogens such as bacteria causing sepsis, and viruses such as HIV and HCV. Monocytes/macrophages constitute the primary source of IL-10 production. These intracellular pathogens infect macrophages and induce host IL-10 production, which may play a significant role in selectively reducing the expression of MHC class II and costimulatory molecules on the surface of antigen presenting cells, leading to reduced antigen presentation and consequently impaired immune responses (Fig. 1).[89] We have demonstrated that peripheral blood mononuclear cells (PBMC) from a subset of patients infected with HSV and exhibiting recurrent genital lesions produce high levels of IL-10. The elevated levels of IL-10 persisted for prolonged periods after a recurrence, while levels of IFN-γ diminished during this period. This suggested that IL-10 exerts a critical influence on recurrence of infection and HSV immunopathogenesis.[90,91] Some viruses such as Epstein-Barr Virus (EBV) and Cytomegalovirus (CMV) may also encode IL-10 homologs to evade host cell defenses.[92,93] The significant role of IL-10 in host immune responses against infectious agents has received tremendous scrutiny and has led to phase I clinical trials. To our knowledge, results of IL-10 administration in clinical trials of human infectious diseases are limited to infections with HIV, HCV and experimental endotoxemia/sepsis. Therefore, we will limit our discussion to the role of IL-10 in the immunopathogenesis of these infections.

Bacterial Sepsis

In adults with sepsis, IL-10 levels were elevated,[94,95] and have been correlated with levels of IL-6, IL-8 and TNF-α.[96] Circulating IL-10 was found to be elevated in adults with shock of both septic and nonseptic origins. However, IL-10 levels were higher in patients with septic shock.[97] Adults with disseminated meningococcemia and shock have high levels of IL-10 in their serum.[96,98] In children, high levels of circulating IL-10 were positively correlated with high levels of IL-12,[99] and with mortality.[100] In meningococcal shock, levels of IL-10 were found to be higher in nonsurvivors than in survivors.[100] Taken together, IL-10 has been found to be increased in patients with sepsis and septic shock in a manner similar to an acute phase reactant.

HIV Infection

The immunosuppressive effects of HIV resulting from the depletion of CD4+ T cell subsets are well documented. In addition, increased levels of IL-10 in HIV infections may have some effects on immunosuppression, particularly in advanced stages of disease. It has been reported that IL-10 concentrations are increased in the serum of HIV-infected patients with ongoing mycobacterial infection,[101] and in patients with Acquired Immunodeficiency Syndrome (AIDS) compared to patients with nonprogressing disease.[102,103] Furthermore, patients undergoing therapy with anti-retroviral regimens displayed a significant decrease in IL-10 producing T cells and serum IL-10 concentrations.[87,101]

Constitutive expression of IL-10 in T cells, including HIV-gag-specific CD4+ T cells, especially from patients with advanced stages of disease has also been observed.[87,104,105] However, variable results have been reported regarding IL-10 production following mitogenic (PHA) stimulation of PBMC from HIV+ individuals. IL-10 production was shown to be enhanced,[106] to be unaltered[107-109] or to be decreased.[105,110,111] IL-10 production was inversely associated with Th activity, as measured by IL-2 production in response to stimulation with recall antigens.[106] Interestingly, the impaired Th functions could be restored with anti-IL-10 antibodies,[105,106,111] suggesting that IL-10 exerts immunosuppressive effects in HIV infection.[105] Furthermore, B cell lines derived from AIDS patients are EBV+ and constitutively secrete high levels of IL-10[112,113] that may play a role in the development of B cell abnormalities in AIDS patients.

We have demonstrated a loss of IL-10 production in T cells but not monocytes, in a subset of HIV+ individuals, which was associated with defective CD28-mediated T cell proliferation.[105] Similar results were obtained in symptomatic rhesus monkeys infected with simian immunodeficiency virus.[114] These observations suggested a differential mechanism for the regulation of IL-10 in T cells and monocytes of HIV-infected individuals. Furthermore, infection of monocytic cells in vitro with HIV induces IL-10 production.[115-117] The molecular mechanism by which HIV infection may differentially induce IL-10 production by T cells and monocytes is not well understood. There is evidence to suggest that the HIV tat protein induces IL-10 expression in T cells and monocytic cells.[118,119] However, gp41/120 caused a rapid increase in IL-10 production by monocytes but not in T, B or NK cells.[120-122] The HIV nef protein has also been suggested to induce IL-10 expression in monocytes.[123,124]

Altered production of IL-10 may be of clinical significance since it has been suggested to play a role in inhibiting HIV replication,[117,125-128] in inducing immune unresponsiveness,[105,106,111,122,129,130] and disease progression.[131] Although IL-10 inhibits HIV replication in T cells and monocytic cells, it has also been shown to enhance replication of HIV in promonocytic cell lines and dendritic cells.[132,133] IL-10 has also been shown to cooperate with TNF-α to activate HIV in monocytic cells.[132,134] The significance of enhanced HIV replication in dendritic cells and other monocytic cells by IL-10 is not clear at present.

HCV Infection

The establishment of chronic HCV infection and poor responsiveness to current therapeutic regimens (IFN-α and ribavirin) may be associated with impaired virus-specific T cell responses, and enhanced levels of immunosuppressive cytokines like IL-10. IL-10 production has been investigated ex vivo in HCV-infected patients and in vitro following stimulation with viral proteins. HCV core and NS3 antigens have been suggested to induce IL-10 production by PBMCs of chronically HCV-infected patients.[84,135-137] IL-10 levels have also been reported to be modulated following stimulation of PBMC from chronically infected patients with PHA or LPS.[138-140] Plasma IL-10 levels were also elevated in patients with chronic infection compared to those that have cleared infection.[84] However, plasma IL-10 levels were not predictive of disease progression or recovery following IFN-α treatment.[141] Furthermore, virus-specific CD4 and CD8 T cells from chronically infected patients have exhibited impaired functional responses to antigens.[142,143] In the CD4 T cells, this has been suggested to be associated with increased IL-10 secretion.[84,144] The clinical significance of increased IL-10 production by antigen- or mitogen-stimulated PBMC of HCV-infected patients is not clear. It is likely that IL-10 production could attenuate virus-specific cell mediated immune responses and promote the establishment of persistent infection. Alternatively, IL-10 may also be beneficial by exerting its anti-inflammatory effects in the liver. Taken together, a balance between these forces may ultimately determine the outcome of infection.

Viral IL-10 Homologs

In addition to influencing the production of cellular IL-10, some viruses encode their own IL-10 homologues (vIL-10). It has been suggested that these viral homologues mimic the biological effects of human IL-10 (hIL-10) and in this way provide a survival advantage for the virus. The first vIL-10 was identified as BCRF1 in the genome of EBV,[92] which displays many of the immunosuppresive activities of its human counterpart in vitro including inhibition of antigen-specific T cell proliferation via the down regulation of MHC class II expression and antigen presentation by macrophages.[145] Viral IL-10 encoded by CMV,[93] despite exhibiting only 27% homology with hIL-10, also exerts potent immuno-suppressive properties.[146] Another example is the parapox Orf virus, which induces cutaneous pustular lesions in sheep, goats as well as humans. It also encodes a vIL-10 that has been recently demonstrated to exert inhibitory effects on inflammatory cytokine production.[147-150]

IL-10 Promoter Polymorphism in Infectious Diseases

Several genetic polymorphisms including that of HLA genes have been proposed to explain differential disease susceptibility and severity in a number of autoimmune disorders. Recently, genetic markers in cytokine genes have been analyzed for their association with disease susceptibility and progression.[151] Since IL-10 expression is of patho-physiological relevance, linkage analysis of the IL-10 promoter haplotype to disease susceptibility has been conducted.[4,152-156] The human IL-10 gene is located on chromosome 1q31-g32 and encodes for 5 exons. The polymorphic IL-10 promoter region contains two informative microsatellites, namely IL-10.G and IL-10.R, which are 1.2 kb and 4 kb upstream of the transcriptional start site, respectively. In addition, the promoter region of the IL-10 gene contains 3 single nucleotide polymorphisms (G/A at -1082, C/T at -819 and C/A at -592), which form three haplotypes, GCC, ACC, and ATA. These haplotypes are known to differ in terms of their effect on IL-10 gene transcription.[155,157-161]

IL-10 promoter polymorphism has been strongly linked to a number of auto-immune diseases, including systemic lupus erythematosus, psoriasis, and rheumatoid arthritis.[154-156,162] Recently, single nucleotide polymorphisms within the IL-10 promoter have been associated with infection including HCV, tuberculosis and leprosy.[163-178] In HCV infection, 85% of patients become chronically infected whereas only 15% clear the virus spontaneously. Interestingly, recovery in HCV infection was associated with high levels of IL-10 production and an increased frequency of the -1082 G/G IL-10 promoter genotype.[163] An association between IL-10 promoter polymorphism (-1082 G/G) and recurrence of HCV infection has also been reported.[168,172,177] Heterozygosity for the -1082 polymorphism was also found to be associated with tuberculosis susceptibility.[165] In leprosy, the frequency of the homozygous -819 T/T genotype was significantly higher among patients than control subjects.[166] In addition, IL-10 promoter polymorphism has been associated with susceptibility to herpes zoster,[171] EBV,[173,176] and experimental human endotoxemia.[175]

Clinical Trials Evaluating the Therapeutic Potential of IL-10 in Patients with Infectious Diseases

Healthy Volunteers

Preliminary studies have been performed in which IL-10 has been administered to healthy adult volunteers.[179-184] Recombinant IL-10 is generally well tolerated by healthy volunteers at doses of 25 μg/kg or less. Higher doses (50 and 100 μg/kg) have been associated with mild-to-moderate flu-like symptoms of fever with chills and myalgias.[183] IL-10 administration resulted in increased numbers of neutrophils, and monocytes. In parallel, platelet counts, CD3+, CD4+ and CD8+ T cell numbers were reduced. IL-10 administration also reduced LPS-stimulated ex vivo production of pro-inflammatory cytokines including TNF-α, IL-1β, IL-6, and IL-8, and PHA- and PMA-stimulated IFN-γ production.[180-185] Studies comparing IL-10 with corticosteroids, both alone and in combination, have demonstrated that even low doses of IL-10 (8 μg/kg) suppressed ex vivo pro-inflammatory cytokine production in a more prolonged and efficacious manner than did corticosteroids.[180] However, IL-10 did not affect the production of endogenous anti-inflammatory molecules such as soluble TNF-receptor p55 or IL-1 receptor antagonist (IL-1ra).[181]

Experimental Human Endotoxemia

IL-10 has been reported to have both pro-inflammatory and anti-inflammatory effects, as well as effects on the fibrinolytic system in experimental human endotoxemia.[186-188] When given to subjects 1 hour after LPS administration, IL-10 was noted to have the pro-inflammatory effects of enhancing ex vivo LPS-induced IFN-γ release, and activation of CTLs and NK cells.[186] Conversely, pretreatment with IL-10 resulted in the anti-inflammatory effects of a reduction in LPS-induced increases in body temperature and ex vivo release of TNF-α, IL-6, IL-8 and IL-1ra,[188] as well as reduced granulocyte accumulation and degranulation.[188] Likewise, IL-10 has been reported to mitigate the pro-inflammatory effects of LPS on the fibrinolytic system.[187]

HIV Infection

Experimental evidence suggests that treatment of HIV-infected monocyte-derived macrophages with IL-10 results in a dose-dependent inhibition of HIV reverse transcriptase activity, and TNF-α-induced HIV replication. Also, exogenous IL-10 may inhibit viral assembly in HIV-infected macrophages. Although the exact role of high IL-10 expression in AIDS patients remains to be fully elucidated, these observations clearly suggested that administration of IL-10 to HIV-infected individuals might confer benefit. To date, three Phase I trials of IL-10 in HIV infection have been reported,[189-191] and in all, IL-10 was well tolerated. In one study, IL-10 transiently reduced HIV replication.[189] In two other studies however, no significant reduction in plasma HIV RNA levels or changes in CD4+ T cell counts were observed.[190,191] These divergent findings highlight the complexity of the immuno-pathogenesis of HIV and the difficulty in selecting patients suitable for IL-10 therapy.

HCV Infection

Therapy for chronic infection with HCV includes IFN-α and/or ribavirin, however, its efficacy is limited. Recent studies demonstrating an association between IL-10 promoter polymorphism and recovery from HCV infection (see above) as well as the potent anti-fibrotic properties of IL-10 prompted its therapeutic evaluation. There have been two pilot studies investigating the safety and efficacy of IL-10 administration in chronically infected HCV patients. McHutchison et al[192] studied the effect of IL-10 treatment on liver alanine amino transferase (ALT) levels and HCV replication in 2 treatment naïve and 14 patients refractory to IFN therapy. IL-10 administration resulted in the normalization of ALT enzyme levels in 50% of patients. However, ALT returned to pretreatment levels 4 wks after cessation of therapy. In a second study, similar promising results were obtained[193] wherein serum ALT levels normalized in 86% of patients. Furthermore, hepatic inflammation and fibrosis decreased in 19 and 14 out of 22 patients, respectively. Overall, it appears that IL-10 improves liver histology and function without any impact on virus replication.

Concluding Remarks

The potent immunoregulatory properties of IL-10 on a wide variety of cell types make this cytokine a potentially therapeutic and broadly effective molecule. IL-10 has shown promising therapeutic effects in many autoimmune and inflammatory disorders including Crohn's disease, rheumatoid arthritis, and psoriasis.[6] Studies of IL-10 manipulation in animal models have revealed a central role for IL-10 in the immunopathogenesis of a number of infectious diseases as well, which has led to clinical trials of IL-10 administration to humans. Overall, the clinical benefit of IL-10 in these diseases appears to be realized under conditions where uncontrolled inflammation constitutes a major component of the disease process. However, the therapeutic administration of IL-10 is unlikely to be beneficial in circumstances where the infectious pathogen has not been controlled. In humans with endotoxemia/sepsis, IL-10 administration has been shown to mediate a beneficial reduction in pro-inflammatory cytokine production, as well as to ameliorate the harmful hemodynamic and coagulopathic effects of endotoxemia. Demonstration of a beneficial anti-inflammatory effect of IL-10 in sepsis is critically dependent on selection of patients in whom the infectious trigger has been controlled and the pro-inflammatory septic cascade is harmful rather than beneficial.

Promising therapeutic effects were observed in HCV infection, demonstrating that IL-10 could inhibit inflammation and fibrosis in the liver. Large, international, placebo controlled trials must be completed to evaluate more clearly the clinical impact of IL-10 therapy in patients with chronic HCV infection. IL-10 administration to HIV+ individuals may be useful in downregulating inflammatory cytokines such as TNF-α and IL-6, which are known to promote virus replication. However, therapeutically IL-10 may be ineffective when viral replication is ongoing and virus loads are high. In fact, IL-10 may be detrimental in this situation, as it has been shown to enhance virus replication in dendritic cells and in cell lines chronically infected with HIV, a situation which may hold true in HIV-infected patients. To date, Phase I trials of administration of IL-10 to HIV+ individuals have met with limited success emphasizing the complex role of IL-10 in HIV immunopathogenesis and HIV replication.

The lack of a clear picture demonstrating the therapeutic efficacy of IL-10 in infectious diseases warrants further investigation. To demonstrate the beneficial effects of IL-10, it is paramount that a proper set of patients are selected for analysis. The clinical value of IL-10 antagonists, such as neutralizing anti-IL-10 or anti-IL-10 receptor antibodies, in diseases which show protective Th1 responses such as lepromatous leprosy, tuberculosis and visceral leishmaniasis should be undertaken. Recently, new IL-10 homologues have been identified which may have better therapeutic value.[194] In addition, little is known about the intracellular signaling pathways implicated in the immunosuppressive and stimulatory effects of IL-10. Understanding of the molecular mechanisms and the signaling molecules involved may help in designing strategies and in identifying new drug targets for mimicking or inhibiting the biological effects of IL-10.

References

1. Lieschke GJ, Burgess AW. Granulocyte colony-stimulating factor and granulocyte-macrophage colony-stimulating factor. N Engl J Med 1992; 327:99-106.
2. Baron S, Tyring SK, Fleischmann WRJ et al. The interferons. Mechanisms of action and clinical applications. JAMA 1991; 266:1375-1383.
3. Davey RTJ, Chaitt DG, Piscitelli SC et al. Subcutaneous administration of interleukin-2 in human immunodeficiency virus type 1-infected persons. J Infect Dis 1997; 175:781-789.
4. Moore KW, de Waal MR, Coffman RL et al. Interleukin-10 and the interleukin-10 receptor. Annu Rev Immunol 2001; 19:683-765.
5. Mocellin S, Panelli MC, Wang E et al. The dual role of IL-10. Trends Immunol 2003; 24:36-43.
6. Asadullah K, Sterry W, Volk HD. Interleukin-10 therapy—review of a new approach. Pharmacol Rev 2003; 55:241-269.
7. Keystone E, Wherry J, Grint P. IL-10 as a therapeutic strategy in the treatment of rheumatoid arthritis. Rheum Dis Clin North Am 1998; 24:629-639.
8. van Deventer SJ, Elson CO, Fedorak RN. Multiple doses of intravenous interleukin 10 in steroid-refractory Crohn's disease. Crohn's Disease Study Group Gastroenterology 1997; 113:383-389.
9. Yang X, Gartner J, Zhu L et al. IL-10 gene knockout mice show enhanced Th1-like protective immunity and absent granuloma formation following Chlamydia trachomatis lung infection. J Immunol 1999; 162:1010-1017.
10. Kuhn R, Lohler J, Rennick D et al. Interleukin-10-deficient mice develop chronic enterocolitis. Cell 1993; 75:263-274.
11. Groux H, Cottrez F, Rouleau M et al. A transgenic model to analyze the immunoregulatory role of IL-10 secreted by antigen-presenting cells. J Immunol 1999; 162:1723-1729.
12. Cua DJ, Groux H, Hinton DR et al. Transgenic interleukin 10 prevents induction of experimental autoimmune encephalomyelitis. J Exp Med 1999; 189:1005-1010.
13. Hagenbaugh A, Sharma S, Dubinett SM et al. Altered immune responses in interleukin 10 transgenic mice. J Exp Med 1997; 185:2101-2110.
14. Li L, Elliott JF, Mosmann TR. IL-10 inhibits cytokine production, vascular leakage, and swelling during T helper 1 cell-induced delayed-type hypersensitivity. J Immunol 1994; 153:3967-3978.
15. Kelly JP, Bancroft GJ. Administration of interleukin-10 abolishes innate resistance to Listeria monocytogenes. Eur J Immunol 1996; 26:356-364.
16. Dai WJ, Kohler G, Brombacher F. Both innate and acquired immunity to Listeria monocytogenes infection are increased in IL-10-deficient mice. J Immunol 1997; 158:2259-2267.
17. Wagner RD, Maroushek NM, Brown JF et al. Treatment with anti-interleukin-10 monoclonal antibody enhances early resistance to but impairs complete clearance of Listeria monocytogenes infection in mice. Infect Immun 1994; 62:2345-2353.
18. Yang X, HayGlass KT, Brunham RC. Genetically determined differences in IL-10 and IFN-gamma responses correlate with clearance of Chlamydia trachomatis mouse pneumonitis infection. J Immunol 1996; 156:4338-4344.
19. Greenberger MJ, Strieter RM, Kunkel SL et al. Neutralization of IL-10 increases survival in a murine model of Klebsiella pneumonia. J Immunol 1995; 155:722-729.
20. van der Poll T, Marchant A, Keogh CV et al. Interleukin-10 impairs host defense in murine pneumococcal pneumonia. J Infect Dis 1996; 174:994-1000.
21. Wang M, Jeng KC, Ping LI. Exogenous cytokine modulation or neutralization of interleukin-10 enhance survival in lipopolysaccharide-hyporesponsive C3H/HeJ mice with Klebsiella infection. Immunology 1999; 98:90-97.
22. van der Poll T, Marchant A, Buurman WA et al. Endogenous IL-10 protects mice from death during septic peritonitis. J Immunol 1995; 155:5397-5401.
23. Murray PJ, Wang L, Onufryk C et al. T cell-derived IL-10 antagonizes macrophage function in mycobacterial infection. J Immunol 1997; 158:315-321.

24. Murray PJ, Young RA. Increased antimycobacterial immunity in interleukin-10-deficient mice. Infect Immun 1999; 67:3087-3095.
25. North RJ. Mice incapable of making IL-4 or IL-10 display normal resistance to infection with Mycobacterium tuberculosis. Clin Exp Immunol 1998; 113:55-58.
26. Denis M, Ghadirian E. IL-10 neutralization augments mouse resistance to systemic Mycobacterium avium infections. J Immunol 1993; 151:5425-5430.
27. Bermudez LE, Champsi J. Infection with Mycobacterium avium induces production of interleukin-10 (IL-10), and administration of anti-IL-10 antibody is associated with enhanced resistance to infection in mice. Infect Immun 1993; 61:3093-3097.
28. Arai T, Hiromatsu K, Nishimura H et al. Endogenous interleukin 10 prevents apoptosis in macrophages during Salmonella infection. Biochem Biophys Res Commun 1995; 213:600-607.
29. Sasaki S, Nishikawa S, Miura T et al. Interleukin-4 and interleukin-10 are involved in host resistance to Staphylococcus aureus infection through regulation of gamma interferon. Infect Immun 2000; 68:2424-2430.
30. Brown JP, Zachary JF, Teuscher C et al. Dual role of interleukin-10 in murine Lyme disease: Regulation of arthritis severity and host defense. Infect Immun 1999; 67:5142-5150.
31. Howard M, Muchamuel T, Andrade S et al. Interleukin 10 protects mice from lethal endotoxemia. J Exp Med 1993; 177:1205-1208.
32. Hasko G, Virag L, Egnaczyk G et al. The crucial role of IL-10 in the suppression of the immunological response in mice exposed to staphylococcal enterotoxin B. Eur J Immunol 1998; 28:1417-1425.
33. Walley KR, Lukacs NW, Standiford TJ et al. Balance of inflammatory cytokines related to severity and mortality of murine sepsis. Infec Immun 1996; 64:4733-4738.
34. Chatelain R, Mauze S, Coffman RL. Experimental Leishmania major infection in mice: Role of IL-10. Parasite Immunol 1999; 21:211-218.
35. Murray HW, Moreira AL, Lu CM et al. Determinants of response to interleukin-10 receptor blockade immunotherapy in experimental visceral leishmaniasis. J Infect Dis 2003; 188:458-464.
36. Noben-Trauth N, Lira R, Nagase H et al. The relative contribution of IL-4 receptor signaling and IL-10 to susceptibility to Leishmania major. J Immunol 2003; 170:5152-5158.
37. Padigel UM, Alexander J, Farrell JP. The role of interleukin-10 in susceptibility of BALB/c mice to infection with Leishmania mexicana and Leishmania amazonensis. J Immunol 2003; 171:3705-3710.
38. Abrahamsohn IA, Coffman RL. Trypanosoma cruzi: IL-10, TNF, IFN-gamma, and IL-12 regulate innate and acquired immunity to infection. Exp Parasitol 1996; 84:231-244.
39. Reed SG, Brownell CE, Russo DM et al. IL-10 mediates susceptibility to Trypanosoma cruzi infection. J Immunol 1994; 153:3135-3140.
40. Holscher C, Mohrs M, Dai WJ et al. Tumor necrosis factor alpha-mediated toxic shock in Trypanosoma cruzi-infected interleukin 10-deficient mice. Infect Immun 2000; 68:4075-4083.
41. Hunter CA, Ellis-Neyes LA, Slifer T et al. IL-10 is required to prevent immune hyperactivity during infection with Trypanosoma cruzi. J Immunol 1997; 158:3311-3316.
42. Uzonna JE, Kaushik RS, Gordon JR et al. Immunoregulation in experimental murine Trypanosoma congolense infection: Anti-IL-10 antibodies reverse trypanosome-mediated suppression of lymphocyte proliferation in vitro and moderately prolong the lifespan of genetically susceptible BALB/c mice. Parasite Immunol 1998; 20:293-302.
43. Neyer LE, Grunig G, Fort M et al. Role of interleukin-10 in regulation of T-cell-dependent and T-cell-independent mechanisms of resistance to Toxoplasma gondii. Infect Immun 1997; 65:1675-1682.
44. van der Heyde HC, Pepper B, Batchelder J et al. The time course of selected malarial infections in cytokine-deficient mice. Exp Parasitol 1997; 85:206-213.
45. Li C, Corraliza I, Langhorne J. A defect in interleukin-10 leads to enhanced malarial disease in Plasmodium chabaudi chabaudi infection in mice. Infect Immun 1999; 67:4435-4442.
46. Tumpey TM, Elner VM, Chen SH et al. Interleukin-10 treatment can suppress stromal keratitis induced by herpes simplex virus type 1. J Immunol 1994; 153:2258-2265.
47. Daheshia M, Kuklin N, Kanangat S et al. Suppression of ongoing ocular inflammatory disease by topical administration of plasmid DNA encoding IL-10. J Immunol 1997; 159:1945-1952.
48. van den BM, Bachmann MF, Kohler G et al. IL-4 and IL-10 antagonize IL-12-mediated protection against acute vaccinia virus infection with a limited role of IFN-gamma and nitric oxide synthetase 2. J Immunol 2000; 164:371-378.
49. Lin MT, Hinton DR, Parra B et al. The role of IL-10 in mouse hepatitis virus-induced demyelinating encephalomyelitis. Virology 1998; 245:270-280.
50. Nishio R, Matsumori A, Shioi T et al. Treatment of experimental viral myocarditis with interleukin-10. Circulation 1999; 100:1102-1108.

51. Tonnetti L, Spaccapelo R, Cenci E et al. Interleukin-4 and -10 exacerbate candidiasis in mice. Eur J Immunol 1995; 25:1559-1565.
52. Romani L, Puccetti P, Mencacci A et al. Neutralization of IL-10 up-regulates nitric oxide production and protects susceptible mice from challenge with Candida albicans. J Immunol 1994; 152:3514-3521.
53. Vazquez-Torres A, Jones-Carson J, Wagner RD et al. Early resistance of interleukin-10 knockout mice to acute systemic candidiasis. Infect Immun 1999; 67:670-674.
54. Del Sero G, Mencacci A, Cenci E et al. Antifungal type 1 responses are upregulated in IL-10-deficient mice. Microbes Infect 1999; 1:1169-1180.
55. Blackstock R, Buchanan KL, Adesina AM et al. Differential regulation of immune responses by highly and weakly virulent Cryptococcus neoformans isolates. Infect Immun 1999; 67:3601-3609.
56. Fierer J, Walls L, Eckmann L et al. Importance of interleukin-10 in genetic susceptibility of mice to Coccidioides immitis. Infect Immun 1998; 66:4397-4402.
57. Cenci E, Mencacci A, Fe dC et al. Cytokine- and T helper-dependent lung mucosal immunity in mice with invasive pulmonary aspergillosis. J Infect Dis 1998; 178:1750-1760.
58. Gerard C, Bruyns C, Marchant A et al. Interleukin 10 reduces the release of tumor necrosis factor and prevents lethality in experimental endotoxemia. J Exp Med 1993; 177:547-550.
59. Berg DJ, Kuhn R, Rajewsky K et al. Interleukin-10 is a central regulator of the response to LPS in murine models of endotoxic shock and the Shwartzman reaction but not endotoxin tolerance. J Clin Invest 1995; 96:2339-2347.
60. Standiford TJ, Strieter RM, Lukacs NW et al. Neutralization of IL-10 increases lethality in endotoxemia. Cooperative effects of macrophage inflammatory protein-2 and tumor necrosis factor. J Immunol 1995; 155:2222-2229.
61. Zheng XX, Steele AW, Nickerson PW et al. Administration of noncytolytic IL-10/Fc in murine models of lipopolysaccharide-induced septic shock and allogeneic islet transplantation. J Immunol 1995; 154:5590-5600.
62. Hickey MJ, Issekutz AC, Reinhardt PH et al. Endogenous interleukin-10 regulates hemodynamic parameters, leukocyte- endothelial cell interactions, and microvascular permeability during endotoxemia. Circ Res 1998; 83:1124-1131.
63. van dP, Marchant A, Buurman WA et al. Endogenous IL-10 protects mice from death during septic peritonitis. J Immunol 1995; 155:5397-5401.
64. Genovese F, Mancuso G, Cuzzola M et al. Role of IL-10 in a neonatal mouse listeriosis model. J Immunol 1999; 163:2777-2782.
65. Belkaid Y, Piccirillo CA, Mendez S et al. CD4+CD25+ regulatory T cells control Leishmania major persistence and immunity. Nature 2002; 420:502-507.
66. Belkaid Y, Hoffmann KF, Mendez S et al. The role of interleukin (IL)-10 in the persistence of Leishmania major in the skin after healing and the therapeutic potential of anti-IL-10 receptor antibody for sterile cure. J Exp Med 2001; 194:1497-1506.
67. Kane MM, Mosser DM. The role of IL-10 in promoting disease progression in leishmaniasis. J Immunol 2001 ;166:1141-1147.
68. Bourreau E, Prevot G, Gardon J et al. High intralesional interleukin-10 messenger RNA expression in localized cutaneous leishmaniasis is associated with unresponsiveness to treatment. J Infect Dis 2001; 1084:1628-1630.
69. Murray HW, Lu CM, Mauze S et al. Interleukin-10 (IL-10) in experimental visceral leishmaniasis and IL-10 receptor blockade as immunotherapy. Infect Immun 2002; 70:6284-6293.
70. Rosas LE, Keiser T, Pyles R et al. Development of protective immunity against cutaneous leishmaniasis is dependent on STAT1-mediated IFN signaling pathway. Eur J Immunol 2003; 33:1799-1805.
71. Sacks D, Noben-Trauth N. The immunology of susceptibility and resistance to Leishmania major in mice. Nat Rev Immunol 2002; 2:845-858.
72. Gazzinelli RT, Wysocka M, Hieny S et al. In the absence of endogenous IL-10, mice acutely infected with Toxoplasma gondii succumb to a lethal immune response dependent on CD4+ T cells and accompanied by overproduction of IL-12, IFN-gamma and TNF- alpha. J Immunol 1996; 157:798-805.
73. Grunig G, Corry DB, Leach MW et al. Interleukin-10 is a natural suppressor of cytokine production and inflammation in a murine model of allergic bronchopulmonary aspergillosis. J Exp Med 1997; 185:1089-1099.
74. Wynn TA, Cheever AW, Williams ME et al. IL-10 regulates liver pathology in acute murine Schistosomiasis mansoni but is not required for immune down-modulation of chronic disease. J Immunol 1998; 160:4473-4480.
75. Bluestone JA, Abbas AK. Natural versus adaptive regulatory T cells. Nat Rev Immunol 2003; 3:253-257.

76. McGuirk P, Mills KH. Pathogen-specific regulatory T cells provoke a shift in the Th1/Th2 paradigm in immunity to infectious diseases.Trends Immunol 2002; 23:450-455.
77. Groux H, O'Garra A, Bigler M et al. A CD4+ T-cell subset inhibits antigen-specific T-cell responses and prevents colitis. Nature 1997; 389:737-742.
78. McGuirk P, McCann C, Mills KH. Pathogen-specific T regulatory 1 cells induced in the respiratory tract by a bacterial molecule that stimulates interleukin 10 production by dendritic cells: A novel strategy for evasion of protective T helper type 1 responses by Bordetella pertussis. J Exp Med 2002; 195:221-231.
79. Gerosa F, Nisii C, Righetti S et al. CD4(+) T cell clones producing both interferon-gamma and interleukin-10 predominate in bronchoalveolar lavages of active pulmonary tuberculosis patients. Clin Immunol 1999; 92:224-234.
80. Pohl-Koppe A, Balashov KE, Steere AC et al. Identification of a T cell subset capable of both IFN-gamma and IL-10 secretion in patients with chronic Borrelia burgdorferi infection. J Immunol 1998; 160:1804-1810.
81. Ganapamo F, Dennis VA, Philipp MT. Early induction of gamma interferon and interleukin-10 production in draining lymph nodes from mice infected with Borrelia burgdorferi. Infect Immun 2000; 68:7162-7165.
82. Plebanski M, Flanagan KL, Lee EA et al. Interleukin 10-mediated immunosuppression by a variant CD4 T cell epitope of Plasmodium falciparum. Immunity 1999; 10:651-660.
83. Hori S, Carvalho TL, Demengeot J. CD25+CD4+ regulatory T cells suppress CD4+ T cell-mediated pulmonary hyperinflammation driven by Pneumocystis carinii in immunodeficient mice. Eur J Immunol 2002; 32:1282-1291.
84. MacDonald AJ, Duffy M, Brady MT et al. CD4 T helper type 1 and regulatory T cells induced against the same epitopes on the core protein in hepatitis C virus-infected persons. J Infect Dis 2002; 185:720-727.
85. Maasho K, Satti I, Nylen S et al. A Leishmania homologue of receptors for activated C-kinase (LACK) induces both interferon-gamma and IL-10 in natural killer cells of healthy blood donors. J Infect Dis 2000; 182:570-578.
86. Bourreau E, Collet M, Prevot G et al. IFN-gamma-producing CD45RA+CD8+ and IL-10-producing CD45RA-CD4+ T cells generated in response to LACK in naive subjects never exposed to Leishmania. Eur J Immunol 2002; 32:510-520.
87. Ostrowski MA, Gu JX, Kovacs C et al. Quantitative and qualitative assessment of human immunodeficiency virus type 1 (HIV-1)-specific CD4+ T cell immunity to gag in HIV-1-infected individuals with differential disease progression: Reciprocal interferon-gamma and interleukin-10 responses. J Infect Dis 2001; 184:1268-1278.
88. Iwashiro M, Messer RJ, Peterson KE et al. Immunosuppression by CD4+ regulatory T cells induced by chronic retroviral infection. Proc Natl Acad Sci USA 2001; 98:9226-9230.
89. Redpath S, Angulo A, Gascoigne NR et al. Murine cytomegalovirus infection down-regulates MHC class II expression on macrophages by induction of IL-10. J Immunol 1999; 162:6701-6707.
90. Singh R, Kumar A, Diaz-Mitoma F. Augmentation of B7 expression by herpes simplex virus antigen. Hum Immunol 2003; 64:780-786.
91. Singh R, Kumar A, Creery WD et al. Dysregulated expression of IFN-gamma and IL-10 and impaired IFN-gamma-mediated responses at different disease stages in patients with genital herpes simplex virus-2 infection. Clin Exp Immunol 2003; 133:97-107.
92. Hsu DH, de Waal MR, Fiorentino DF et al. Expression of interleukin-10 activity by Epstein-Barr virus protein BCRF1. Science 1990; 250:830-832.
93. Kotenko SV, Saccani S, Izotova LS et al. Human cytomegalovirus harbors its own unique IL-10 homolog (cmvIL-10). Proc Natl Acad Sci USA 2000; 97:1695-1700.
94. Marchant A, Deviere J, Byl B et al. Interleukin-10 production during septicaemia. Lancet 1994; 343:707-708.
95. Sherry RM, Cue JI, Goddard JK et al. Interleukin-10 is associated with the development of sepsis in trauma patients. J Traum 1996; 40(4):613-616.
96. Lehmann AK, Halstensen A, Sornes S et al. High levels of interleukin 10 in serum are associated with fatality in meningococcal disease. Infect Immun 1995; 63:2109-2112.
97. Marchant A, Alegre ML, Hakim A et al. Clinical and biological significance of interleukin-10 plasma levels in patients with septic shock. J Clin Immunol 1995; 15:266-273.
98. Derkx B, Marchant A, Goldman M et al. High levels of interleukin-10 during the initial phase of fulminant meningococcal septic shock. J Infect Dis 1995; 171:229-232.
99. Hazelzet JA, Kornelisse RF, van der Pouw Kraan TC et al. Interleukin 12 levels during the initial phase of septic shock with purpura in children: Relation to severity of disease. Cytokine 1997; 9:711-716.

100. Kornelisse RF, Hazelzet JA, Savelkoul HF et al. The relationship between plasminogen activator inhibitor-1 and proinflammatory and counterinflammatory mediators in children with meningococcal septic shock. J Infect Dis 1996; 173:1148-1156.
101. Stylianou E, Aukrust P, Kvale D et al. IL-10 in HIV infection: Increasing serum IL-10 levels with disease progression—down-regulatory effect of potent anti-retroviral therapy. Clin Exp Immunol 1999; 116:115-120.
102. Muller F, Aukrust P, Lien E et al. Enhanced interleukin-10 production in response to Mycobacterium avium products in mononuclear cells from patients with human immunodeficiency virus infection. J Infect Dis 1998; 177:586-594.
103. Muller F, Aukrust P, Nordoy I et al. Possible role of interleukin-10 (IL-10) and CD40 ligand expression in the pathogenesis of hypergammaglobulinemia in human immunodeficiency virus infection: Modulation of IL-10 and Ig production after intravenous Ig infusion. Blood 1998; 92:3721-3729.
104. Diaz-Mitoma F, Kumar A, Karimi S et al. Expression of IL-10, IL-4 and interferon-gamma in unstimulated and mitogen-stimulated peripheral blood lymphocytes from HIV-seropositive patients. Clin Exp Immunol 1995; 102:31-39.
105. Kumar A, Angel J, Daftarian MP et al. Differential production of IL-10 by T cells and monocytes of HIV-infected individuals: association of IL-10 production with CD28-mediated immune responsiveness. Clin Exp Immunol 1998; 114: 78-86.
106. Clerici M, Wynn TA, Berzofsky JA et al. Role of interleukin-10 in T helper cell dysfunction in asymptomatic individuals infected with the human immunodeficiency virus. J Clin Invest 1994; 93:768-775.
107. Graziosi C, Pantaleo G, Gantt KR et al. Lack of evidence for the dichotomy of TH1 and TH2 predominance in HIV- infected individuals. Science 1994; 265:248-252.
108. Chehimi J, Starr SE, Frank I et al. Impaired interleukin 12 production in human immunodeficiency virus- infected patients. J Exp Med 1994; 179:1361-1366.
109. Zhang M, Gong J, Iyer DV et al. T cell cytokine responses in persons with tuberculosis and human immunodeficiency virus infection. J Clin Invest 1994; 94:2435-2442.
110. Meyaard L, Hovenkamp E, Keet IP et al. Single cell analysis of IL-4 and IFN-gamma production by T cells from HIV-infected individuals: Decreased IFN-gamma in the presence of preserved IL-4 production. J Immunol 1996; 157:2712-2718.
111. Daftarian MP, Diaz-Mitoma F, Creery WD et al. Dysregulated production of interleukin-10 (IL-10) and IL-12 by peripheral blood lymphocytes from human immunodeficiency virus-infected individuals is associated with altered proliferative responses to recall antigens. Clin Diagn Lab Immunol 1995; 2:712-718.
112. Benjamin D, Knobloch TJ, Dayton MA. Human B-cell interleukin-10: B-cell lines derived from patients with acquired immunodeficiency syndrome and Burkitt's lymphoma constitutively secrete large quantities of interleukin-10. Blood 1992; 80:1289-1298.
113. Masood R, Zhang Y, Bond MW et al. Interleukin-10 is an autocrine growth factor for acquired immunodeficiency syndrome-related B-cell lymphoma. Blood 1995; 85:3423-3430.
114. Brodie SJ, Sasseville VG, Reimann KA et al. Macrophage function in simian AIDS. Killing defects in vivo are independent of macrophage infection, associated with alterations in Th phenotype, and reversible with IFN-gamma. J Immunol 1994; 153:5790-5801.
115. Akdis CA, Blaser K. IL-10-induced anergy in peripheral T cell and reactivation by microenvironmental cytokines: Two key steps in specific immunotherapy. FASEB J 1999; 13:603-609.
116. Yoo J, Chen H, Kraus T et al. Altered cytokine production and accessory cell function after HIV-1 infection. J Immunol 1996; 157:1313-1320.
117. Kootstra NA, van 't W, Huisman HG et al. Interference of interleukin-10 with human immunodeficiency virus type 1 replication in primary monocyte-derived macrophages. J Virol 1994; 68:6967-6975.
118. Sharma V, Knobloch TJ, Benjamin D. Differential expression of cytokine genes in HIV-1 tat transfected T and B cell lines. Biochem Biophys Res Commun 1995; 208:704-713.
119. Blazevic V, Heino M, Lagerstedt A et al. Interleukin-10 gene expression induced by HIV-1 Tat and Rev in the cells of HIV-1 infected individuals. J Acquir Immune Defic Syndr Hum Retrovirol 1996; 13:208-214.
120. Barcova M, Kacani L, Speth C et al. gp41 envelope protein of human immunodeficiency virus induces interleukin (IL)-10 in monocytes, but not in B, T, or NK cells, leading to reduced IL-2 and interferon-gamma production. J Infect Dis 1998; 177:905-913.
121. Borghi P, Fantuzzi L, Varano B et al. Induction of interleukin-10 by human immunodeficiency virus type 1 and its gp120 protein in human monocytes/macrophages. J Virol 1995; 69:1284-1287.
122. Schols D, De Clercq E. Human immunodeficiency virus type 1 gp120 induces anergy in human peripheral blood lymphocytes by inducing interleukin-10 production. J Virol 1996; 70:4953-4960.

123. Brigino E, Haraguchi S, Koutsonikolis A et al. Interleukin 10 is induced by recombinant HIV-1 Nef protein involving the calcium/calmodulin-dependent phosphodiesterase signal transduction pathway. Proc Natl Acad Sci USA 1997; 94:3178-3182.
124. Creery D, Angel JB, Aucoin S et al. Nef protein of human immunodeficiency virus and lipopolysaccharide induce expression of CD14 on human monocytes through differential utilization of interleukin-10. Clin Diagn Lab Immunol 2002; 9:1212-1221.
125. Akridge RE, Oyafuso LK, Reed SG. IL-10 is induced during HIV-1 infection and is capable of decreasing viral replication in human macrophages. J Immunol 1994; 53:5782-5789.
126. Saville MW, Taga K, Foli A et al. Interleukin-10 suppresses human immunodeficiency virus-1 replication in vitro in cells of the monocyte/macrophage lineage. Blood 1994; 3:3591-3599.
127. Kollmann TR, Pettoello-Mantovani M, Katopodis NF et al. Inhibition of acute in vivo human immunodeficiency virus infection by human interleukin 10 treatment of SCID mice implanted with human fetal thymus and liver. Proc Natl Acad Sci USA 1996; 3:3126-3131.
128. Patterson BK, Czerniewski M, Andersson J et al. Regulation of CCR5 and CXCR4 expression by type 1 and type 2 cytokines: CCR5 expression is downregulated by IL-10 in CD4-positive lymphocytes. Clin Immunol 1999; 1:254-262.
129. Clerici M, Hakim FT, Venzon DJ et al. Changes in interleukin-2 and interleukin-4 production in asymptomatic, human immunodeficiency virus-seropositive individuals. J Clin Invest 1993; 1:759-765.
130. Landay AL, Clerici M, Hashemi F et al. In vitro restoration of T cell immune function in human immunodeficiency virus-positive persons: Effects of interleukin (IL)-12 and anti-IL-10. J Infect Dis 1996; 73:1085-1091.
131. Clerici M, Balotta C, Salvaggio A et al. Human immunodeficiency virus (HIV) phenotype and interleukin-2/ interleukin-10 ratio are associated markers of protection and progression in HIV infection. Blood 1996; 8:574-579.
132. Angel JB, Saget BM, Wang MZ et al. Interleukin-10 enhances human immunodeficiency virus type 1 expression in a chronically infected promonocytic cell line (U1) by a tumor necrosis factor alpha-independent mechanism. J Interferon Cytokine Res 1995; 5:575-584.
133. Ancuta P, Bakri Y, Chomont N et al. Opposite effects of IL-10 on the ability of dendritic cells and macrophages to replicate primary CXCR4-dependent HIV-1 strains. J Immunol 2001; 66:4244-4253.
134. Barcellini W, Rizzardi GP, Marriott JB et al. Interleukin-10-induced HIV-1 expression is mediated by induction of both membrane-bound tumour necrosis factor (TNF)-alpha and TNF receptor type 1 in a promonocytic cell line. AIDS 1996; 0:835-842.
135. Kakumu S, Okumura A, Ishikawa T et al. Serum levels of IL-10, IL-15 and soluble tumour necrosis factor-alpha (TNF-alpha) receptors in type C chronic liver disease. Clin Exp Immunol 1997; 09:458-463.
136. Kakumu S, Okumura A, Ishikawa T et al. Production of interleukins 10 and 12 by peripheral blood mononuclear cells (PBMC) in chronic hepatitis C virus (HCV) infection. Clin Exp Immunol 1997; 08:138-143.
137. Dolganiuc A, Kodys K, Kopasz A et al. Hepatitis C virus core and nonstructural protein 3 proteins induce pro- and anti-inflammatory cytokines and inhibit dendritic cell differentiation. J Immunol 2003; 70:5615-5624.
138. Sarih M, Bouchrit N, Benslimane A. Different cytokine profiles of peripheral blood mononuclear cells from patients with persistent and self-limited hepatitis C virus infection. Immunol Lett 2000; 4:117-120.
139. Amaraa R, Mareckova H, Urbanek P et al. Production of interleukins 10 and 12 by activated peripheral blood monocytes/macrophages in patients suffering from chronic hepatitis C virus infection with respect to the response to interferon and ribavirin treatment. Immunol Lett 2002; 83:209-214.
140. Woitas RP, Petersen U, Moshage D et al. HCV-specific cytokine induction in monocytes of patients with different outcomes of hepatitis C. World J Gastroenterol 2002; 8:562-566
141. Bozkaya H, Bozdayi AM, Aslan N et al. Circulating IL-2 and IL-10 in chronic active hepatitis C with respect to the response to IFN treatment. Infection 2000; 28:309-313.
142. Ulsenheimer A, Gerlach JT, Gruener NH et al. Detection of functionally altered hepatitis C virus-specific CD4 T cells in acute and chronic hepatitis C. Hepatology 2003; 37:1189-1198.
143. Wedemeyer H, He XS, Nascimbeni M et al. Impaired effector function of hepatitis C virus-specific CD8+ T cells in chronic hepatitis C virus infection. J Immunol 2002; 169:3447-3458.
144. Sobue S, Nomura T, Ishikawa T et al. Th1/Th2 cytokine profiles and their relationship to clinical features in patients with chronic hepatitis C virus infection. J Gastroenterol 2001; 36:544-551.

145. de Waal MR, Haanen J, Spits H et al. Interleukin 10 (IL-10) and viral IL-10 strongly reduce antigen-specific human T cell proliferation by diminishing the antigen-presenting capacity of monocytes via downregulation of class II major histocompatibility complex expression. J Exp Med 1991; 174:915-924.
146. Spencer JV, Lockridge KM, Barry PA et al. Potent immunosuppressive activities of cytomegalovirus-encoded interleukin-10. J Virol 2002; 76:1285-1292.
147. Haig DM, McInnes CJ. Immunity and counter-immunity during infection with the parapoxvirus orf virus. Virus Res 2002; 88:3-16.
148. Imlach W, McCaughan CA, Mercer AA et al. Orf virus-encoded interleukin-10 stimulates the proliferation of murine mast cells and inhibits cytokine synthesis in murine peritoneal macrophages. J Gen Virol 2002; 83:1049-1058.
149. Fleming SB, McCaughan CA, Andrews AE et al. A homolog of interleukin-10 is encoded by the poxvirus orf virus. J Virol 1997; 71:4857-4861.
150. Lateef Z, Fleming S, Halliday G et al. Orf virus-encoded interleukin-10 inhibits maturation, antigen presentation and migration of murine dendritic cells. J Gen Virol 2003; 84:1101-1109.
151. Bidwell J, Keen L, Gallagher G et al. Cytokine gene polymorphism in human disease: on-line databases, supplement 1. Genes Immun 2001; 2:61-70.
152. Anaya JM, Correa PA, Herrera M et al. Interleukin 10 (IL-10) influences autoimmune response in primary Sjogren's syndrome and is linked to IL-10 gene polymorphism. J Rheumatol 2002; 29:1874-1876.
153. de Jong BA, Westendorp RG, Eskdale J et al. Frequency of functional interleukin-10 promoter polymorphism is different between relapse-onset and primary progressive multiple sclerosis. Hum Immunol 2002; 63:281-285.
154. Asadullah K, Eskdale J, Wiese A et al. Interleukin-10 promoter polymorphism in psoriasis. J Invest Dermatol 2001; 116:975-978.
155. Eskdale J, McNicholl J, Wordsworth P et al. Interleukin-10 microsatellite polymorphisms and IL-10 locus alleles in rheumatoid arthritis susceptibility. Lancet 1998; 352:1282-1283
156. Eskdale J, Wordsworth P, Bowman S et al. Association between polymorphisms at the human IL-10 locus and systemic lupus erythematosus. Tissue Antigens 1997; 49:635-639.
157. Eskdale J, Keijsers V, Huizinga T et al. Microsatellite alleles and single nucleotide polymorphisms (SNP) combine to form four major haplotype families at the human interleukin-10 (IL-10) locus. Genes Immun 1999; 1:151-155.
158. Eskdale J, Gallagher G, Verweij CL et al. Interleukin 10 secretion in relation to human IL-10 locus haplotypes. Proc Natl Acad Sci USA 1998; 95:9465-9470.
159. De SK, Venkateshan CN, Seth P et al. Adenovirus-mediated human immunodeficiency virus-1 Nef expression in human monocytes/macrophages and effect of Nef on downmodulation of Fcgamma receptors and expression of monokines. Blood 1998; 91:2108-2117.
160. Eskdale J, Kube D, Tesch H et al. Mapping of the human IL10 gene and further characterization of the 5' flanking sequence. Immunogenetics 1997; 46:120-128.
161. Turner DM, Williams DM, Sankaran D et al. An investigation of polymorphism in the interleukin-10 gene promoter. Eur J Immunogenet 1997; 24:1-8.
162. Llorente L, Zou W, Levy Y et al. Role of interleukin 10 in the B lymphocyte hyperactivity and autoantibody production of human systemic lupus erythematosus. J Exp Med 1995; 181:839-844.
163. Lio D, Caruso C, Di Stefano R et al. IL-10 and TNF-alpha polymorphisms and the recovery from HCV infection. Hum Immunol 2003; 64:674-680.
164. Schaaf BM, Boehmke F, Esnaashari H et al. Pneumococcal septic shock is associated with the interleukin-10-1082 gene promoter polymorphism. Am J Respir Crit Care Med 2003; 168:476-480.
165. Lopez-Maderuelo D, Arnalich F, Serantes R et al. Interferon-gamma and interleukin-10 gene polymorphisms in pulmonary tuberculosis. Am J Respir Crit Care Med 2003; 167:970-975.
166. Santos AR, Suffys PN, Vanderborght PR et al. Role of tumor necrosis factor-alpha and interleukin-10 promoter gene polymorphisms in leprosy. J Infect Dis 2002; 186:1687-1691.
167. Delgado JC, Baena A, Thim S et al. Ethnic-specific genetic associations with pulmonary tuberculosis. J Infect Dis 2002; 186:1463-1468.
168. Constantini PK, Wawrzynowicz-Syczewska M, Clare M et al. Interleukin-1, interleukin-10 and tumour necrosis factor-alpha gene polymorphisms in hepatitis C virus infection: An investigation of the relationships with spontaneous viral clearance and response to alpha-interferon therapy. Liver 2002; 22:404-412.
169. Miyazoe S, Hamasaki K, Nakata K et al. Influence of interleukin-10 gene promoter polymorphisms on disease progression in patients chronically infected with hepatitis B virus. Am J Gastroenterol 2002; 97:2086-2092.

170. Konenkov VI, Smol'nikova MV. Polymorphism of promotor sites of interleukins-4 and -10 and tumor necrosis factor-alpha genes in HIV-infected patients. Bull Exp Biol Med 2002; 133:389-391.
171. Haanpaa M, Nurmikko T, Hurme M. Polymorphism of the IL-10 gene is associated with susceptibility to herpes zoster. Scand J Infect Dis 2002; 34:112-114.
172. Vidigal PG, Germer JJ, Zein NN. Polymorphisms in the interleukin-10, tumor necrosis factor-alpha, and transforming growth factor-beta1 genes in chronic hepatitis C patients treated with interferon and ribavirin. J Hepatol 2002; 36:271-277.
173. Wu MS, Huang SP, Chang YT et al. Tumor necrosis factor-alpha and interleukin-10 promoter polymorphisms in Epstein-Barr virus-associated gastric carcinoma. J Infect Dis 2002; 185:106-109.
174. Kumarvelu J, Shanmugasundaram GK, Unwalla H et al. Genetic analyses of the promoter region of interleukin-10 gene in different species of monkeys: Implications for HIV/AIDS progression. Genes Immun 2001; 2:404-407.
175. Fijen JW, Tulleken JE, Hepkema BG et al. The influence of tumor necrosis factor-alpha and interleukin-10 gene promoter polymorphism on the inflammatory response in experimental human endotoxemia. Clin Infect Dis 2001; 33:1601-1603.
176. Helminen ME, Kilpinen S, Virta M et al. Susceptibility to primary Epstein-Barr virus infection is associated with interleukin-10 gene promoter polymorphism. J Infect Dis 2001; 184:777-780.
177. Yee LJ, Tang J, Gibson AW et al. Interleukin 10 polymorphisms as predictors of sustained response in antiviral therapy for chronic hepatitis C infection. Hepatology 2001; 33:708-712.
178. Edwards-Smith CJ, Jonsson JR, Purdie DM et al. Interleukin-10 promoter polymorphism predicts initial response of chronic hepatitis C to interferon alfa. Hepatology 1999; 0:526-530.
179. Chakraborty A, Blum RA, Mis SM et al. Pharmacokinetic and adrenal interactions of IL-10 and prednisone in healthy volunteers. J Clin Pharmacol 1999; 9:624-635.
180. Chakraborty A, Blum RA, Cutler DL et al. Pharmacoimmunodynamic interactions of interleukin-10 and prednisone in healthy volunteers. Clin Pharmacol Ther 1999; 5:304-318.
181. Chernoff AE, Granowitz EV, Shapiro L et al. A randomized, controlled trial of IL-10 in humans. Inhibition of inflammatory cytokine production and immune responses. J Immunol 1995; 54:5492-5499.
182. Fuchs AC, Granowitz EV, Shapiro L et al. Clinical, hematologic, and immunologic effects of interleukin-10 in humans. J Clin Immunol 1996; 291-303.
183. Huhn RD, Radwanski E, O'Connell SM et al. Pharmacokinetics and immunomodulatory properties of intravenously administered recombinant human interleukin-10 in healthy volunteers. Blood 1996; 7:699-705.
184. Huhn RD, Radwanski E, Gallo J et al. Pharmacodynamics of subcutaneous recombinant human interleukin-10 in healthy volunteers. Clin Pharmacol Ther 1997; 2:171-180.
185. Radwanski E, Chakraborty A, Van Wart S et al. Pharmacokinetics and leukocyte responses of recombinant human interleukin-10. Pharm Res 1998; 5:1895-1901.
186. Lauw FN, Pajkrt D, Hack CE et al. Proinflammatory effects of IL-10 during human endotoxemia. J Immunol 2000; 65:2783-2789.
187. Pajkrt D, van der Poll T, Levi M et al. Interleukin-10 inhibits activation of coagulation and fibrinolysis during human endotoxemia. Blood 1997; 9:2701-2705.
188. Pajkrt D, Camoglio L, Tiel-van Buul MC et al. Attenuation of proinflammatory response by recombinant human IL-10 in human endotoxemia: effect of timing of recombinant human IL-10 administration. J Immunol 1997; 58:3971-3977.
189. Weissman D, Ostrowski M, Daucher J A et al. Interleukin-10 Decreases HIV Plasma Viral Load: Results of a Phase I Clinical Trial. Abstract #37, Abstracts of the 4th Conference on Retroviruses and Opportunistic Infections, 1998:71.
190. Ahrendt L D, Witek J, Dean LM et al. A Phase I Study of Interleukin-10 (IL-10) in HIV-infected Individuals. Abstract #613, Abstracts of the 5th Conference on Retroviruses and Opportunistic Infections, 1999:195.
191. Angel JB, Jacobson MA, Skolnik PR et al. A multicenter, randomized, double-blind, placebo-controlled trial of recombinant human interleukin-10 in HIV-infected subjects. AIDS 2000; 14:2503-2508.
192. McHutchison JG, Giannelli G, Nyberg L et al. A pilot study of daily subcutaneous interleukin-10 in patients with chronic hepatitis C infection. J Interferon Cytokine Res 1999; 19:1265-1270.
193. Nelson DR, Lauwers GY, Lau JY et al. Interleukin 10 treatment reduces fibrosis in patients with chronic hepatitis C: A pilot trial of interferon nonresponders. Gastroenterology. 2000; 118:655-660.
194. Volk H, Asadullah K, Gallagher G et al. IL-10 and its homologs: important immune mediators and emerging immunotherapeutic targets. Trends Immunol 2001; 22:414-417.

Chapter 5

Interleukin-10 in Allergic Disease

Peter J. Barnes

Abstract

IL-10 plays a key regulatory role in allergic diseases. It is produced by many of the inflammatory cells involved in allergic inflammation, including macrophages, regulatory T lymphocytes, dendritic cells, mast cells and eosinophils. IL-10 suppresses allergic inflammation by inhibiting the expression of inflammatory cytokines, Th2 cell-derived cytokines, chemokines and inflammatory mediator enzymes. In addition, it suppresses antigen presentation and increases the production of endogenous anti-inflammatory molecules. There is increasing evidence for defective production of IL-10 in allergic diseases, including asthma and rhinitis and this is associated with disease severity. This may lead to amplification of the inflammatory response in allergic diseases. It may be determined by polymorphisms of the IL-10 promoter linked to low endogenous IL-10 production. IL-10 itself may be a therapeutic approach to allergic disease, but because of side effects stimulation of endogenous IL-10 production may be a more useful approach. Corticosteroids restore the impaired IL-10 secretion in asthma and specific immunotherapy increases IL-10 production by regulatory T cells. Novel therapies in the future might include drugs that selectively activate IL-10 signal transduction pathways.

Introduction

There is increasing interest in the role of IL-10 in allergic diseases, particularly asthma, as is it an important anti-inflammatory mediator in allergic diseases, but it may also have therapeutic potential and lead to novel approaches to the treatment of allergic diseases in the future. IL-10 is a unique cytokine with a wide spectrum of anti-inflammatory effects and a profile of activity that suggests that it may be a powerful inhibitor of allergic responses.[1-3]

Recently several related cytokines have been identified, including IL-19, IL-20, IL-22, IL-24 and IL-26, making up a new IL-10 family of cytokines.[4] Some of these cytokines may activate IL-10 receptors, but little is known about their function or their role in allergic diseases. The IL-10 relate cytokines are widely expressed in immune cells making it likely that they pay a role in the inflammatory response.[5]

Secretion on IL-10 Allergic Disease

IL-10 is produced by several cell types, including monocytes, macrophages, T-lymphocytes, dendritic cells, eosinophils and mast cells, all of which are involved in allergic diseases.[6]

Macrophages

Macrophages appear to be the predominant source of IL-10 in the lungs. Human alveolar macrophages spontaneously release IL-10 and this is increased by inflammatory stimuli, such as interleukin(IL)-1β and endotoxin.[7,8] Alveolar macrophages from asthmatic patients release significantly less IL-10 at baseline and in response to inflammatory stimuli than cells from normal subjects.[8] A similar pattern is observed with monocytes from peripheral blood and this

Interleukin-10, edited by Francesco M. Marincola. ©2006 Eurekah.com.

defect in IL-10 secretion is more marked in patients with severe compared to mild asthma, suggesting that this deficit may relate to the severity of asthma and the intensity of airway inflammation.[9]

T Lymphocytes

IL-10 was first discovered as a cytokine synthesis inhibitor secreted from Th2 cells that suppressed Th1 cells.[10] Classically IL-10 production is described as a property of Th2 cells and is often classified as a Th2 cytokine,[11] but it is now apparent that several subtypes of T lymphocytes may secrete IL-10. Recent studies indicate that a subtype of $CD4^+$ T cells termed type 1 regulatory T cells (Tr1) are a major source of IL-10.[12] Tr1 cells produce high concentrations of IL-10 and transforming growth factor-β (TGF-β) and play a major role in suppressing both Th1 and Th2 cells. Differentiation of T cells to the Tr1 subtype appears to be regulated predominantly by IL-10 secretion from dendritic cells.[13] Tr1 cells through the release of IL-10 have a powerful inhibitory effect on allergic sensitization and may produce T cell tolerance.[14]

Dendritic Cells

Dendritic cells play a critical role in setting the balance between Th1 and Th2 immunity and this is mediated partly through the secretion of IL-10. Myeloid dendritic cells have a higher capacity than plasmacytoid dendritic cells to capture allergens, such as the allergenic protein from house dust mite Der p1. In normal individuals Der p1 pulsing results in IL-10 release from myeloid dendritic cells and this induces IL-10 producing $CD4^+$ (Tr1) cells.[15] In allergic subjects, however, these dendritic cells fail to produce IL-10 and do not induce IL-10 production in $CD4^+$ cells. This may contribute to the impaired suppression of Th2 cells in allergic diseases.

Eosinophils

Human eosinophils have the capacity to secrete IL-10 and this is released by activation through high affinity IgE receptors.[16]

Mast Cells

Human mast cells have the capacity to produce IL-10 and this is stimulated by exposure to allergen, via the high affinity IgE receptor, and by endotoxin acting via toll-like receptors (TLR2, TLR4).[17,18] IL-10 may therefore have an autoregulatory effect on mast cells to inhibit release of inflammatory mediators.

Structural Cells

Structural cells of the respiratory tract are increasingly recognized to produce cytokines in allergic diseases. IL-10 immunoreactivity is weakly detectable in airway epithelial cells and is significantly reduced in biopsies obtained from asthmatic patients.[19] Our own studies have been unable to demonstrate secretion of IL-10 from human airway epithelial cells with a variety of stimulants (Lim S, Barnes PJ: unpublished observations).

Effects of IL-10

Mechanisms of Action

IL-10 acts on specific IL-10 receptors that have now been cloned, although the signal transduction pathways and kinases that lead to the widespread anti-inflammatory actions of this cytokine are not yet well understood.[6] Activation of the transcription factor nuclear factor-κB (NF-κB) is important for switching on multiple inflammatory genes in allergic diseases.[20,21] Many of the actions of IL-10 that are relevant to allergic diseases can be explained by an inhibitory effect on NF-κB, as a result of reduced activity of inhibitor of NF-κB kinases (IKK), as well as reducing binding of NF-κB to its recognition sites in the promoter regions of inflammatory genes.[22,23] But this cannot account for all of the effects of IL-10. For example, IL-10 is very

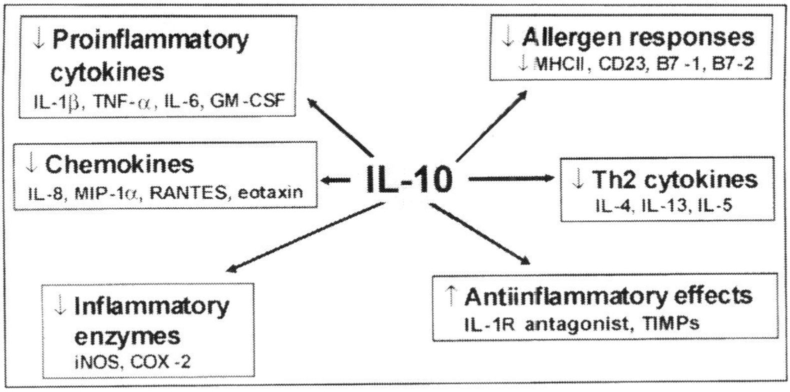

Figure 1. IL-10 has a broad spectrum of anti-allergic effects.

effective at inhibiting IL-5 transcription which is independent of NF-κB.[24] In mice many effects of IL-10 appear to be mediated by an inhibitory effect on phosphodiesterase (PDE)-4,[25] but this does not appear to be the case in human cells.[26,27] An antibody to IL-10 increases the release of cytokines from monocytes after activation, indicating an important endogenous negative feedback role of IL-10 in these cells, but this blocking antibody does not reduce the anti-inflammatory effect of PDE4 inhibitors.

Effects on Inflammatory Mediator Secretion

IL-10 has a very broad spectrum of anti-inflammatory and anti-allergic effects (Fig. 1). It inhibits the synthesis of proinflammatory cytokines (IL-1β, TNF-α, IL-6, GM-CSF), and the Th2 cell-derived cytokines IL-4 and IL-5[24,28] (Fig. 2). IL-10 similarly inhibits the secretion of multiple chemokines, such as MIP-1α, RANTES, IL-8 and eotaxin.[29-32] IL-10 also inhibits the expression of the inflammatory enzymes inducible nitric oxide synthase (iNOS) and inducible cyclooxygenase (COX-2) in macrophages.[33,34]

IL-10 markedly suppresses the secretion of inflammatory cytokines from human alveolar macrophages in vitro, suggesting that it may act as an autoregulator of inflammatory gene expression in alveolar macrophages that also produce it.[29,35,36] IL-10 released from alveolar macrophages tightly regulates the secretion of IL-12, thus playing a critical role in setting the balance between Th1 and Th2 cells.[37]

Inhibitory Effects on Cells

IL-10 inhibits the proliferation of CD4+ T lymphocytes by inhibiting IL-2 release. It reduces the expression of major histocompatibility (MHC) Class II molecules, the co-stimulatory molecules B7-1 and B7-2 and low affinity IgE receptors (CD23) in antigen-presenting cells, thus effectively blocking allergen presentation by mononuclear cells and dendritic cells to T-cells.[38] In addition, IL-10 increases the expression of several anti-inflammatory proteins, including IL-1 receptor antagonist[39] and tissue inhibitor of matrix metalloproteinases.[40]

IL-10 inhibits the release of cytokines from several cell types, including inflammatory cells, epithelial cells and proliferating airway smooth muscle cells.[31,32] It is very effective in inhibiting eosinophilic inflammation in animal models and this may be through its combined inhibitory effects on IL-5 synthesis, the release of eosinophil chemotactic chemokines (such as eotaxin and RANTES) and through a reduction in eosinophil survival, probably as a result of decreased GM-CSF secretion.[41] IL-10 also has an inhibitory effect on IgE-dependent activation of human lung mast cells, with inhibition of TNF-α and IL-8.[42] Surprisingly (and in contrast to results in rodent mast cells) IL-10 also inhibits the release of histamine, suggesting an inhibitory effect on degranulation as well as cytokine synthesis.

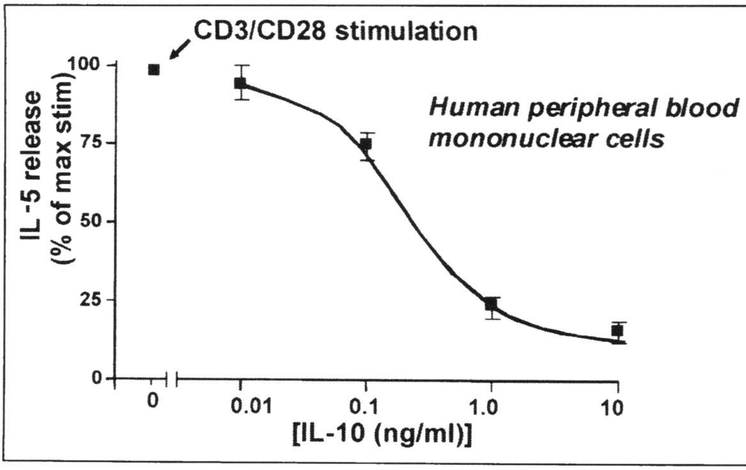

Figure 2. Inhibition of interleukin-5 (IL-5) release from human peripheral blood mononuclear cells activated by anti-CD3 and anti-CD28 by increasing concentrations of IL-10. Adapted from reference 24.

A blocking antibody to IL-10 increases the release of cytokines from monocytes and macrophages, suggesting that IL-10 serves as an endogenous feedback inhibitory mechanism to damp down the inflammatory response.[38] This has also been demonstrated in vivo in a murine model of allergic inflammation, where an IL-10 blocking antibody increases the airway inflammatory response to allergen.[43] Similarly, IL-10 gene knock-out mice have an increased eosinophilic inflammatory response to allergen.[44] The kinetics of IL-10 production show a late secretion, which is not maximal until 24 hours after stimulation, whereas the inflammatory genes suppressed by IL-10 are much more rapidly expressed (6-12 hours). This suggests that IL-10 may function as a late "braking" mechanism that prevents persistence of the inflammatory response. Surprisingly, IL-10 knock-out mice fail to develop airway hyperresponsiveness after allergen sensitization, indicating that IL-10 has some unexpected inhibitory action on airway hyperresponsiveness, at variance with previous studies.[45] However, when these animals are infected with respiratory syncytial virus they show a skewed Th2 response with exaggerated eosinophilic inflammation.[46]

Role in Allergic Diseases

There is increasing evidence that IL-10 secretion may be defective in patients with asthma.[1,47] Lower concentrations of IL-10 are found in bronchoalveolar lavage fluid of asthmatic patients than in normal control subjects[48] and there is a reduction in the number of macrophages expressing IL-10 in induced sputum.[49] There is a reduced secretion of IL-10 from alveolar macrophages obtained by bronchoalveolar lavage from asthmatic patients compared to normal controls, and this is at the level of gene expression.[8] This reduced expression of IL-10 is correlated with an increased production of proinflammatory cytokines, such as TNF-α and GM-CSF, and the chemokine MIP-1α from these cells. This suggests that a defect in IL-10 synthesis may be result in exaggerated and more prolonged inflammatory responses in asthmatic airways (Fig. 3). Furthermore, since IL-10 appears to act as a feedback inhibitor of antigen presentation by mononuclear cells, this may also account for the observation that macrophages from asthmatic patients are less effective at inhibiting T cell proliferative responses.[50] The production of IL-10 from peripheral blood monocytes is reduced in allergic compared to normal subjects after influenza A infection, suggesting an impaired ability to mount an anti-inflammatory response in allergy.[51] Similarly, IL-10 production from peripheral blood mononuclear cells is reduced in children who develop atopy.[52] IL-10 levels are reported to be normal in patients with mild

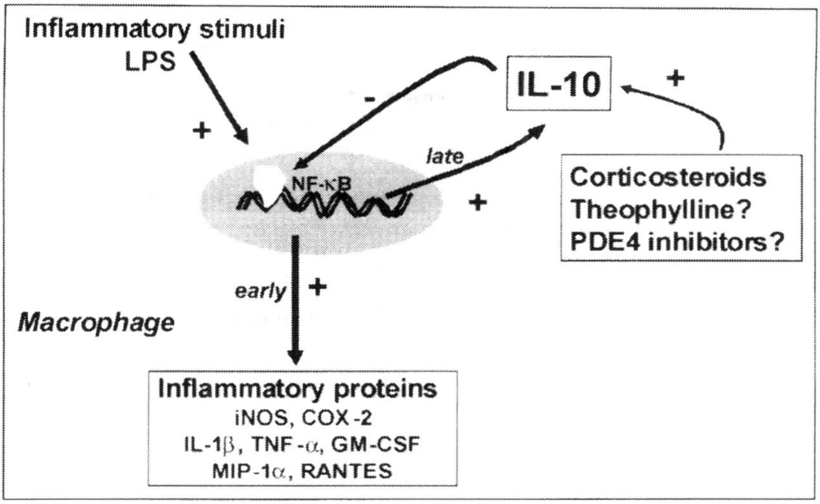

Figure 3. Macrophages rapidly release many inflammatory proteins in response to activating stimuli, such as lipopolysaccharide (LPS), many of which are regulated by nuclear factor-κB (NF-κB). The same activating stimuli increase release of IL-10, but this occurs later and results in inhibition of these inflammatory proteins through inhibition of NF-κB. In allergic disease there is defective production of IL-10, so that the inflammatory response is less suppressed leading to amplification of and prolongation of inflammation.

asthma, however[53] and IL-10 release from peripheral blood monocytes is increased during the late response to allergen[54] and in bronchoalveolar lavage after allergen challenge.[53] Nasal allergen challenge in patients with seasonal allergic rhinitis reduces the concentrations of IL-10 in nasal lavage, however, in line with most of the studies in asthma.[55] In atopic dermatitis an increase in IL-10 expression in the epidermis has been described.[56]

Genetic Mechanisms

The gene for IL-10 has been mapped to chromosome #1 and several polymorphisms in the 5'-promoter region of the IL-10 gene have been identified that are associated with altered synthesis of IL-10 in response to inflammatory stimuli.[57] These polymorphisms have not usually been associated with the prevalence of asthma, but a haplotype that results in reduced IL-10 synthesis is found significantly more often in patients with severe asthma, who require high does of inhaled or oral corticosteroids for control.[58] This is consistent with the greater reduction in IL-10 expression in monocytes of patients with severe compared to mild asthma.[9] A similar association of this haplotypes has also been reported in rheumatoid arthritis with a more severe form of the disease.[59] This suggests that IL-10 may play a key role in determining disease severity and that this may be genetically determined. A polymorphism in the IL-10 promoter has also been associated with increased levels of total IgE[60] and with the number of circulating eosinophils, suggesting that its effect in asthma is mediated through the allergic process.[61] A recent study has reported an association of a promoter polymorphism with asthma prevalence,[62] although this has been negated by another study which confirmed an association between IL-10 polymorphisms and total IgE and eosinophil counts.[63]

Therapeutic Implications

The potent immunosuppressive and anti-inflammatory action of IL-10 have suggested that it may be useful therapeutically in the treatment of allergic diseases.[64-67]

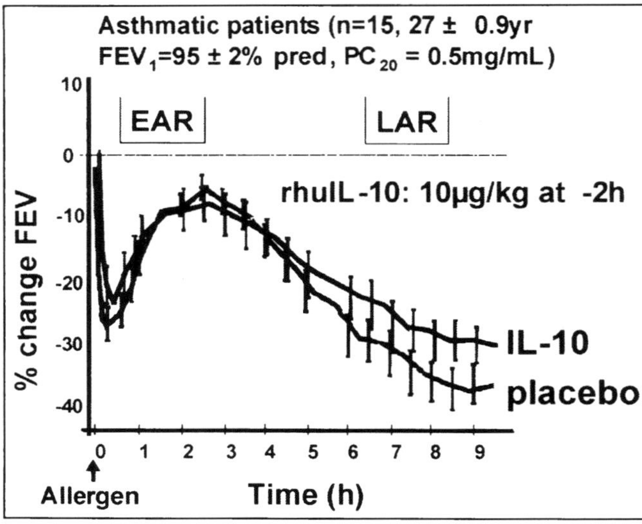

Figure 4. The effect of a single injection of human recombinant IL-10 on allergen challenge in patients with mild asthma. No significant difference compared to placebo was found for the early asthmatic response (EAR) or the late asthmatic response (LAR).

IL-10 Therapy

There are theoretical reasons why IL-10 itself may not be suitable as a treatment, since it suppresses the production of IL-12 and IFN-γ, thus favoring the predominance of Th2 cells. However, its powerful anti-inflammatory effects appear to more than compensate for any tendency to increase eosinophilia, so that in animal studies it is very effective at blocking eosinophilic inflammation. Recombinant IL-10 (given intranasally) is effective in suppressing allergen inflammation in a murine model.[43] Recombinant human IL-10 has been given to healthy volunteers by subcutaneous injection and is reasonably well tolerated, with a reduction in LPS-induced IL-1β and TNF-α release from whole blood ex vivo.[68] In the future it may be possible to deliver IL-10 by a transgene adenovirus vector which results more prolonged production of IL-10[69] and this approach inhibits allergic sensitization in mice.[70] Experimentally allergen-specific $CD4^+$ cells genetically engineered to produce IL-10 protect against the development of eosinophilic inflammation in the lungs and airway hyperresponsiveness in mice challenged with the allergen (ovalbumin),[71] validating the approach of giving IL-10 locally to the lungs. We have studied the effect of a single injection of IL-10 on allergen-induced bronchoconstriction in asthmatic patients, but found no protective effect (Lim S, Barnes PJ: unpublished observations), although it is likely that the duration of treatment was too short (Fig. 4).

Increased IL-10 Synthesis

Another approach to increase IL-10 production is to stimulate its endogenous synthesis and this has the attraction of restoring the defect in IL-10 secretion found in allergic diseases. Treatment with inhaled corticosteroids restores the abnormally low production of IL-10 from alveolar macrophages and concomitantly reduces the release of inflammatory cytokines.[8] It is not certain whether this is a direct effect of corticosteroids on IL-10 synthesis, or whether it is mediated indirectly. In T lymphocytes corticosteroids increase IL-10 production and this is impaired by pretreating the cells with IL-2 and IL-4 which have been shown to induce steroid resistance. This effect that is mediated via activation of p38 mitogen-activated protein kinase.[72] The increased production of IL-10 may account for some of the anti-inflammatory actions of corticosteroids.[73] Inhaled corticosteroids and oral anti-leukotriene therapy have been reported to increase the circulating concentration of IL-10 in children with asthma.[74]

Therapeutic concentrations of theophylline have been reported to increase IL-10 production from monocytes and this may underlie the recently recognized immunomodulatory and anti-inflammatory action of this drug.[75] Theophylline may be effective by inhibition of phosphodiesterases (PDE), thus increasing intracellular cyclic AMP concentrations. In mononuclear cells PDE inhibitors increase IL-10 release,[76] although this does not appear to be the mechanism for inhibiting TNF-α release, as discussed above.[26,27] We have demonstrated that therapeutic concentrations of theophylline in asthmatic patients, while reducing the numbers of airway eosinophils, do not increase IL-10 production by alveolar macrophages, in bronchoalveolar lavage or by blood monocyte or lymphocytes.[77] This makes it unlikely that therapeutic doses of theophylline produce their anti-inflammatory effects by increasing IL-10 secretion.

Specific Immunotherapy

Specific immunotherapy to specific allergens induces allergen-specific anergy in peripheral T lymphocytes and this appears to be due to IL-10 production by Tr1 cells, followed by its production from monocytes.[78] Endogenous IL-10 switches off the production of specific IgE and IgG4. IL-10 acts directly T cells by suppressing the CD28 co-stimulation signal transduction via tyrosine phosphorylation of CD28.[79] Immunotherapy results in increased numbers of Tr1 cells that secrete both IL-10 and TGF-β, resulting in suppression of Th2-driven allergic inflammation.[80] Grass pollen immunotherapy in seasonal rhinitis results in increased IL-10 production and this is expressed predominantly in $CD4^+CD25^+$ T cells that are a variant of regulatory T cells.[81]

Certain bacterial products stimulate Th1 immunity and may therefore have a therapeutic role in reversing the Th2 immunity found in allergic disease. *Lactobacillus* added to the diet of children with atopic dermatitis improves symptoms and this is accompanied by an increased secretion of IL-10 from peripheral blood monocytes.[82]

Novel Approaches

Another possibility for therapy in the future is the development of other agonists for the IL-10 receptor, or drugs that activate the unique but so far unidentified signal transduction pathways activated by this cytokine.

References

1. Borish L. IL-10: Evolving concepts. J Allergy Clin Immunol 1998; 101:293-297.
2. Barnes PJ. IL-10: A key regulator of allergic disease. Clin Exp Allergy 2001; 31:667-669.
3. Bellinghausen I, Knop J, Saloga J. The role of interleukin 10 in the regulation of allergic immune responses. Int Arch Allergy Immunol 2001; 126:97-101.
4. Kotenko SV. The family of IL-10-related cytokines and their receptors: Related, but to what extent? Cytokine Growth Factor Rev 2002; 13:223-240.
5. Wolk K, Kunz S, Asadullah K et al. Immune cells as sources and targets of the IL-10 family members? J Immunol 2002; 168:5397-5402.
6. Moore KW, de Waal MR, Coffman RL et al. Interleukin-10 and the interleukin-10 receptor. Annu Rev Immunol 2001; 19:683-765.
7. Toossi Z, Hirsch CS, Hamilton BD et al. Decreased production of TGF-beta 1 by human alveolar macrophages compared with blood monocytes. J Immunol 1996; 156:3461-3468.
8. John M, Lim S, Seybold J et al. Inhaled corticosteroids increase IL-10 but reduce MIP-1α, GM-CSF and IFN-γ release from alveolar macrophages in asthma. Am J Respir Crit Care Med 1998; 157:256-262.
9. Tomita K, Lim S, Hanazawa T et al. Attenuated production of intracellular IL-10 and IL-12 in monocytes from patients with severe asthma. Clin Immunol 2002; 102:258-266.
10. Fiorentino DF, Bond MW, Mosmann TR. Two types of mouse T helper cell. IV. Th2 clones secrete a factor that inhibits cytokine production by Th1 clones. J Exp Med 1989; 170:2081-2095.
11. Romagnani S. The Th1/Th2 paradigm. Immunol Today 1997; 18:263-266.

12. Roncarolo MG, Bacchetta R, Bordignon C et al. Type 1 T regulatory cells. Immunol Rev 2001; 182:68-79.
13. Levings MK, Bacchetta R, Schulz U et al. The role of IL-10 and TGF-β in the differentiation and effector function of T regulatory cells. Int Arch Allergy Immunol 2002; 129:263-276.
14. Akbari O, Freeman GJ, Meyer EH et al. Antigen-specific regulatory T cells develop via the ICOS-ICOS-ligand pathway and inhibit allergen-induced airway hyperreactivity. Nat Med 2002; 8:1024-1032.
15. Charbonnier AS, Hammad H, Gosset P et al. Der p 1-pulsed myeloid and plasmacytoid dendritic cells from house dust mite-sensitized allergic patients dysregulate the T cell response. J Leukoc Biol 2003; 73:91-99.
16. Kayaba H, Dombrowicz D, Woerly G et al. Human eosinophils and human high affinity IgE receptor transgenic mouse eosinophils express low levels of high affinity IgE receptor, but release IL-10 upon receptor activation. J Immunol 2001; 167:995-1003.
17. Ishizuka T, Okayama Y, Kobayashi H et al. Interleukin-10 is localized to and released by human lung mast cells. Clin Exp Allergy 1999; 29:1424-1432.
18. Varadaradjalou S, Feger F, Thieblemont N et al. Toll-like receptor 2 (TLR2) and TLR4 differentially activate human mast cells. Eur J Immunol 2003; 33:899-906.
19. Bosson J, Stenfors N, Bucht A et al. Ozone-induced bronchial epithelial cytokine expression differs between healthy and asthmatic subjects. Clin Exp Allergy 2003; 33:777-782.
20. Barnes PJ, Karin M. Nuclear factor-κB: A pivotal transcription factor in chronic inflammatory diseases. New Engl J Med 1997; 336:1066-1071.
21. Barnes PJ, Adcock IM. Transcription factors and asthma. Eur Respir J 1998; 12:221-234.
22. Wang P, Wu P, Siegel MI et al. Interleukin(IL)-10 inhibits nuclear factor κB activation in human monocytes. IL-10 and IL-4 suppress cytokine synthesis by different mechanisms. J Biol Chem 1995; 270:9558-9563.
23. Schottelius AJ, Mayo MW, Sartor RB et al. Interleukin-10 signaling blocks inhibitor of κB kinase activity and nuclear factor kappaB DNA binding. J Biol Chem 1999; 274:31868-31874.
24. Staples KJ, Bergmann M, Barnes PJ et al. Stimulus-specific inhibition of IL-5 by cAMP-elevating agents and IL-10 reveals differential mechanisms of action. Biochem Biophys Res Commun 2000; 273:811-815.
25. Kambayashi T, Jacob CO, Zhou D et al. Cyclic nucleotide phosphodiesterase type IV participates in the regulation of IL-10 and in the subsequent inhibition of TNF-α and IL-6 release by endotoxin-stimulated macrophages. J Immunol 1995; 155:4909-4916.
26. Seldon PM, Barnes PJ, Giembycz MA. Interleukin-10 does not mediate the inhibitory effect of PDE4 inhibitors and other cAMP-elevating drugs on lipopolysaccharide-induced tumor necrosis factor-α generation from human peripheral blood monocytes. Cell Biochem Biophys 1998; 28:179-201.
27. Seldon PM, Giembycz MA. Suppression of granulocyte/macrophage colony-stimulating factor release from human monocytes by cyclic AMP-elevating drugs: Role of interleukin-10. Br J Pharmacol 2001; 134:58-67.
28. Schandene L, Alonso-Vega C, Willems F et al. B7/CD28-dependent IL-5 production by human resting T cells is inhibited by IL-10. J Immunol 1994; 152:4368-4374.
29. Berkman N, John M, Roesens G et al. Inhibition of induced expression of macrophage inflammatory protein-1α in human blood monocytes and alveolar macrophages by interleukin-10: Differential sensitivities in human blood monocytes and alveolar macrophages. J Immunol 1995; 155:4412-4418.
30. Berkman N, Robichaud A, Krishnan VL et al. Expression of RANTES in human airway epithelial cells: effect of corticosteroids and interleukins-4, 10 and 13. Immunol 1996; 87:599-603.
31. John M, Jose PJ, Lim S et al. Expression and release of interleukin-8 by human airway smooth muscle cells: Inhibition by Th2 cytokines and corticosteroids. Am J Respir Cell Mol Biol 1998; 18:84-90.
32. Chung KF, Patel HJ, Fadlon EJ et al. Induction of eotaxin expression and release from human airway smooth muscle cells by IL-1β and TNFα: Effects of IL-10 and corticosteroids. Br J Pharmacol 1999; 127:1145-1150.
33. Cunha FQ, Moncada S, Liew FY. Interleukin-10 (IL-10) inhibits the induction of nitric oxide synthase by interferon-gamma in murine macrophages. Biochem Biophys Res Commun 1992; 182:1155-1159.
34. Niiro H, Otsuka T, Tanabe T et al. Inhibition by interleukin-10 of inducible cyclooxygenase expression in lipopolysaccharide-stimulated monocytes: its underlying mechanism in comparison with interleukin-4. Blood 1995; 85:3736-3745.

35. Armstrong L, Jordan N, Millar A. Interleukin 10 (IL-10) regulation of tumour necrosis factor α (TNF-α) from human alveolar macrophages and peripheral blood monocytes. Thorax 1996; 51:143-149.
36. Raychaudhuri B, Fisher CJ, Farver CF et al. Interleukin 10 (IL-10)-mediated inhibition of inflammatory cytokine production by human alveolar macrophages. Cytokine 2000; 12:1348-1355.
37. Isler P, de Rochemonteix BG, Songeon F et al. Interleukin-12 production by human alveolar macrophages is controlled by the autocrine production of interleukin-10. Am J Respir Cell Mol Biol 1999; 20:270-278.
38. de Waal Malefyt R, Haanen J, Spits H et al. Interleukin 10 (IL-10) and viral IL-10 strongly reduce antigen-specific human T cell proliferation by diminishing the antigen-presenting capacity of monocytes via downregulation of class II major histocompatibility complex expression. J Exp Med 1991; 174:915-924.
39. Kline JN, Fisher PA, Monick MM et al. Regulation of interleukin-1 receptor antagonist by Th1 and Th2 cytokines. Am J Physiol 1995; 269:L92-98.
40. Lacraz S, Nicod LP, Chicheportiche R et al. IL-10 inhibits metalloproteinase and stimulates TIMP-1 production in human mononuclear phagocytes. J Clin Invest 1995; 96:2304-2310.
41. Takanaski S, Nonaka R, Xing Z et al. Interleukin 10 inhibits lipopolysaccharide-induced survival and cytokine production by human peripheral blood eosinophils. J Exp Med 1994; 180:711-715.
42. Royer B, Varadaradjalou S, Saas P et al. Inhibition of IgE-induced activation of human mast cells by IL-10. Clin Exp Allergy 2001; 31:694-704.
43. Zuany-Amorim C, Haile S, Leduc D et al. Interleukin-10 inhibits antigen-induced cellular recruitment into the airways of sensitized mice. J Clin Invest 1995; 95:2644-2651.
44. Tournoy KG, Kips JC, Pauwels RA. Endogenous interleukin-10 suppresses allergen-induced airway inflammation and nonspecific airway responsiveness. Clin Exp Allergy 2000; 30:775-783.
45. Makela MJ, Kanehiro A, Borish L et al. IL-10 is necessary for the expression of airway hyperresponsiveness but not pulmonary inflammation after allergic sensitization. Proc Natl Acad Sci USA 2000; 97:6007-6012.
46. Makela MJ, Kanehiro A, Dakhama A et al. The failure of interleukin-10-deficient mice to develop airway hyperresponsiveness is overcome by respiratory syncytial virus infection in allergen-sensitized/challenged mice. Am J Respir Crit Care Med 2002; 165:824-831.
47. Barnes PJ, Lim S. Inhibitory cytokines in asthma. Mol Medicine Today 1998; 4:452-458.
48. Borish L, Aarons A, Rumbyrt J et al. Interleukin-10 regulation in normal subjects and patients with asthma. J Allergy Clin Immunol 1996; 97:1288-1296.
49. Takanashi S, Hasegawa Y, Kanehira Y et al. Interleukin-10 level in sputum is reduced in bronchial asthma, COPD and in smokers. Eur Respir J 1999; 14:309-314.
50. Spiteri MA, Knight RA, Jeremy JY et al. Alveolar macrophage-induced suppression of peripheral blood mononuclear cell responsiveness is reversed by in vitro allergen exposure in bronchial asthma. Eur Resp J 1994; 7:1431-438.
51. Gentile DA, Patel A, Ollila C et al. Diminished IL-10 production in subjects with allergy after infection with influenza A virus. J Allergy Clin Immunol 1999; 103:1045-1048.
52. van der Velden, V, Laan MP, Baert MR et al. Selective development of a strong Th2 cytokine profile in high-risk children who develop atopy: Risk factors and regulatory role of IFN-γ, IL-4 and IL-10. Clin Exp Allergy 2001; 31:997-1006.
53. Colavita AM, Hastie AT, Musani AI et al. Kinetics of IL-10 production after segmental antigen challenge of atopic asthmatic subjects. J Allergy Clin Immunol 2000; 106:880-886.
54. Lim S, John M, Seybold J et al. Increased interleukin-10 and macrophage inflammatory protein-1α release from blood monocytes ex vivo during late-phase response to allergen in asthma. Allergy 2000; 55:489-495.
55. Beeh KM, Beier J, Kornmann O et al. A single nasal allergen challenge increases induced sputum inflammatory markers in non-asthmatic subjects with seasonal allergic rhinitis: Correlation with plasma interleukin-5. Clin Exp Allergy 2003; 33:475-482.
56. Ohmen JD, Hanifin JM, Nickoloff BJ et al. Overexpression of IL-10 in atopic dermatitis. Contrasting cytokine patterns with delayed-type hypersensitivity reactions. J Immunol 1995; 154:1956-1963.
57. Turner DM, Williams DM, Sankaran D et al. An investigation of polymorphism in the interleukin-10 gene promoter. Eur J Immunogenet 1997; 24:1-8.
58. Lim S, Crawley E, Woo P et al. Haplotype associated with low interleukin-10 production in patients with severe asthma. Lancet 1998; 352:113.
59. Hajeer AH, Lazarus M, Turner D et al. IL-10 gene promoter polymorphisms in rheumatoid arthritis. Scand J Rheumatol 1998; 27:142-145.

60. Hobbs K, Negri J, Klinnert M et al. Interleukin-10 and transforming growth factor-β promoter polymorphisms in allergies and asthma. Am J Respir Crit Care Med 1998; 158:1958-1962.
61. Immervoll T, Loesgen S, Dutsch G et al. Fine mapping and single nucleotide polymorphism association results of candidate genes for asthma and related phenotypes. Hum Mutat 2001; 18:327-336.
62. Hang LW, Hsia TC, Chen WC et al. Interleukin-10 gene-627 allele variants, not interleukin-I beta gene and receptor antagonist gene polymorphisms, are associated with atopic bronchial asthma. J Clin Lab Anal 2003; 17:168-173.
63. Karjalainen J, Hulkkonen J, Nieminen MM et al. Interleukin-10 gene promoter region polymorphism is associated with eosinophil count and circulating immunoglobulin E in adult asthma. Clin Exp Allergy 2003; 33:78-83.
64. Pretolani M, Goldman M. IL-10: A potential therapy for allergic inflammation? Immunol Today 1997; 18:277-280.
65. Asadullah K, Sterry W, Volk HD. Interleukin-10 therapy-review of a new approach. Pharmacol Rev 2003; 55:241-269.
66. Barnes PJ. Cytokine modulators for allergic diseases. Curr Opin Allergy Clin Immunol 2001; 1:555-560.
67. Barnes PJ. Cytokine modulators as novel therapies for asthma. Ann Rev Pharmacol Toxicol 2002; 42:81-98.
68. Huhn RD, Radwanski E, Gallo J et al. Pharmacodynamics of subcutaneous recombinant human interleukin-10 in healthy volunteers. Clin Pharmacol Ther 1997; 62:171-180.
69. Xing Z, Ohkawara Y, Jordana M et al. Adenoviral vector-mediated interleukin-10 expression in vivo: intramuscular gene transfer inhibits cytokine responses in endotoxemia. Gene Ther 1997; 4:140-149.
70. Stampfli MR, Cwiartka M, Gajewska BU et al. Interleukin-10 gene transfer to the airway regulates allergic mucosal sensitization in mice. Am J Respir Cell Mol Biol 1999; 21:586-596.
71. Oh JW, Seroogy CM, Meyer EH et al. CD4 T-helper cells engineered to produce IL-10 prevent allergen-induced airway hyperreactivity and inflammation. J Allergy Clin Immunol 2002; 110:460-468.
72. Irusen E, Matthews JG, Takahashi A et al. p38 Mitogen-activated protein kinase-induced glucocorticoid receptor phosphorylation reduces its activity: Role in steroid-insensitive asthma. J Allergy Clin Immunol 2002; 109:649-657.
73. Barnes PJ, Adcock IM. How do corticosteroids work in asthma? Ann Intern Med 2003; 139:359-370.
74. Stelmach I, Jerzynska J, Kuna P. A randomized, double-blind trial of the effect of glucocorticoid, anti-leukotriene and beta-agonist treatment on IL-10 serum levels in children with asthma. Clin Exp Allergy 2002; 32:264-269.
75. Mascali JJ, Cvietusa P, Negri J et al. Anti-inflammatory effects of theophylline: modulation of cytokine production. Ann Allergy Asthma Immunol 1996; 77:34-38.
76. Eigler A, Siegmund B, Emmerich U et al. Anti-inflammatory activities of cAMP-elevating agents: enhancement of IL-10 synthesis and concurrent suppression of TNF production. J Leukoc Biol 1998; 63:101-107.
77. Oliver B, Tomita K, Keller A et al. Low-dose theophylline does not exert its anti-inflammatory effects in mild asthma through upregulation of interleukin-10 in alveolar macrophages. Allergy 2001; 56:1087-1090.
78. Akdis CA, Blesken T, Akdis M et al. Role of interleukin 10 in specific immunotherapy. J Clin Invest 1998; 102:98-106.
79. Joss A, Akdis M, Faith A et al. L-10 directly acts on T cells by specifically altering the CD28 co-stimulation pathway. Eur J Immunol 2000; 30:1683-1690.
80. Jutel M, Akdis M, Budak et al. IL-10 and TGF-β cooperate in the regulatory T cell response to mucosal allergens in normal immunity and specific immunotherapy. Eur J Immunol 2003; 33:1205-1214.
81. Francis JN, Till SJ, Durham SR. Induction of IL-10+CD4+CD25+ T cells by grass pollen immunotherapy. J Allergy Clin.Immunol. 2003;111:1255-1261.
82. Pessi T, Sutas Y, Hurme M et al. Interleukin-10 generation in atopic children following oral Lactobacillus rhamnosus GG. Clin Exp Allergy 2000; 30:1804-1808.

Chapter 6

The Role of IL-10 in Autoimmune Pathology

Andrew W. Gibson, Jeffrey C. Edberg, Jianming Wu and Robert P. Kimberly

Abstract

The association of high levels of IL-10 with autoimmune diseases suggests that this cytokine plays an important role in the disease pathogenesis. Family studies of first-degree relatives and analysis of twins indicate that heritable genetic factors underlie inter-individual differences in quantitative IL-10 production. Genome wide scans have shown an association between the chromosomal region containing the IL-10 locus and autoimmune diseases, and studies of the IL-10 gene have identified both single nucleotide polymorphisms (SNP) and short tandem repeat polymorphisms (STRP) in the promoter. Susceptibility to and manifestations of autoimmune diseases have reported associations with SNPs, SNP haplotypes and STRPs in the IL-10 promoter region, but results have varied. Since the total numbers of affecteds studied in different ethnic groups have been small, further studies with greater power will be necessary to uncover the genetic basis of dysregulated IL-10 production in autoimmune diseases.

Introduction

Autoimmune diseases are characterized by the presence of dysregulated cytokine expression that plays a role in the maintenance of autoreactive lymphocytes. Genetic polymorphisms that alter the function or expression of these cytokine genes are likely candidates as factors either predisposing to disease or modulating disease manifestations, and are the subject of intense research.

Interleukin-10 (IL-10), a cytokine that exhibits multiple modulatory effects on the immune system, has altered expression in many autoimmune diseases. The expression of IL-10 appears to be influenced by polymorphisms in its promoter, which may be associated with autoimmune disease. However, the definition of which of the many currently identified promoter polymorphisms are involved in disease pathogenesis, and whether new, as yet unidentified polymorphisms might also play a role in IL-10 dysregulation, is still a subject of continued investigation.

Interleukin-10 is produced by $CD4^+$ and $CD8^+$ T cells, activated B lymphocytes, monocytes, macrophages, and keratinocytes.[1,2] As an anti inflammatory cytokine IL-10 down-regulates the expression of Th1 cytokines, MHC class II and costimulatory molecules on macrophages. However, IL-10 also stimulates FcγR expression on the same cells,[1,2] and has been shown to prolong B cell survival, to induce B cell differentiation, and to enhance B cell proliferation and antibody production.[1-5] The effects of IL-10 on B cells, particularly on the stimulation and survival of autoreactive B cells are believed to be of great importance in autoimmune diseases.[2,5] Additionally, IL-10 may play an important role in influencing the balance of Th1 versus Th2 cytokines, which can influence the progression of autoimmune diseases.[6-8]

Interleukin-10, edited by Francesco M. Marincola. ©2006 Eurekah.com.

Association of IL-10 Production with Autoimmune Diseases

A number of studies of both mouse and human models of autoimmune diseases have documented altered IL-10 serum levels, suggesting a direct link between IL-10 levels and disease.[1,2] Evidence from mouse models suggest that IL-10 production may play a protective role in organ-specific autoimmune disease and alter the balance between pathogenic Th1 cells and protective anti-inflammatory Th2 cells.[2] However, while some studies show that administration of IL-10 could result in improvement of the disease phenotype, other studies have shown that IL-10-deficient mice fail to develop an autoimmune syndrome, and the expression of an IL-10 transgene resulted in earlier disease onset or exacerbated disease rather than in protection.[2]

High serum levels of IL-10 have been documented in human autoimmune diseases. In systemic lupus erythematosus (SLE) patients, high IL-10 levels have been shown to correlate with disease activity,[9,10] and studies indicate that cultured PBMC from lupus patients spontaneously produce high levels of IL-10.[1] Production of high levels of IL-10 has also been demonstrated in synovial T-lymphocytes of rheumatoid arthritis patients, in the serum of systemic sclerosis, Kawasaki disease, and ALPS patients, as well as in the cultured cells of polymyositis and dermatomyositis patients.[2] High IL-10 levels have been found in the blister fluid of pemphigus vulgaris and bullous pemphigoid patients and in the mucosal T cells in ulcerative colitis.[2,11-16] Increased expression of IL-10 mRNA by PBMCs and/or various other tissues was associated with Sjogren's syndrome, Grave's disease, myasthenia gravis, psoriasis and autoimmune lymphoproliferative syndrome (ALPS).[1,2,8,16,17] Taken together, these studies indicate that dysregulated expression of IL-10 plays a significant role in autoimmune diseases. While the disease association with IL-10 levels is clear, the degree to which IL-10 production precedes disease onset, as opposed to resulting from disease activity, is less clear. In addition, while autoimmune disease progression may play a role in altering the levels of IL-10 and other cytokines, evidence from family studies suggest that, to a large extent, heritable genetic differences determine IL-10 production and thus contribute to autoimmune disease phenotype.[18,19] The association of high levels of IL-10 mRNA and protein with early stages of Grave's disease and Hashimoto's thyroiditis also suggests a role for IL-10 in the early stages of some diseases.[8]

IL-10 Promoter Polymorphisms and IL-10 Production

IL-10 production is regulated at the transcriptional level, and studies suggest that heritable genetic differences that associate with inter-individual differences in gene expression may differentially regulate IL-10 production.[19-21] There is also evidence that suggests the existence of post-transcriptional regulation by RNA-destabilizing, AU-rich elements within the 3'-untranslated of IL-10 mRNA.[22]

Inter-individual differences in IL-10 production is heritable.[23] First degree relatives of SLE patients produce high levels of IL-10 compared with unrelated controls,[24] and first-degree relatives of nonsurvivors of fatal meningococcal disease produce significantly lower levels of IL-10 than relatives of survivors.[23,25] The implication of a heritable genetic basis for IL-10 production is also supported by the concordance of IL-10 production in monozygotic twins, which suggests that genetics could account for up to 75% of IL-10 production.[23] Therefore, high IL-10 production associated with autoimmune diseases may represent a risk factor for disease susceptibility or severity, and the genetic polymorphisms that regulate IL-10 production represent genetic risk factors for autoimmune diseases.[1,9,10,18,19,24,26]

Promoter-reporter studies have identified several positive and negative regulatory promoter sequences within the 1.3 kb region upstream of the transcription start site, and a transcription factor binding site search of the immediate 4kb promoter region has identified more than 74 potential binding sites.[27,28] Three functional NF-κB and a Stat3 binding site, and four cAMP-responsive elements have been identified in the 2kb promoter region of the human IL-10 promoter.[29-31] However, none of these elements appear to be polymorphic. Functional Sp1 and Sp3 transcription factor binding sites have similarly been identified in the murine IL-10 promoter.[32]

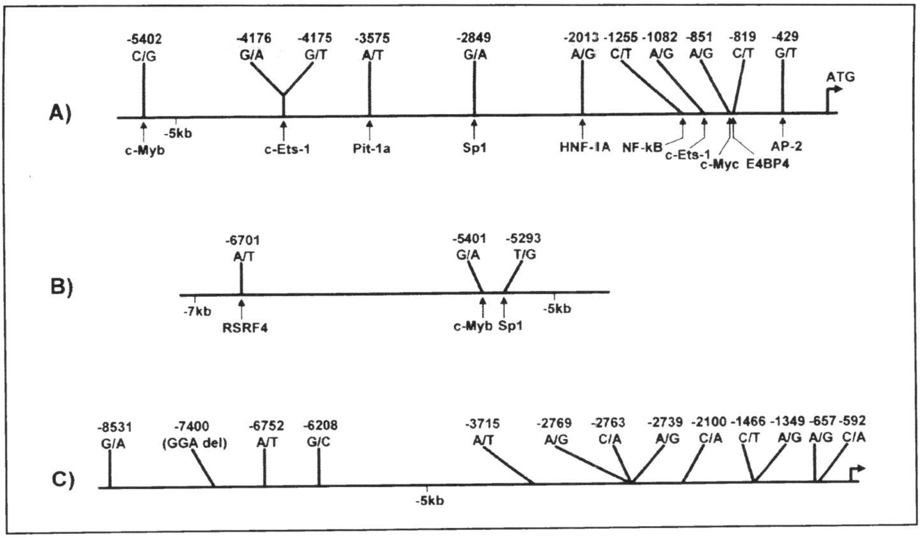

Figure 1. Schematic representation of the IL-10 promoter region: A) SNPs identified within putative transcription factor binding sites using the TESS search tool (http://www.cbil.upenn.edu/tess/); B) Novel SNPs identified in the distal promoter (AWG, unpublished results); C) Additional identified SNPs not lying in currently recognized transcription factor binding sites.

The human IL-10 promoter is highly polymorphic. Two (CA) short tandem repeat polymorphisms (STRP), IL-10.R and IL-10.G, have been identified at -4 kb and -1.1 kb, respectively, and three single nucleotide polymorphisms (SNP), -1082G/A, -819C/T and -597C/A, which form three predominant haplotypes (GCC, ACC, ATA), had been identified previously in the 1.0 kb promoter region.[33-36] A fourth haplotype, GTA, was seen only in a Chinese population.[37] The G allele at -1082 and haplotypes containing this allele have been associated with high IL-10 production, while the A allele and the ATA haplotype have been associated with low IL-10 production.[35,38]

Resequencing within the 8kb promoter region by various groups has identified at least 23 additional SNPs (Fig. 1),[19,39,40] and four novel SNPs have been identified recently in the 3' untranslated region (3' UTR).[41] Significant differences in the allele frequencies of some promoter SNPs have been seen in different ethnic or geographic populations, underscoring the need for appropriate stratification in disease association studies (Table 1).[42] Haplotypes defined by proximal and distal SNPs[19,35,43] and by SNPs and STRP alleles associate with IL-10 production.[39,44] However, because of variable, inconsistent associations between IL-10 production and different SNP alleles, SNP haplotypes and STRP alleles, it is currently unclear which promoter polymorphisms are causal in determining differences in IL-10 production. Data from our group, which showed an association between distal SNP haplotypes and IL-10 production suggest that SNPs in the distal promoter may play a role in IL-10 production.[19]

IL-10 Promoter Polymorphisms and Automimmune Disease

Genome wide scans have shown linkage of disease susceptibility to the region on chromosome 1 that encompasses the IL-10 locus. The IL-10 gene has been localized at 1q31-1q32 region, and several genome scans have identified regions of chromosome 1 (1q21-23, 1q31-32, 1q42-44) which may contain susceptibility loci for SLE and other autoimmune diseases.[45-51]

Using case-control approaches to look at the distribution of promoter SNP or STRP alleles, several studies have found an association between IL-10 promoter polymorphisms and autoimmune disease susceptibility or disease manifestations (Tables 2 and

Table 1. SNP allele frequencies in different ethnic groups

	-3575 A/T	-2849 G/A	-2776 A/G	-2763 C/A	-2100 C/A	-2013 G/A	-1466 C/T	-1349 A/G	-1082 A/G	-819 C/T	-592 C/A	-429 G/T
Cauc	0.98A	0.72G	0.95A	0.63C	0.99C	0.97G	1.00C	0.50A	0.49A*	0.79C*	0.79C*	1.00G
	0.02T	0.28A	0.05G	0.37A	0.01A	0.03A	0.00T	0.50G	0.51G*	0.21T*	0.21A*	0.00T
Af-Am	0.99A	0.75G	0.88A	0.60C	0.99C	1.00G	0.79C	0.48A	0.46A	0.85G	0.67C	0.77G
	0.01T	0.25A	0.12G	0.40A	0.01A	0.00A	0.21T	0.52G	0.54G	0.15T	0.33T	0.13T
Chin	-	-	-	-	-	-	-	-	0.94A†	0.33C†	0.33C†	-
	-	-	-	-	-	-	-	-	0.06G†	0.67T†	0.67A†	-

Cauc: Caucasians; Af-Am: African Americans; Chin: Chinese. *Data taken from Turner et al, 1997.[35] † Data taken from Mok et al, 1998.[55] All other data taken from Gibson et al, 2001.[19]

3).[19,37,38,41,43,52-65,74] In each autoimmune disease examined, some studies have failed to find associations between promoter polymorphisms and susceptibility or manifestations.[37,43,61,63-73] For example, Lazarus et al (1997) have found an association between -1082G, -1082G-containing haplotypes and the presence of anti-Ro autoantibodies and nephritis in Caucasian SLE.[53] A study by Crawley et al (1999) also used a Caucasian group, but failed to find a similar association, and Mok et al (1998) have found an association between the ATA haplotype and lupus nephritis in a Chinese population.[37,68] Therefore, currently it is not clear which of the 26 identified SNPs, which haplotypes or which STRP alleles are relevant in the disease process. In reviewing these studies several observations are clear: (1) the total number of affecteds that have been studied is small (Table 2); (2) different genetic markers, either SNPs or STRPs, have been screened; (3) different ethnic populations have been used in different studies; and (4) different genotyping techniques have been used, often without delineation of error rates. Since there are clear differences in the allele frequencies in different ethnic and geographic groups, appropriate stratification by ethnicity will be important. In support of this, our group has found an association between the promoter SNP (-2763A/T) in African Americans with SLE [19] but not in a Dutch Caucasian cohort of 98 affecteds compared with 128 controls

Table 2. Association studies of IL-10 polymorphisms and autoimmune disease susceptibility

Disease	Positive Association Reported No. of Studies (Total Affecteds Studied)	No. Association Identified No. of Studies (Total Affecteds Studied)
SLE	6* (455 Cauc; 200 Af-Am; 230 Mex)[19,41,52,53,55,58]	3 (180 Cauc; 330 Mex; 88 Chin)[37,67,72]
RA	1 (251 Cauc; 61 Af-Am) [54]	5 (811 Cauc; 138Af-Am) [43,64,66,69,73]
MS	2* (463 Cauc) [59,60]	2 (366 Cauc)[70,71]
Psoriasis	2 (352 Cauc)[57,62]	1 (248 Cau) [63]
AS	-	1 (468 Cauc) [61]
GBS	1 (87 Cauc) [65]	-

*Different IL-10 markers reported as associated with disease susceptibility in each of these studies. Systemic lupus erythematosus (SLE), rheumatoid arthritis (RA), multiple sclerosis (MS), ankylosing spondylitis (AS), Guillian-Barre syndrome (GBS). Cauc: Caucasians; Af-Am: African Americans; Chin: Chinese.

Table 3. Association studies of IL-10 polymorphisms and autoimmune disease manifestations

Disease	Positive Association Reported No. of Studies (Total Affecteds Studied)	No Association Identified No. of Studies (Total Affecteds Studied)
SLE	4* (224 Cauc; 88 Chin) [37,52,53,56]	2 (300 Cauc) [68,72]
RA	3 (547 Cauc) [38,43,64]	2 (269 Cauc) [66,69]
MS	-	2 (366 Cauc) [70,71]
Psoriasis	1 (248 Cauc) [63]	
AS	1 (468 Cauc) [61]	-
GBS	-	1 (87 Cauc) [65]

*Different disease manifestations reported as associated with IL-10 polymorphisms in each of these studies. Systemic lupus erythematosus (SLE), rheumatoid arthritis (RA), multiple sclerosis (MS), ankylosing spondylitis (AS), Guillian-Barre syndrome (GBS). Cauc: Caucasians; Af-Am: African Americans; Chin: Chinese.

(-2763AA = 0.37 in affecteds vs 0.36 in controls; Gibson and Huizinga, unpublished results). Others have failed to find a similar association in an Italian SLE group.[41]

Variable associations between IL-10 STRP alleles and autoimmune disease have also been documented. In different studies, different IL-10G STRP alleles associated with SLE in a UK Caucasian cohort, an Italian Caucasian cohort and in Mexican Americans, while IL-10R alleles associated with RA in two different UK Caucasian cohorts, and in an African American group.[41,52,54,55,58] Data from our group did not show an association between the IL-10R STRP and SLE in three ethnic groups (Caucasians, African Americans or Mexican Americans). However, we found significant association between SLE and IL-10.G8 and IL-10.G9 in Cauca-

Table 4. IL-10G allele frequencies in African American, Caucasian and Mexican American SLE populations

Alleles			African Am.		Mexican Am.		Caucasian Am.	
Size (bp)	Eksdale (1995)	Mehrian (1998)	NL (2n = 192	SLE 2n = 184	NL 2n = 88	SLE 2n = 144	NL 2n = 206	SLE 2n = 132)
192	IL10G-3	115	0.005	0.000	0.000	0.000	0.000	0.000
194	IL10G-4	117	0.005	0.011	0.000	0.000	0.000	0.000
196	IL10G-5	119	0.005	0.000	0.000	0.000	0.000	0.000
198	IL10G-6	121	0.005	0.043	0.023	0.017	0.019	0.023
200	IL10G-7	123	0.042	0.022	0.068	0.076	0.034	0.023
202	**IL10G-8**	**125**	[a] **0.177**	**0.283**	0.250	0.306 [b]	**0.165**	**0.311**
204	**IL10G-9**	**127**	[a] **0.260**	**0.152**	0.091	0.069 [b]	**0.296**	**0.083**
206	IL10G-10	129	0.156	0.239	0.136	0.097	0.073	0.114
208	**IL10G-11**	**131**	[b]**0.198**	**0.082**	0.091	0.083	0.073	0.114
210	IL10G-12	133	0.047	0.109	0.250	0.174	0.146	0.220
212	IL10G-13	135	0.063	0.043	0.023	0.132	0.126	0.083
214	IL10G-14	137	0.036	0.016	0.068	0.042	0.068	0.023
216	IL10G-15	139	0.005	0.000	0.000	0.014	0.000	0.008

Different allele nomenclatures are shown for comparison. [a] $p < 0.02$; [b] $p < 0.002$

sians, and IL-10.G8, IL-10.G9 and IL-10.G11 in African Americans (Table 4). We have not replicated the association in Mexican Americans observed by Mehrian and colleagues, but our cohort numbers were not adequately powered to provide a definitive result (Table 4).

Of particular interest is a recent study that examined distal SNPs in the -8kb promoter region, and extended haplotypes in the 4kb region in an Italian SLE cohort.[41] While the authors found no association between SLE and distal SNPs or SNPs in the 3'UTR, a meta-analysis of four IL-10.G STRP association studies found significant association with STRPs containing larger numbers of (CA) repeats. IL-10.G alleles containing 21 (CA) repeats or more (IL-10.G9 and up), and extended SNP-STRP haplotypes that included these longer repeat alleles were associated significantly with SLE.[41] This study suggests that the disease association might not be with a particular STRP allele, but with a larger stretch of (CA) repeats. The biologic basis for this observation is not clear at present.

In summary, associations between IL-10 promoter polymorphisms, IL-10 production and autoimmune diseases have been documented. However, while association studies have yielded inconsistent results, the total number of affecteds from each of the various ethnic groups studied has not been large. Additionally, adequate numbers of studies have not been done in all ethnic groups. Recent resequencing efforts have uncovered a more complex promoter structure, and additional SNPs are being discovered in the promoter and in the 3'UTR[41,75] (and Gibson, unpublished results). This raises the possibility that other, more complex SNP associations with disease phenotypes might be uncovered in future studies.

References

1. Lalani I, Bhol K, Ahmed AR. Interleukin-10: Biology, role in inflammation and autoimmunity. Ann Allergy Asthma Immunol 1997; 79:469-483.
2. Moore KW, de Waal Malefyt R, Coffman RL et al. Interleukin-10 and the interleukin-10 receptor. Annu Rev Immunol 2001; 19:683-765.
3. te Velde AA, de Waal Malefijt R, Huijbens RJ et al. IL-10 stimulates monocyte Fc gamma R surface expression and cytotoxic activity. Distinct regulation of antibody-dependent cellular cytotoxicity by IFN-gamma, IL-4, and IL-10. J Immunol 1992; 149:4048-4052.
4. Levy Y, Brouet JC. Interleukin-10 prevents spontaneous death of germinal center B cells by induction of the bcl-2 protein. J Clin Invest 1994; 93:424-428.
5. Llorente L, Zou W, Levy Y et al. Role of interleukin 10 in the B lymphocyte hyperactivity and autoantibody production of human systemic lupus erythematosus. J Exp Med 1995; 181:839-844.
6. Katsikis PD, Cohen SB, Londei M et al. Are CD4+ Th1 cells pro-inflammatory or anti-inflammatory? The ratio of IL-10 to IFN-gamma or IL-2 determines their function. Int Immunol 1995; 7:1287-1294.
7. Hagenbaugh A, Sharma S, Dubinett SM et al. Altered immune responses in interleukin 10 transgenic mice. J Exp Med 1997; 185:2101-2110.
8. Mirakian R, Hammond LJ, Bottazzo GF. TH1 and TH2 cytokine control of thyrocyte survival in thyroid autoimmunity. Nat Immunol 2001; 2:371.
9. Mongan AE, Ramdahin S, Warrington RJ. Interleukin-10 response abnormalities in systemic lupus erythematosus. Scand J Immunol 1997; 46:406-412.
10. Houssiau FA, Lefebvre C, Vanden Berghe M et al. Serum interleukin 10 titers in systemic lupus erythematosus reflect disease activity. Lupus 1995; 4:393-395.
11. Llorente L, Richaud-Patin Y, Fior R et al. In vivo production of interleukin-10 by nonT cells in rheumatoid arthritis, Sjogren's syndrome, and systemic lupus erythematosus. A potential mechanism of B lymphocyte hyperactivity and autoimmunity. Arthritis Rheum 1994; 37:1647-1655.
12. Cush JJ, Splawski JB, Thomas R et al. Elevated interleukin-10 levels in patients with rheumatoid arthritis. Arthritis Rheum 1995; 38:96-104.
13. Cohen SB, Katsikis PD, Chu CQ et al. High level of interleukin-10 production by the activated T cell population within the rheumatoid synovial membrane. Arthritis Rheum 1995; 38:946-952.
14. Isomaki P, Luukkainen R, Saario R et al. Interleukin-10 functions as an antiinflammatory cytokine in rheumatoid synovium. Arthritis Rheum 1996; 39:386-395.
15. Furuzawa-Carballeda J, Alcocer-Varela J. Interleukin-8, interleukin-10, intercellular adhesion molecule-1 and vascular cell adhesion molecule-1 expression levels are higher in synovial tissue from patients with rheumatoid arthritis than in osteoarthritis. Scand J Immunol 1999; 50:215-222.

16. Melgar S, Yeung MM, Bas A et al. Over-expression of interleukin 10 in mucosal T cells of patients with active ulcerative colitis. Clin Exp Immunol 2003; 134:127-137.
17. Lopatin U, Yao X, Williams RK et al. Increases in circulating and lymphoid tissue interleukin-10 in autoimmune lymphoproliferative syndrome are associated with disease expression. Blood 15 2001; 97(10):3161-3170.
18. Harley JB, Moser KL, Gaffney PM et al. The genetics of human systemic lupus erythematosus. Curr Opin Immunol 1998; 10:690-696.
19. Gibson AW, Edberg JC, Wu J et al. Novel single nucleotide polymorphisms in the distal IL-10 promoter affect IL-10 production and enhance the risk of systemic lupus erythematosus. J Immunol 2001; 166:3915-3922.
20. Le T, Leung L, Carroll WL et al. Regulation of interleukin-10 gene expression: Possible mechanisms accounting for its upregulation and for maturational differences in its expression by blood mononuclear cells. Blood 1997; 89:4112-4119.
21. Knolle PA, Uhrig A, Protzer U et al. Interleukin-10 expression is autoregulated at the transcriptional level in human and murine Kupffer cells. Hepatology 1998; 27:93-99.
22. Powell MJ, Thompson SA, Tone Y et al. Posttranscriptional regulation of IL-10 gene expression through sequences in the 3'-untranslated region. J Immunol 2000; 165:292-296.
23. Westendorp RG, Langermans JA, Huizinga TW et al. Genetic influence on cytokine production in meningococcal disease. Lancet 1997; 349:1912-1913.
24. Llorente L, Richaud-Patin Y, Couderc J et al. Dysregulation of interleukin-10 production in relatives of patients with systemic lupus erythematosus. Arthritis Rheum 1997; 40:1429-1435.
25. van Dissel JT, van Langevelde P, Westendorp RG et al. Anti-inflammatory cytokine profile and mortality in febrile patients. Lancet 1998; 351:950-953.
26. Hagiwara E, Gourley MF, Lee S et al. Disease severity in patients with systemic lupus erythematosus correlates with an increased ratio of interleukin-10:interferon-gamma-secreting cells in the peripheral blood. Arthritis Rheum 1996; 39:379-385.
27. Kube D, Platzer C, von Knethen A et al. Isolation of the human interleukin 10 promoter. Characterization of the promoter activity in Burkitt's lymphoma cell lines. Cytokine 1995; 7:1-7.
28. Eskdale J, Kube D, Tesch H et al. Mapping of the human IL10 gene and further characterization of the 5' flanking sequence. Immunogenetics 1997; 46:120-128.
29. Mori N, Prager D. Activation of the interleukin-10 gene in the human T lymphoma line HuT 78: Identification and characterization of NF-kappa B binding sites in the regulatory region of the interleukin-10 gene. Eur J Haematol 1997; 59:162-170.
30. Platzer C, Fritsch E, Elsner T et al. Cyclic adenosine monophosphate-responsive elements are involved in the transcriptional activation of the human IL-10 gene in monocytic cells. Eur J Immunol 1999; 29:3098-3104.
31. Benkhart EM, Siedlar M, Wedel A et al. Role of Stat3 in lipopolysaccharide-induced IL-10 gene expression. J Immunol 2000; 165:1612-1617.
32. Tone M, Powell MJ, Tone Y et al. IL-10 gene expression is controlled by the transcription factors Sp1 and Sp3. J Immunol 2000; 165:286-291.
33. Eskdale J, Gallagher G. A polymorphic dinucleotide repeat in the human IL-10 promoter. Immunogenetics 1995; 42:444-445.
34. Eskdale J, Kube D, Gallagher G. A second polymorphic dinucleotide repeat in the 5' flanking region of the human IL10 gene. Immunogenetics 1996; 45:82-83.
35. Turner DM, Williams DM, Sankaran D et al. An investigation of polymorphism in the interleukin-10 gene promoter. Eur J Immunogenet 1997; 24:1-8.
36. Rieth H, Assohou AC, Mormann M et al. A new allelic variation within the 5'-flanking region of the interleukin-10 gene. Eur J Immunogenet 2003; 30:191-193.
37. Mok CC, Lanchbury JS, Chan DW et al. Interleukin-10 promoter polymorphisms in Southern Chinese patients with systemic lupus erythematosus. Arthritis Rheum 1998; 41:1090-1095.
38. Crawley E, Kay R, Sillibourne J et al. Polymorphic haplotypes of the interleukin-10 5' flanking region determine variable interleukin-10 transcription and are associated with particular phenotypes of juvenile rheumatoid arthritis. Arthritis Rheum 1999; 42:1101-1108.
39. Eskdale J, Keijsers V, Huizinga T et al. Microsatellite alleles and single nucleotide polymorphisms (SNP) combine to form four major haplotype families at the human interleukin-10 (IL-10) locus. Genes Immun 1999; 1:151-155.
40. D'Alfonso S, Rampi M, Rolando V et al. New polymorphisms in the IL-10 promoter region. Genes Immun 2000; 1:231-233.
41. D'Alfonso S, Giordano M, Mellai M et al. Association tests with systemic lupus erythematosus (SLE) of IL10 markers indicate a direct involvement of a CA repeat in the 5' regulatory region. Genes Immun 2002; 3:454-463.

42. Moraes MO, Santos AR, Schonkeren JJ et al. Interleukin-10 promoter haplotypes are differently distributed in the Brazilian versus the Dutch population. Immunogenetics 2003; 54:896-899.
43. Hajeer AH, Lazarus M, Turner D et al. IL-10 gene promoter polymorphisms in rheumatoid arthritis. Scand J Rheumatol 1998; 27:142-145.
44. Eskdale J, Gallagher G, Verweij CL et al. Interleukin 10 secretion in relation to human IL-10 locus haplotypes. Proc Natl Acad Sci USA 1998; 95:9465-9470.
45. Tsao BP, Cantor RM, Kalunian KC et al. Evidence for linkage of a candidate chromosome 1 region to human systemic lupus erythematosus. J Clin Invest 1997; 99:725-731.
46. Cornelis F, Faure S, Martinez M et al. New susceptibility locus for rheumatoid arthritis suggested by a genome-wide linkage study. Proc Natl Acad Sci USA 1998; 95:10746-10750.
47. Gaffney PM, Kearns GM, Shark KB et al. A genome-wide search for susceptibility genes in human systemic lupus erythematosus sib-pair families. Proc Natl Acad Sci USA 1998; 95:14875-14879.
48. Moser KL, Neas BR, Salmon JE et al. Genome scan of human systemic lupus erythematosus: Evidence for linkage on chromosome 1q in African-American pedigrees. Proc Natl Acad Sci USA 1998; 95:14869-14874.
49. Shai R, Quismorio Jr FP, Li L et al. Genome-wide screen for systemic lupus erythematosus susceptibility genes in multiplex families. Hum Mol Genet 1999; 8:639-644.
50. Capon F, Novelli G, Semprini S et al. Searching for psoriasis susceptibility genes in Italy: Genome scan and evidence for a new locus on chromosome 1. J Invest Dermatol 1999; 112:32-35.
51. Johanneson B, Lima G, von Salome J et al. A major susceptibility locus for systemic lupus erythemathosus maps to chromosome 1q31. Am J Hum Genet 2002; 71:1060-1071.
52. Eskdale J, Wordsworth P, Bowman S et al. Association between polymorphisms at the human IL-10 locus and systemic lupus erythematosus. Tissue Antigens 1997; 49:635-639.
53. Lazarus M, Hajeer AH, Turner D et al. Genetic variation in the interleukin 10 gene promoter and systemic lupus erythematosus. J Rheumatol 1997; 24:2314-2317.
54. Eskdale J, McNicholl J, Wordsworth P et al. Interleukin-10 microsatellite polymorphisms and IL-10 locus alleles in rheumatoid arthritis susceptibility. Lancet 1998; 352:1282-1283.
55. Mehrian R, Quismorio Jr FP, Strassmann G et al. Synergistic effect between IL-10 and bcl-2 genotypes in determining susceptibility to systemic lupus erythematosus. Arthritis Rheum 1998; 41:596-602.
56. Rood MJ, Keijsers V, van der Linden MW et al. Neuropsychiatric systemic lupus erythematosus is associated with imbalance in interleukin 10 promoter haplotypes. Ann Rheum Dis 1999; 58:85-89.
57. Asadullah K, Eskdale J, Wiese A et al. Interleukin-10 promoter polymorphism in psoriasis. J Invest Dermatol 2001; 116:975-978.
58. D'Alfonso S, Rampi M, Bocchio D et al. Systemic lupus erythematosus candidate genes in the Italian population: Evidence for a significant association with interleukin-10. Arthritis Rheum 2000; 43:120-128.
59. de Jong BA, Westendorp RG, Eskdale J et al. Frequency of functional interleukin-10 promoter polymorphism is different between relapse-onset and primary progressive multiple sclerosis. Hum Immunol 2002; 63(4):281-285.
60. Martinez Doncel A, Rubio A, Arroyo R et al. Interleukin-10 polymorphisms in Spanish multiple sclerosis patients. J Neuroimmunol 2002; 131:168-172.
61. Goedecke V, Crane AM, Jaakkola E et al. Interleukin 10 polymorphisms in ankylosing spondylitis. Genes Immun 2003; 4:74-76.
62. Hensen P, Asadullah K, Windemuth C et al. Interleukin-10 promoter polymorphism IL10.G and familial early onset psoriasis. Br J Dermatol 2003; 149:381-385.
63. Kingo K, Koks S, Silm H et al. IL-10 promoter polymorphisms influence disease severity and course in psoriasis. Genes Immun 2003; 4:455-457.
64. Lard LR, van Gaalen FA, Schonkeren JJ et al. Association of the -2849 interleukin-10 promoter polymorphism with autoantibody production and joint destruction in rheumatoid arthritis. Arthritis Rheum 2003; 48:1841-1848.
65. Myhr KM, Vagnes KS, Maroy TH et al. Interleukin-10 promoter polymorphisms in patients with Guillain-Barre syndrome. J Neuroimmunol 2003; 139:81-83.
66. Coakley G, Mok CC, Hajeer AH et al. Interleukin-10 promoter polymorphisms in rheumatoid arthritis and Felty's syndrome. Br J Rheumatol 1998; 37:988-991.
67. Alarcon-Riquelme ME, Lindqvist AK, Jonasson I et al. Genetic analysis of the contribution of IL10 to systemic lupus erythematosus. J Rheumatol 1999; 26:2148-2152.
68. Crawley E, Woo P, Isenberg DA. Single nucleotide polymorphic haplotypes of the interleukin-10 5' flanking region are not associated with renal disease or serology in Caucasian patients with systemic lupus erythematosus. Arthritis Rheum 1999; 42:2017-2018.

69. Cantagrel A, Navaux F, Loubet-Lescoulie P et al. Interleukin-1beta, interleukin-1 receptor antagonist, interleukin-4, and interleukin-10 gene polymorphisms: Relationship to occurrence and severity of rheumatoid arthritis. Arthritis Rheum 1999; 42:1093-1100.
70. Pickard C, Mann C, Sinnott P et al. Interleukin-10 (IL10) promoter polymorphisms and multiple sclerosis. J Neuroimmunol 1999; 101:207-210.
71. Maurer M, Kruse N, Giess R et al. Genetic variation at position -1082 of the interleukin 10 (IL10) promotor and the outcome of multiple sclerosis. J Neuroimmunol 2000; 104:98-100.
72. Dijstelbloem HM, Hepkema BG, Kallenberg CG et al. The R-H polymorphism of Fcgamma receptor IIa as a risk factor for systemic lupus erythematosus is independent of single-nucleotide polymorphisms in the interleukin-10 gene promoter. Arthritis Rheum 2002; 46:1125-1126.
73. MacKay K, Milicic A, Lee D et al. Rheumatoid arthritis susceptibility and interleukin 10: A study of two ethnically diverse populations. Rheumatology (Oxford) 2003; 42:149-153.
74. Lin M, Storer B, Martin P et al. Relation of an interleukin-10 promoter polymorphism to Graft-versus-Host Disease and survival after hematopoietic-cell transplantation. N Engl J Med 2003; 349:2201-2210.
75. Kube D, Rieth H, Eskdale J et al. Structural characterisation of the distal 5' flanking region of the human interleukin-10 gene. Genes Immun 2001; 2:181-190.

CHAPTER 7

Role of Interleukin-10 in Transplantation

Thomas Ritter and Hans-Dieter Volk

Abstract

Organ transplantation has been developed as ultimative option in patients with end-stage organ diseases. New immunosuppressive regimens and improved perioperative management did improve the short-term results during the last decade, however, insufficient control of chronic rejection and side effects of continuous immunosuppression limit the long-term success of this approach. The immunosuppressive drugs used today suppress both unspecific inflammatory reactions and specific T-cell activation. Unfortunately, these powerful drugs also prevent the activation of the endogenous counterregulatory processes that control the immune response (e.g., regulatory T cells, anti-inflammatory cytokines). Thus, targeting more selective the inflammatory process would be of outstanding interest.

IL-10 might be an ideal candidate for this approach as it expresses both direct anti-inflammatory and immunoregulatory properties. Therefore, IL-10 has been administered in various models of cell and organ transplantation using different ways of application which will be discussed in the following chapter.

Introduction

Organ transplantation has been developed as ultimative option in patients with end-stage organ diseases. New immunosuppressive regimens and improved perioperative management did improve the short-term results during the last decade, however, insufficient control of chronic rejection and side effects of continuous immunosuppression limit the long-term success of this approach. Although most of the proteins expressed in the particular organ of the donor and recipient are identical, the polymorphisms in some genes, particularly in the major histocompatibility complex (MHC) are highly immunogenic. To minimize the risk of rejection, donor/recipient pairs are selected according to MHC typing where it is possible (kidney, bone marrow). During the last years we had to learn, however, that the MHC mismatch determines only part of the immunogenicity. Much more important is the activation of donor antigen-presenting cells by "dangerous" signals, e.g., following ischemia-reperfusion injury. This explains why kidneys from living-unrelated donors with high MHC mismatch have a better half-life than well MHC-matched organs from cadaver donors because brain death and prolonged ischemic time trigger a lot of dangerous signals. The rejection of allogeneic organs is always accompanied by an infiltration of the graft with inflammatory cells (mainly T cells and macrophages) and the production of proinflammatory cytokines, e.g., Interferon-gamma (IFN-γ) or Tumour Necrosis Factor-alpha (TNF-α), and cytotoxic effector molecules, e.g., granzymes, perforine, granulysin. The immunosuppressive drugs used today suppress both unspecific inflammatory reactions (e.g., steroids) and specific T-cell activation (calcineurin inhibitors, T-cell antibodies, anti-proliferative drugs). Unfortunately, these powerful drugs also prevent the activation of the endogenous counterregulatory processes that control the immune response (e.g., regulatory T cells, anti-inflammatory cytokines). Thus, targeting more selective the inflammatory process would be of outstanding interest.

Interleukin-10, edited by Francesco M. Marincola. ©2006 Eurekah.com.

Table 1. Summary of IL-10 studies in transplantation

Treatment	Type	Model/Organ/Animal	References
Donor	Protein	IRI/ liver/ pig	Donckier et al, 2003[6]
Donor organ	Plasmid	Allo/ heart/ mouse	DeBruyne et al, 1998[22]
Donor organ	Plasmid	Allo/ lung/ rabbit	Hong et al, 2002[23]
Donor organ	Plasmid	Allo/ lung/ rat	Itano et al, 2001[24]
Donor organ	Plasmid	Allo/ cornea/ mouse	König-Merediz et al, 2000[25]
Donor organ	Retrovirus	Allo/ heart/ rat	Qin et al, 1995[27]
Donor organ	Adenovirus	Allo/ heart/ rat	Brauner et al, 1997[28]
Donor organ	Adenovirus	Allo/ heart/ rat	Qin et al, 1997[29]
Donor organ	Adenovirus	Allo/ heart/ rat	David et al, 2000[31]
Donor organ	Adenovirus	Allo/ lung/ rat	Fischer et al, 2001[32]
Donor organ	Adenovirus	Allo/ cornea/ sheep	Klebe et al, 2001[33]
Donor organ	Adenovirus	Allo/ kidney/ rat	Kato et al, 1999[34]
Recipient	Protein	Allo/ heart/ mouse	Li et al, 1999[7]
Recipient	Protein	Allo/ heart/ mouse	Li et al, 1999[7]
Recipient	Protein	Allo/ cornea/ rat	Torres et al, 1999[10]
Recipient	Transgenic mice	Chronic/ heart/ mouse	Fischbein et al, 2003[26]
Recipient	AAV	Syngeneic/ islets/ NOD-mouse	Goudy et al, 2001[35]
Recipient	AAV	Syngeneic/ islets/ NOD-mouse	Yang et al, 2002[36]
Recipient	AAV	Syngeneic/ islets/ NOD-mouse	Zhang et al, 2003[37]

IL-10 might be an ideal candidate for this approach as it expresses both direct anti-inflammatory and immunoregulatory properties. Therefore, IL-10 has been administered in various models of cell and organ transplantation (Table 1, Fig. 1). A review of published data as well as recent findings from our own group will be discussed in this chapter.

Interleukin-10 Protein Therapy

Due to its anti-inflammatory and immunomodulatory effects recombinant IL-10 protein has been applied in various animal models of immune dysregulation, e.g., in autoimmune diseases (rheumatoid arthritis, experimental autoimmune encephalitis, diabetes, inflammatory bowel disease or psoriasis, for review see ref. 1). In 1995 the first study using recombinant IL-10 in patients has been performed. It showed that systemic application of IL-10 has no serious side effects and is well tolerated by the patients.[2] Recent data from our group demonstrated its efficacy and safety in psoriasis patients particularly in preventing of relapse (for review see ref. 3).

IL-10 is generally seen as a anti-inflammatory and immunomodulatory cytokine, however, recent data suggest that IL-10 might also express pro-inflammatory properties. It can stimulate NK cells and B cells, particularly at higher doses. Kinetic studies in patients have shown that at a daily dose >10 μg/kg IL-10 shows increasing evidence for inflammatory effects, even IFN-γ is up-regulated (and our own observations).[4] These biphasic properties might explain at least in part contradictory results in experimental transplant models described in the following chapters.

Rationale of IL-10 Application in Transplantation

In principle, IL-10 has been used in experimental transplantation in three different ways:

- donor pretreatment with recombinant IL-10 or systemic gene therapy with the aim to reduce the ischemia/reperfusion injury and the graft immunogenicity by deactivating donor intragraft macrophages and antigen-presenting cells.

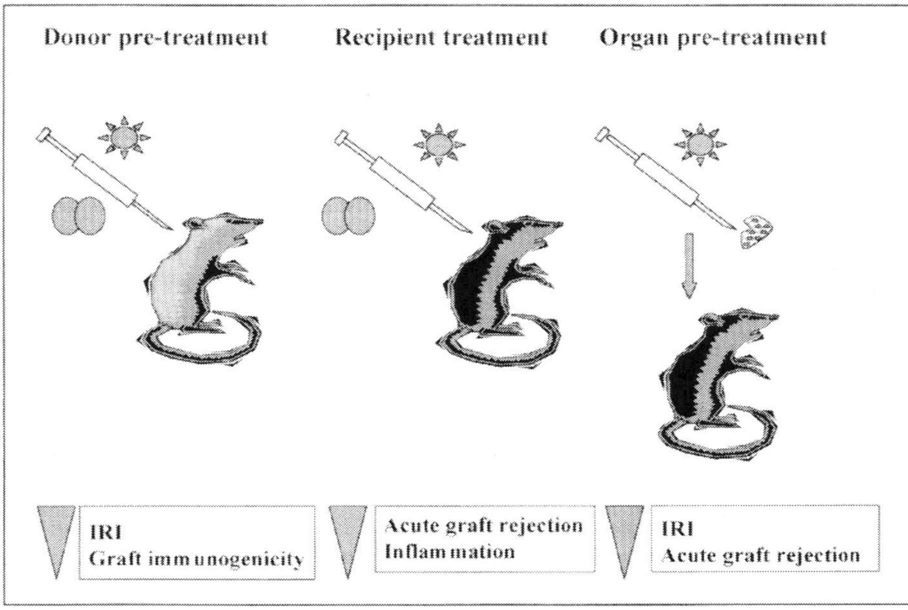

Figure 1. The different ways of IL-10 application in transplantation are shown. Abbreviations: IRI= ischemia-/reperfusion injury; [gene therapy symbol]: gene therapy; [IL-10 protein symbol]: IL-10 protein; [triangle symbol]: reduction of; [transplant symbol]: transplant.

- recipient treatment with recombinant IL-10 or systemic gene therapy with the aim to prevent acute graft rejection by deactivating inflammatory reactions and inducing regulatory T cells.
- gene therapy resulting in intragraft over-expression of IL-10 with the aim to combat both ischemia/reperfusion injury and T-cell mediated graft rejection by combination of the effects described above.

Protective IL-10 can also be induced indirectly, e.g., by application of cAMP elevating drugs like prostaglandins or 15-deoxyspergualin[5] but these approaches will not be discussed in this review.

Interleukin-10 Protein Therapy in Transplantation

With the knowledge of the anti-inflammatory and immunomodulatory activity of IL-10 and the availability of recombinant IL-10, several experimental transplant studies were performed to investigate the protective capacity of IL-10.

Donor Pretreatment

The prevention of ischemia-/reperfusion injury (IRI) is a serious problem in transplantation by maintaining the integrity of the transplant. IRI is a complex phenomenon which damages the transplant already shortly after harvesting the graft from animals or patients. It leads to the production of toxic oxygen radicals, up-regulation of adhesion molecules and production of proinflammatory cytokines e.g., by resident antigen presenting cells (APC) in the liver (Kupffer cells) e.g., TNF-α. Donckier et al[6] could show in a preclinical pig liver transplant model that preconditioning of the donor with IL-10 at the beginning of graft harvest followed by 5 hours of cold ischemia led to significant reduction of liver transaminases (early indicator of liver damage) in the transplant donor. However, no differences regarding the infiltration of cells and necrosis could be observed when compared to untreated controls.[6]

Recipient Treatment

Recently, the effects of systemic IL-10 administration on transplant survival have been investigated in a vascularized mouse heart transplant model. Whereas pretreatment of the donor with recombinant IL-10 (0.2-200 μg/day) leads to prolongation of allogeneic transplant survival, perioperative or post-transplant administration was not effective or even led to accelerated graft rejection.[7] Pretreatment of graft recipients with IL-10 led to reduced alloantigen specific T cells in the spleen and to reduced graft infiltration.[7] Similar data concerning transplant survival have also been obtained in a rat allogeneic liver transplant model. Administration of recombinant human IL-10 at lower dosis (10-20 μg/kg/day) led to prolonged graft survival and significantly decreased serum levels of cytokine-induced neutrophil chemoattractant protein. In contrast, high dose IL-10 application shortened transplant survival.[8] Less promising results have been reported in a rat model of corneal transplantation. This procedure is up to now the most frequently performed transplantation of human tissue and therefore of considerable interest.[9] Torres et al[10] could show that both local (subconjunctivally) and systemic (intraperitoneal) injection of IL-10 at various concentrations at the day before or around transplantation does not lead to prolonged transplant survival.

In summary, these data indicate that extensive IL-10 pretreatment with careful dosage is required to be effective in organ transplantation.

Clinical Applications

So far in the clinical organ transplant situation, IL-10 was used only in a pilot trial for preventing the OKT 3 monoclonal antibody (mab) mediated side effects.[11] OKT3 is a mab targeting T cells. Its powerful immunosuppressive capacity is mostly used for reversal of steroid-resistant acute rejections. Cross-linking of CD3 molecules by OKT3 mab, however, initially activates T cells resulting in a cytokine release syndrome (e.g., release of TNF-α) which can harm the patient. Pretreatment of patients with recombinant IL-10 reduces the release of TNF-α induced by OKT3 and prevents most of the side effects.[11]

It has been reported that patients with spontaneously elevated IL-10 serum levels prior to receiving allogenic bone marrow transplantation have a reduced risk of graft versus host disease (GVHD) and improved survival.[12-14] It was speculated that IL-10 induces alloreactive regulatory T cells. This was the rationale to generate regulatory T cells by ex vivo allostimulation in the presence of IL-10 (see chapter on regulatory T cells).[15]

Interleukin-10 Gene Therapy

At the beginning of the early 90s gene therapy was introduced into medicine. The major difference to other therapeutic options is, that the drug is produced by the patients themselves due to the introduction of genetically-modified viruses or cells/tissues/organs into the patients. Early studies with gene therapy in men have dealt with the treatment of severe immunodeficiencies (SCID) or with the treatment of cancer.[16] Many different ways of genetic modification have been investigated so far but until now the "break-through" for gene therapy as a serious therapy option is still far away, despite some promising results obtained recently.[17] This is mainly due to still unsolved problems concerning the suitable gene transfer vehicle (e.g., the viral vector), the efficiency of gene transfer and problems related to immune reactions of the treated patient against the vector or the transferred therapeutic gene.

In this chapter the effort of gene therapy by using different gene transfer systems to modify the transplant and/or the recipient will be discussed as well as the genetic modification of lymphoid cells to produce IL-10. To our knowledge, there is no clinical study using IL-10 gene therapy in transplantation, therefore all experiments described here are performed in preclinical models.

In principle, there a two ways to introduce genetic material into cells or tissues: gene transfer with recombinant viruses or techniques based on nonviral gene transfer methods. Nonviral gene transfer include physical and chemical approaches to transfer recombinant DNA into cells/tissues (for review see ref. 18). For viral gene transfer different families of viruses have been investigated (for review see refs. 16,19).

Table 2. Vectors for gene transfer

Viral	Stable/ Transient	Infection of Nondividing Cells	Packaging Capacity (kb)	Advantage	Disadvantage
Retrovirus (MoMuLV)	Stable	No	Up to 8	Integration Many cell types mutagenesis	Instability Insertional
Lentivirus (HIV)	Stable	Yes	Up to 10	Transduces both quies. and prolif. cells	Safety concerns by HIV-based vectors
Adenovirus	Transient	Yes	Up to 8	Transduces both quies. and prolif. cells High titer	Immunogenic
Herpes simplex virus	Transient	Yes	> 30!	Tropism for neural cells	Safety concerns
Adeno-associated virus (AAV)	Stable (?)	Yes	Up to 5	Low immunogenic	Production of high titer stocks
Nonviral	Stable/ Transient	Transfection of nondividing cells			
Naked DNA	Transient	Yes	-		
Liposome	Transient	Yes	-		
DNA conjugates	Transient	Yes	-		

Nonviral IL-10 Gene Transfer

Several techniques have been developed so far for the transfer of genetic material into cells or tissues without using viral vectors as gene transfer vehicle. These techniques include electroporation, "gene gun"-mediated DNA transfer, liposomal particles or simple injection of "naked" DNA (Table 2). In contrast to viral vectors, these gene transfer techniques generally do not induce an immune response by the recipient, although their application is limited due to their low efficiency in vivo.[18] Despite this low transduction efficiency there are some reports in the literature showing successful gene transfer with plasmid-DNA encoding for IL-10 in transplantation. It is important to note that most of the gene transfer studies are using the Epstein-Barr virus (EBV) encoded IL-10 homologue viral IL-10 (here referred to as vIL-10), since it was shown that vIL-10 has similar immunosuppressive effects as cellular (human or mouse) IL-10 but no growth stimulatory activities on B cells of cellular IL-10.[20,21]

Donor Organ Pretreatment (ex Vivo)

In 1998, DeBruyne et al[22] reported on the prevention of allograft rejection due to vIL-10 DNA-liposome-mediated intragraft gene transfer in a nonvascularized mouse heart transplant model. Enhanced graft survival was specific for vIL-10 gene transfer since treatment with an antisense-plasmid for IL-10 or anti-IL-10 mab abrogated the effect. In addition, vIL-10 gene transfer inhibited the generation of donor-specific T cells and led to a decrease in IgM and IgG alloantibody production. Similar results using novel lipid/DNA formulations have been recently reported in a rabbit heterotopic heart transplant model.[23] Beside the transfer of plasmid-DNA in heart transplants, studies have also been performed in a model of rat lung transplantation. Allograft gas exchange and vascular rejection score was significantly better in vIL-10 treated grafts than in transplants receiving control plasmid.[24] In a model of allogeneic cornea transplantation, topical administration of plasmid-DNA on the cornea at day 10 after transplantation by gene gun technology has been investigated. In contrast to the experiments described before cIL-10 gene transfer had no significant effect on allograft survival.[25]

Recipient Treatment (in Vivo)

Interesting results have been published by Fischbein et al[26] studying transgenic, IL-10 secreting mice as recipients in a model of chronic heart rejection. They could show that in IL-10 transgenic mice the number of graft-infiltrating T lymphocytes and the percentage of interleukin-2/interferon-gamma producing T lymphocytes were markedly reduced. In addition, the over-expression of interleukin-10 resulted in the decline of graft-infiltrating macrophages.[26]

The results described above indicate that many factors (gene transfer efficiency, source of IL-10, localization of gene transfer, systemic vs. local expression) may influence the survival rate of allogeneic transplants after genetic modification.

Gene Transfer with Viral Vectors

In contrast to gene transfer technologies using nonviral strategies, viral vectors are generally characterized by their high transduction efficiency of different cell types and tissues, although they do not show the same transduction efficiency of a given cell type.[16] Therefore the selection of a gene therapy vector is critically dependent on the cell type or tissue being the target for gene therapy. However, virus-mediated toxicity and the induction of an immune response against the vector are still major problems which have to be solved before gene therapy with viral vectors will be considered as therapeutic option e.g., in nonlethal diseases.

Different virus families have been investigated for their potential as gene therapy vector (see Table 1). These include vectors mainly derived from Retroviruses, Adenoviruses, Adeno-associated viruses and Herpesviruses (Herpes-Simplex virus type 1 (HSV-1), Cytomegalo-virus (CMV), Epstein-Barr virus (EBV)). Below advantages and disadvantages of viral vectors derived from different families will be briefly described. A common feature of all recombinant viral vectors is the deletion of essential genes within the viral genome. This is an important prerequisite due to biosafety issues of viral vectors and necessary to generate enough space within the viral genome for the insertion of the therapeutic gene. These manipulations lead to the generation of replication-deficient viral particles which subsequently need for their in-vitro replication either a packaging line providing the deleted genes in trans or a helper virus.[16,19]

Adenoviral Vectors

Beside retroviral vectors, adenoviruses are the most frequently used carriers for the transfer of genetic information in human gene therapy trials. More than 50 serotypes belonging to different subgroups are known. So far no severe human diseases are known to be associated with adenoviruses, although they may induce tumours in rodents. A severe complication in the application of recombinant adenoviruses is the generation of both cellular and humoral immune responses against transduced cells or tissues.[16,19]

Retroviral Vectors

Retroviruses are single stranded RNA viruses surrounded by a lipid envelope. Upon entry into the target cell, the RNA is released, reverse transcribed into double-stranded DNA which is stably integrated into the host DNA. In contrast to retroviral vectors based on HIV or SIV genomes (lentiviruses), retroviruses belonging to the subgroup of oncoretroviruses (e.g., MoMuLV) are strictly dependent on cell division and break-down of the nuclear membrane for integration in the cellular DNA.[16] A major disadvantage of retroviral vectors is characterized by their integration into transcriptionally-active sites within the cellular DNA which may lead to undesired proliferation of transduced cells.

Adeno-Associated Virus Vectors

Since it has been realized that the immune response induced by the injection of recombinant adenovirus is a major disadvantage in Ad-mediated gene therapy protocols, other vector systems have been investigated for their potential as gene therapy vector. One promising candidate is the Adeno-associated-virus (AAV) originally identified as contamination in adenovirus preparations. AAV belongs to the family of parvoviruses and contains single stranded DNA. Six serotypes (AAV 1-6) with different cell specificity have been identified so far without pathologic signs upon infection in humans. In contrast to adenovirus vectors AAV does not seem to induce a strong immune response (no potentially immunogenic viral protein is remaining) making them suitable as gene therapy vector.

In addition to the virus families described above, the family of Herpes viruses (HSV-1, CMV, EBV) have been exploited for their capacity as gene therapy vectors.

Gene Transfer of IL-10 by Viral Vectors

Donor Organ Pretreatment (ex Vivo)

Most of the animal studies using viral vectors for IL-10 gene transfer have been performed in the heart transplant model. In 1995, Qin et al[27] reported on the successful gene transfer in a mouse nonvascularized heart transplant model. In this study mouse hearts were injected with a retroviral vector encoding for vIL-10. Expression of vIL-10 from transduced grafts was able to prolong allogeneic graft survival. Similar data have been obtained when recombinant adenovirus expressing vIL-10 has been used as gene transfer vehicle, although there are different ways of vector application.[28,29] Due to its immunosuppressive effect on graft-infiltrating cells it was shown that gene transfer of IL-10 inhibits immune responses to both alloantigens and the adenoviral antigens.[29] Interestingly, mouse IL-10 was less effective than vIL-10 in this model. The authors explain this by the immunostimulatory properties of endogenous IL-10 on NK cells and mast cells which is not observed with vIL-10.[30] Unfortunately, our own data suggest that this is not the case in the human situation. vIL-10 and human endogenous IL-10 have quite similar effects in-vitro, however, vIL-10 with some lower activity because of the lower affinity to the human IL-10 receptor (Sabat et al, in preparation).

Prolongation of allogeneic graft survival could as well be obtained when rat IL-10 was used for gene transfer.[31] IL-10 gene transfer was as well investigated in models of lung, cornea and kidney transplantation. Successful gene transfer of IL-10 has been as well obtained in models of lung[32] and cornea transplantation.[33] Data from our group using a full mismatched rat renal allo-transplant model indicate that combined intra-graft gene transfer of IL-4 and IL-10 significantly prolongs survival, whereas IL-4 gene transfer had no effect.[34]

Recipient Treatment (in Vivo)

Islet transplantation represents a potential cure for type 1 diabetes although autoimmune and allogeneic immune responses limit the success in clinical transplantation. Gene transfer in islet transplantation has been performed by several groups. In these studies the major focus was on the genetic modification of the transplant recipient by AAV vectors expressing IL-10 (e.g., by intramuscular injection) rather than gene transfer into the pancreatic islets themselves. The reason for this might be that islets are extremely sensitive to inflammatory responses and virus-mediated gene transfer might lead to donor-derived inflammation which in turn may mediate destruction of the islets. It could be shown in these models that IL-10 expression prevents diabetes in nonobese diabetic (NOD) mice.[35-37]

Role of ex Vivo Gene-Modified, IL-10 Producing Cells in Transplantation

As already discussed above, a major limitation in gene therapy experiments is the generation of immune responses against the viral vector. Therefore other applications have been investigated for their use in gene therapy. An interesting possibility to prevent transplant rejection represents the ex vivo modification of antigen presenting cells (APC) or T cells. These cells could be used as transport vehicle of therapeutic genes (e.g., IL-10) to environments where antigen-presentation takes place (lymph nodes, spleen) as well as to the transplant.

First experiments have been described in 1998, showing the ex vivo modification of myeloid dendritic cells (DC) with a retrovirus encoding for vIL-10. These vIL-10 producing DCs have decreased expression of major histocompatibility (MHC) class II antigens and costimulatory molecules, thereby reducing their capacity to generate alloantigen-specific T-cells.[38] Recently, human DCs transduced with an adenovirus construct expressing vIL-10 have been analysed for their effect on skin graft rejection in humanized NOD-scid chimeric mice. In animals receiving vIL-10 modified DCs skin rejection was reduced when compared to animals receiving control DCs.[39] However, contradictory results when using DC expressing vIL-10 have been reported. Adenoviral over-expression of murine IL-10 enhanced the allostimulatory activity of DCs in mixed lymphocyte reactions. Consequently, injection of IL-10 DCs in allogeneic mice accelerated graft rejection.[40] These data indicate that the genetic modification of DCs with immunosuppressive molecules seems to be an interesting approach although further careful analysis is required.

Beside DCs, genetic modification of T cells has been investigated for their immunosuppressive potential. Since T cells can be only transduced with high efficiency using retroviral vectors, all studies published so far are dealing with the generation of retrovirus-mediated IL-10 expressing T cells. The majority of these studies are designed to inhibit autoimmune colitis[40a] or airway-mediated diseases and inflammation.[41,42] Ex vivo generated IL-10 producing T cells showed inhibitory effects both in-vitro as well in vivo. We have analysed the potential of alloantigen-specific T cells engineered to express vIL-10 in an allogeneic heart transplant model. Although vIL-10 expressing T cells show inhibitory effects on naïve T cells in-vitro (proliferation, production of IFN-γ) injection of vIL-10 producing T cells at the day of transplantation is unable to prevent graft rejection in high responder graft recipients (Brandt et al, submitted). This indicates that strong inflammatory responses as obtained after transplantation of MHC class I/II disparate grafts can not be inhibited by IL-10 producing cells.

Regulatory T Cells Expressing IL-10 and Their Role in Transplantation

For 20 years it is known from transplant experiments that a cell population exists within the T cell compartment which is able to exert suppressive/regulatory functions after immunosuppressive treatment.[43] Further experiments have clearly shown that tolerogenic cells which have been generated under the umbrella of monoclonal antibody induced tolerance can be transferred into naïve animals by adoptive transfer.[44,45] Since then, many groups have investigated the role of regulatory T cells in autoimmune diseases and transplantation (for review see refs. 15,46-52). It becomes more and more clear that different cell types within the regulatory cell compartment may have regulatory/suppressive effects dependent on the expression of "regulatory" molecules. The majority of these cell populations express IL-10 which is a clear indication that IL-10 plays an important role in the generation/maintenance of a regulatory cell phenotype. However, in many cases these cells have been generated in vivo during tolerance inducing protocols e.g., with monoclonal antibodies.[53] It would be of great importance to establish methods for the generation of regulatory T cells in-vitro. First attempts have been reported by Groux et al,[54] who demonstrated that CD4+ T cells subset expressing IL-10 inhibits antigen-specific T cell responses and prevents colitis upon adoptive transfer. Recently it has been shown that incubation of mixed lymphocyte reactions (MLR) with IL-10 and Transforming Growth Factor-β (TGF-β) induce alloreactive T cells with regulatory function in a GVHD-model.[55] In addition to the experiments described above, experiments have been performed the generate IL-10 expressing regulatory cells by treatment with Vitamin D3 and Dexamethason. This leads to the generation of regulatory cells expressing IL-10.[56]

As it is now possible to generate and expand in vitro alloreactive regulatory T cells—Tregl—that are able to suppress alloreactive effector T cells in-vitro[15] there is the opportunity to apply Treg in clinical situations. Presently, Roncarolo and colleagues are performing a pilot trial to test the immunoregulatory capacity of these IL-10 driven Tregl cells for preventing acute and chronic graft-versus-host disease in haploidentical bone marrow transplantation.

Conclusions

The anti-inflammatory and immunomodulatory capacity of IL-10 makes it an ideal candidate for targeted immunotherapy in transplantation. There are a lot of promising experimental data using systemic or intragraft IL-10 delivery by administration of recombinant protein or various gene vectors. Most promising are the data with IL-10 dependent regulatory T cells. However, in some models IL-10 delivery does not show any protective effects or even deteriorates the graft.[57] These contradictory results demonstrate that we are far from understanding the action of IL-10 in detail. This explains why clinical applications of IL-10 in transplantation are limited so far.

References

1. Moore KW, de Waal Malefyt R, Coffman RL et al. Interleukin-10 and the interleukin-10 receptor. Annu Rev Immunol 2001; 19:683-765, (Review).
2. Chernoff AE, Granowitz EV, Shapiro L et al. A randomized, controlled trial of IL-10 in humans. Inhibition of inflammatory cytokine production and immune responses. J Immunol 1995; 15:154(10):5492-9.
3. Asadullah K, Sterry W, Volk HD. Interleukin-10 therapy—review of a new approach. Pharmacol Rev 2003; 55(2):241-69, (Review).
4. Tilg H, van Montfrans C, van den Ende A et al. Treatment of Crohn's disease with recombinant human interleukin 10 induces the proinflammatory cytokine interferon gamma. Gut 2002; 50(2):191-5.
5. Hutchings A, Wu J, Asiedu C et al. The immune decision toward allograft tolerance in nonhuman primates requires early inhibition of innate immunity and induction of immune regulation. Transpl Immunol 2003; 11(3-4):335-44.
6. Donckier V, Loi P, Closset J et al. Preconditioning of donors with interleukin-10 reduces hepatic ischemia-reperfusion injury after liver transplantation in pigs. Transplantation 2003; 75(6):902-4.
7. Li W, Fu F, Lu L et al. Differential effects of exogenous interleukin-10 on cardiac allograft survival: Inhibition of rejection by recipient pretreatment reflects impaired host accessory cell function. Transplantation 1999; 68(9):1402-9.
8. Zou XM, Yagihashi A, Hirata K et al. Downregulation of cytokine-induced neutrophil chemoattractant and prolongation of rat liver allograft survival by interleukin-10. Surg Today 1998; 28:184-191.
9. Eye Bank Association of America (1998). Eye Bank Association of America Annual Report: 1997-1998.
10. Torres PF, de Vos AF, Martins B et al. Interleukin 10 treatment does not prolong experimental corneal allograft survival. Ophthalmic Res 1999; 31:297-303.
11. Wissing KM, Morelon E, Legendre C et al. A pilot trial of recombinant human interleukin-10 in kidney transplant recipients receiving OKT3 induction therapy. Transplantation 1997; 64:999-1006.
12. Bacchetta R, Bigler M, Touraine JL et al. High levels of interleukin 10 production in vivo are associated with tolerance in SCID patients transplanted with HLA mismatched hematopoietic stem cells. J Exp Med 1994; 179(2):493-502.
13. Baker KS, Roncarolo MG, Peters C et al. High spontaneous IL-10 production in unrelated bone marrow transplant recipients is associated with fewer transplant-related complications and early deaths. Bone Marrow Transplant 1999; 23(11):1123-9.
14. Holler E, Roncarolo MG, Hintermeier-Knabe R et al. Prognostic significance of increased IL-10 production in patients prior to allogeneic bone marrow transplantation. Bone Marrow Transplant 2000; 25(3):237-41.
15. Roncarolo MG, Bacchetta R, Bordignon C et al. Type 1 T regulatory cells. Immunol Rev 2001; 182:68-79, (Review).
16. Kay MA, Glorioso JC, Naldini L. Viral vectors for gene therapy: The art of turning infectious agents into vehicles of therapeutics. Nat Med 2001; 7(1):33-40, (Review).
17. Cavazzana-Calvo M, Hacein-Bey S, de Saint Basile G et al. Gene therapy of human severe combined immunodeficiency (SCID)-X1 disease. Science 2000; 28:288(5466):669-72.

18. Nishikawa M, Huang L. Nonviral vectors in the new millennium: Delivery barriers in gene transfer. Hum Gene Ther 2001; 12(8):861-70, (Review).
19. Walther W, Stein U. Viral vectors for gene transfer: A review of their use in the treatment of human diseases. Drugs 2000; 60:249-71, (Review).
20. de Waal Malefyt R, Abrams J, Bennett B et al. Interleukin 10(IL-10) inhibits cytokine synthesis by human monocytes: An autoregulatory role of IL-10 produced by monocytes. J Exp Med 1991a; 174(5):1209-20.
21. de Waal Malefyt R, Haanen J, Spits H et al. Interleukin 10 (IL-10) and viral IL-10 strongly reduce antigen-specific human T cell proliferation by diminishing the antigen-presenting capacity of monocytes via downregulation of class II major histocompatibility complex expression. J Exp Med 1991b; 174(4):915-24.
22. DeBruyne LA, Li K, Chan SY et al. Lipid-mediated gene transfer of viral IL-10 prolongs vascularized cardiac allograft survival by inhibiting donor-specific cellular and humoral immune responses. Gene Ther 1998; 5(8):1079-87.
23. Hong YS, Laks H, Cui G et al. Localized immunosuppression in the cardiac allograft induced by a new liposome-mediated IL-10 gene therapy. J Heart Lung Transplant 2002; 21(11):1188-200.
24. Itano H, Mora BN, Zhang W et al. Lipid-mediated ex vivo gene transfer of viral interleukin 10 in rat lung allotransplantation. J Thorac Cardiovasc Surg 2001; 122(1):29-38.
25. Konig Merediz SA, Zhang EP, Wittig B et al. Hoffmann F. Ballistic transfer of minimalistic immunologically defined expression constructs for IL4 and CTLA4 into the corneal epithelium in mice after orthotopic corneal allograft transplantation. Graefes Arch Clin Exp Ophthalmol 2000; 238(8):701-7.
26. Fischbein MP, Yun J, Laks H et al. Regulated interleukin-10 expression prevents chronic rejection of transplanted hearts. J Thorac Cardiovasc Surg 2003; 126(1):216-23.
27. Qin L, Chavin KD, Ding Y et al. Multiple vectors effectively achieve gene transfer in a murine cardiac transplantation model. Immunosuppression with TGF-beta 1 or vIL-10. Transplantation 1995; 27:59(6):809-16.
28. Brauner R, Nonoyama M, Laks H et al. Intracoronary adenovirus-mediated transfer of immunosuppressive cytokine genes prolongs allograft survival. J Thorac Cardiovasc Surg 1997; 114(6):923-33.
29. Qin L, Ding Y, Pahud DR et al. Adenovirus-mediated gene transfer of viral interleukin-10 inhibits the immune response to both alloantigen and adenoviral antigen. Hum Gene Ther 1997; 8(11):1365-74.
30. Ding Y, Qin L, Kotenko SV et al. A single amino acid determines the immunostimulatory activity of interleukin 10. J Exp Med 2000; 191(2):213-24.
31. David A, Chetritt J, Guillot C et al. Interleukin-10 produced by recombinant adenovirus prolongs survival of cardiac allografts in rats. Gene Ther 2000; 7(6):505-10.
32. Fischer S, Liu M, MacLean AA et al. In vivo transtracheal adenovirus-mediated transfer of human interleukin-10 gene to donor lungs ameliorates ischemia-reperfusion injury and improves early posttransplant graft function in the rat. Hum Gene Ther 2001; 12:1513-1526.
33. Klebe S, Sykes PJ, Coster DJ et al. Prolongation of sheep corneal allograft survival by ex vivo transfer of the gene encoding interleukin-10. Transplantation 2001; 71:1214-1220.
34. Kato H, Fieblinger C, Ke B et al. Synergistic effects of cytokine gene transfer in high responder rat renal allograft recipients. Transplantation 1999; 67:S570.
35. Goudy K, Song S, Wasserfall C et al. Adeno-associated virus vector-mediated IL-10 gene delivery prevents type 1 diabetes in NOD mice. Proc Natl Acad Sci USA 2001; 98:13913-13918.
36. Yang Z, Chen M, Wu R et al. Suppression of autoimmune diabetes by viral IL-10 gene transfer. J Immunol 2002; 168:6479-6485.
37. Zhang YC, Pileggi A, Agarwal A et al. Adeno-associated virus-mediated IL-10 gene therapy inhibits diabetes recurrence in syngeneic islet cell transplantation of NOD mice. Diabetes 2003; 52:708-716.
38. Takayama T, Nishioka Y, Lu L et al. Retroviral delivery of viral interleukin-10 into myeloid dendritic cells markedly inhibits their allostimulatory activity and promotes the induction of T-cell hyporesponsiveness. Transplantation 1998; 66:1567-1574.
39. Coates PT, Krishnan R, Kireta S et al. Human myeloid dendritic cells transduced with an adenoviral interleukin-10 gene construct inhibit human skin graft rejection in humanized NOD-scid chimeric mice. Gene Ther 2001; 8:1224-1233.
40. Lee WC, Qiani S, Wan Y et al. Contrasting effects of myeloid dendritic cells transduced with an adenoviral vector encoding interleukin-10 on organ allograft and tumour rejection. Immunology 2000; 101:233-241.
40a. Van Montfraus C, Rodriguez Pena MS, Prouk I et al. Prevention of colitis by interleukin-10 transduced T lymphocytes in the SCID mice transfer model. Gastroenterology 2002; 123:(6):1865-1876.

41. Murakami Y, Akahoshi T, Kawai S et al. Antiinflammatory effect of retrovirally transfected interleukin-10 on monosodium urate monohydrate crystal-induced acute inflammation in murine air pouches. Arthritis Rheum 2002; 46:2504-2513.
42. Oh JW, Seroogy CM, Meyer EH et al. CD4 T-helper cells engineered to produce IL-10 prevent allergen-induced airway hyperreactivity and inflammation. J Allergy Clin Immunol 2002; 110:460-468.
43 Hall BM, Jelbart ME, Gurley KE et al. Specific unresponsiveness in rats with prolonged cardiac allograft survival after treatment with cyclosporine. Mediation of specific suppression by T helper/ inducer cells. J Exp Med 1985; 162:1683-1694.
44. Qin S, Cobbold SP, Pope H et al. "Infectious" transplantation tolerance. Science 1993; 259:974-977.
45. Onodera K, Lehmann M, Akalin E et al. Induction of "infectious" tolerance to MHC-incompatible cardiac allografts in CD4 monoclonal antibody-treated sensitized rat recipients. J Immunol 1996; 157:1944-1950.
46. Zhai Y, Kupiec-Weglinski JW. What is the role of regulatory T cells in transplantation tolerance? Curr Opin Immunol 1999; 11:497-503, (Review).
47. Waldmann H, Cobbold S. Regulating the immune response to transplants. A role for CD4+ regulatory cells? Immunity 2001; 14:399-406, (Review).
48. Battaglia M, Blazar BR, Roncarolo MG. The puzzling world of murine T regulatory cells. Microbes Infect 2002; 4:559-66, (Review).
49. Karim M, Bushell AR, Wood KJ. Regulatory T cells in transplantation. Curr Opin Immunol 2002; 14:584-591.
50. Levings MK, Bacchetta R, Schulz U et al. The role of IL-10 and TGF-beta in the differentiation and effector function of T regulatory cells. Int Arch Allergy Immunol 2002; 129:263-276.
51. Shevach EM. CD4+ CD25+ suppressor T cells: More questions than answers. Nat Rev Immunol 2002; 2:389-400.
52. Wood KJ, Sakaguchi S. Regulatory T cells in transplantation tolerance. Nat Rev Immunol 2003; 3:199-210.
53. Hara M, Kingsley CI, Niimi M et al. IL-10 is required for regulatory T cells to mediate tolerance to alloantigens in vivo. J Immunol 2001; 166(6):3789-96.
54. Groux H, O'Garra A, Bigler M et al. A CD4+ T-cell subset inhibits antigen-specific T-cell responses and prevents colitis. Nature 1997; 389:737-742.
55. Chen ZM, O'Shaughnessy MJ, Gramaglia I et al. IL-10 and TGF-beta induce alloreactive CD4+CD25- T cells to acquire regulatory cell function. Blood 2003; 101:5076-5083.
56. Barrat FJ, Cua DJ, Boonstra A et al. In vitro generation of interleukin 10-producing regulatory CD4(+) T cells is induced by immunosuppressive drugs and inhibited by T helper type 1 (Th1)- and Th2-inducing cytokines. J Exp Med 2002; 4:195(5):603-16.
57. Nussler NC, Muller AR, Weidenbach H et al. IL-10 increases tissue injury after selective intestinal ischemia/reperfusion. Ann Surg 2003; 238(1):49-58.

Chapter 8A

Role of IL-10 in Reactive Arthritis and Ankylosing Spondylitis

Joachim Sieper and Martin Rudwaleit

Introduction

Reactive arthritis (ReA) and ankylosing spondylitis (AS) are part of the so called spondyloarthropathies which also include arthritis/spondylitis with inflammatory bowel disease (IBD) and arthritis/spondylitis with psoriasis. The main link between each other is the association with HLA-B27, the same pattern of peripheral joint involvement with an asymmetrical arthritis predominantly of the lower limbs, and the possible occurence of sacroiliitis, spondylitis, enthesitis and uveitis. ReA and AS attracted special interest in the past because in ReA the triggering bacteria is known and AS is the most frequent disease and the disease with the worst outcome among the SpA. Furthermore, especially the HLA-B27-positive patients with ReA, with arthritis/spondylitis associated with psoriasis or with arthritis/spondylitis associated with IBD develop in about 20-50% of cases the full blown picture of AS in the course of their disease.

Although a substantial proportion of ReA patients will develop AS later on there is no evidence at this moment that persistent bacteria play a role in the pathogenesis of AS. It can be rather assumed that bacteria associated with ReA trigger somehow an autoimmune response in AS, possibly through cross-reactivity. Thus, these diseases offer the rather unique possibility to clarify the induction of autoimunnity by bacterial infection.[1]

ReA occurs between a few days and up to 6 weeks after a preceding bacterial infections of the gut with enterobacteriae such as yersinia, salmonella, shigella or campylobacter or of the urogential tract with Chlamydia trachomatis in about 2-6% of the infected patients. Similar to other SpA an asymmetrical arthritis with a predominant involvement of the lower limbs is typical. In the majority of patients the disease is self limiting with about 50% of patients without symptoms after 6 months, however, up to 20% of patients with symptoms longer than 12 months have been reported. The persistence of bacteria or bacterial antigen seems to be the most important trigger for acute and chronic ReA. It has been clearly shown over the past 10 years that bacteria or bacterial products from ReA-inducing bacteria such as Chlamydia trachomatis, Yersinia, Salmonella, and Shigella can be detected in synovial fluid and synovial membrane by various techniques. For Chlamydia trachomatis, chlamydial DNS and RNS have been found making it likely that live Chlamydia are present. This is different for enteric ReA. Yersinia- or Salmonella-DNA in the ReA joint has not been detected (with few exceptions) but constantly elevated Yersinia- or Salmonella-specific IgA antibodies (indicating recent contact with the bacterium because of the short half life of IgA antibodies), the detection of pieces from Yersinia or Salmonella in peripheral blood MNC over years in patients with chronic ReA, and animal models of Yersinia-induced arthritis indicate that enterobacteria also persist alive in vivo, probably outside the joint, at sites such as gut mucosa or lymph nodes.[2]

Interleukin-10, edited by Francesco M. Marincola. ©2006 Eurekah.com.

Role of Cytokines in the Pathogenesis of ReA

On this background the question has been raised why these bacteria persist in some patients but not in others and why some patients develop chronic courses of their arthritis. The ReA-associated bacteria are obligate (such as Chlamydia) or facultative intracellular bacteria. In previous research it could be shown that T-helper 1 cytokines such TNFα and IFNγ are crucial for an effective elimination of these bacteria while T-helper 2 cytokines such as IL-4 or T-helper 3 cytokines such as IL-10 might inhibit effective elimination.[3,4] In the last years we have investigated quite intensively these cytokines in peripheral blood, synovial fluid and synovial membrane of patients with ReA in comparison with other arthritides such as rheumatoid arthritis. We and others could indeed show that there is a relative deficiency of Th1-cytokines in ReA, especially of TNFα but also of IFNγ, both locally in synovial fluid and synovial membrane,[5-7] and systemically in peripheral blood.[8] Furthermore, we could demonstrate a correlation between a low TNFα production in peripheral blood and a longer duration of arthritic symptoms.[8] Thus, a relative lack of Th1-cytokines seems to be relevant for the occurrence and persistence of ReA, probably mediated by a persistence of bacteria. Immunopathology would then be caused by a hypersensitive immune response against persitent bacterial antigens.

IL-10 in Reactive Arthritis

IL-10 has received increasing interest in the last years as a potentially immunosuppressive cytokine. While upregulation of such a cytokine would be wanted in assumed autoimmune diseases such as rheumatoid arthritis we hypothesized that it might contribute to bacterial persistence in ReA, possibly by downregulation of the Th1-cytokines IFNγ and TNFα.

To investigate this question further we performed a serie of experiments. First, we could show that Chlamydia trachomatis specific CD4+ T cells clones derived from the synovial fluid from patients with Chlamydia-induced ReA were capable to secrete besides IFNγ also IL-10 upon in vitro stimulation with whole Chlamydia.[9] Subsequently, synovial fluid mononuclear cells (MNC) from 9 ReA patients due to Chlamydia trachomatis (4 patients), Yersinia enterocolitica (4 patients), or Salmonella enteritidis (1 patient) were stimulated in vitro with the responsible bacterium and cytokines were measured in the supernatant by ELISA.[6] SF MNC produced high amounts of IFNγ and IL-10, relatively less TNFα, and no detectable IL-4. These amounts were significantly higher than cytokine production by SF MNC stimulated with LPS and than cytokines secreted by PB MNC derived from healthy controls upon stimulation with the same whole bacteria. Interestingly, the amount of IL-10 was relatively the highest compared to IFNγ and TNFα resulting in a IL-10/IFNγ ratio of 2 (±1.9) and in a IL-10/TNFα ratio of 14 (±11) (Fig. 1). These ratios were strikingly high compared with the results found in the joints of 6 patients with Lyme arthritis (after stimulation of these SF MNC with Borrelia burgdorferi), in which the mean ratio was 0.6 (±0.4) for IL-10/IFNγ and 1.2 (±1.1) for IL-10/TNFα (Fig. 1).[6] Other researchers could also show in patients with Chlamydia-induced and other ReA that a relative high amount of IL-10 mRNA, as measured by PCR,[10] and a low amount of IFNγ, as detected by immunohistology,[7] are present. When peripheral blood MNC from ReA were stimulated in vitro with mitogens a significant lower amount of TNFα was found in ReA compared to rheumatoid arthritis. In the same study, IL-10 levels were found to be higher in ReA than in RA, although this difference was not significant.[8]

In contrast to the ReA-associated bacteria, a Th2 response seems to be necessary for the elimination of Borrelia burgdorferi.[11] Thus, these data indicate that the presence of the "wrong" Th2/Th3-response in the joint of ReA-patients and the "wrong" Th1-response in the joint of Lyme arthritis patients contributes to the persistence of these bacteria.

We further investigated whether IL-10 would have a regulatory role in the production of IFNγ and TNFα produced by SF-derived MNC. We could indeed show that IL-10 is most likely responsible for the inhibition of IFNγ and TNFα secretion within the ReA joint, because neutralizing endogenous IL-10 in culture (SF MNC stimulated with the responsible bacterium in vitro) with anti-IL-10 resulted in enhanced seceretion of IFNγ and TNFα. The addition of exogenous IL-10 further reduced secretion of both cytokines.[6] Interestingly, the enhanced

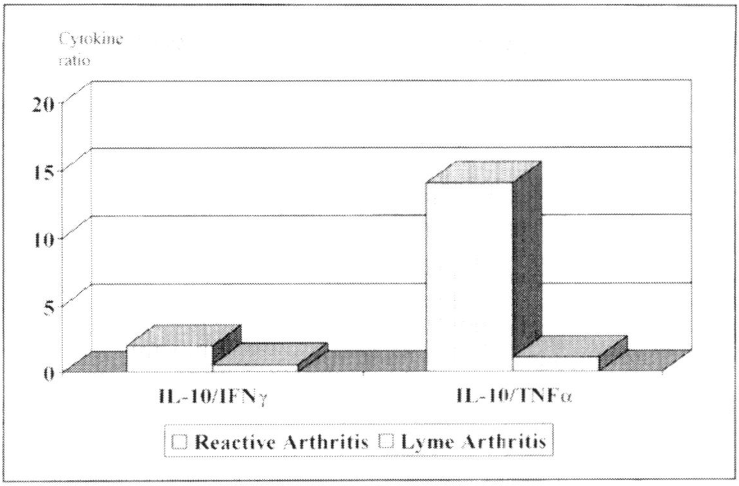

Figure 1. Ratio of IL-10/TNFα and IL-10/IFNγ- secreted by synovial fluid mononuclear cells stimulated with the triggering bacteria in case of reactive arthritis or with Borrelia burgdorferi in case of Lyme arthritis for 24 hours amd measured in the supernatant by ELISA (modified from Yin et al, 1997).[6]

effect of anti-IL-10 could be reduced by adding neutralizing antibody against IL-12, indicating that IL-10 suppressed IFNγ and TNFα secretion by inhibiting IL-12 synthesis. It has been shown before that the Th1 inhibitory effect of IL-10 might be mediated by suppression of IL-12.[12]

While in these studies we used whole bacteria for in vitro stimulation of MNC we next asked the question whether the cytokine secretion pattern, including IL-10, is dependent on which immunodominant antigen is used for stimulation of CD4+ T cells. We performed a detailed analysis of the response of T cell clones derived from synovial fluid from a patient with Yersinia-induced ReA which were specific for the heat shock protein (hsp) 60 from Yersinia.[13] Yersinia hsp60 specific CD4+ T cells were raised which were specific for different hsp60-derived antigens. Four clones specific for the same peptide were used for further analysis of the cytokine secretion pattern. Most interestingly, only one clone produced, in addition to IFNγ, significant amounts of IL-10. A longer peptide induced a greater production of IL-10 than the 12 amino acid long core peptide and an altered peptide (single AA substitution at position 10) induced the lowest secretion of IL-10.[13]

While in this study antigen specific T cell clones were used we applied subsequently the technique of antigen specific flow cytometry to investigate to capacity of T cell subpopulations to produce IL-10 upon antigenic stimulus.[14] For this method whole SF cells or whole PB are stimulated with antigen in vitro for 6 hours. For the last 4 hours the secretion inhibitor brefeldin A is added for accumulation of intracellular cytokines and the cytokine positive cells are quantified by flow cytometry. Besides antibodies specific for the IFNγ and IL-10 we used also antibodies binding to the cell surface marker CD4 allowing us to quantify the percentage of CD4+ T cells secreting IL-10 or IFNγ alone or together. In 2 patients with Chlamydia-induced ReA the percentage of IL-10-positive CD4+ T cells was between 0.10 and 0.23% while the percentage of IFNγ-positive CD4+ T cells was between 0.5 and 2.5% after in vitro stimulation with the Chlamydia-specific 'major outer membrane protein' (MOMP) or hsp 60 (Fig. 2) Interestingly, there were nearly no IFNγ/IL-10-double-positive T cells (Fig. 2).

These data clearly indicate that CD4+ T cells (and maybe also others, such as macrophages, which were not further differentiated in our experiments using whole MNC) produce IL-10 upon stimulation both with whole ReA-associated bacteria and with immunodominant bacteria derived from these bacteria. The production of IL-10 seems to be different for different T cell clones recognizing the same antigen and seems also to be dependent on the antigen.

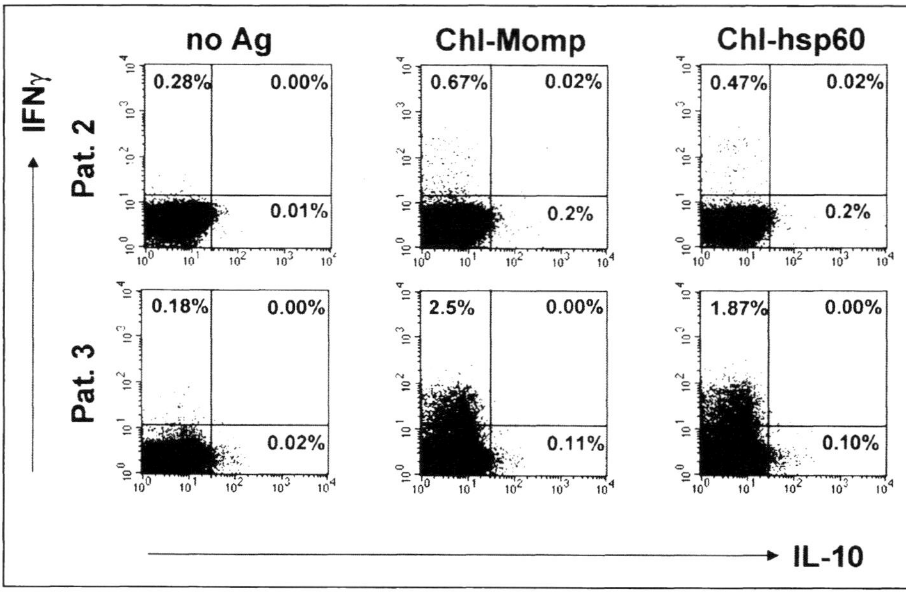

Figure 2. Stimulation of whole synovial fluid with different recombinant proteins from Chlamydia trachomatis. After staining for T cell surface markers and intracellular cytokines, a gate for CD4+ T cells was set. Double staining was performed for IL-10 and IFNγ. The percentage of the IFNγ single positive cells (left upper quadrant), the IFNγ/IL-10 double positive cells (right upper quadrant), and the IL-10 single-positive cells (right lower quadrant) of the CD4+/CD69+ T cell subpopulation are indicated. Reproduced with permission from Thiel et al, 2000. ©2000 Wiley Publishers.[14]

This flexibility of the IL-10-response offers the possibility for therapeutic manipulations. Therefore, an effective elimination of bacteria might be inhibited by IL-10 which can be overcome by exogenuous IL-12.

Subsequently, colleagues from Germany and Finland addressed the question whether differences in the production of IL-10 in ReA patients might be genetically determined. In this study 85 finish ReA patients and 62 HLA-B27-positive Finish controls were investigated. From genomic DNA, IL-10 microsatellites G and R and IL-10 promotor polymorphisms at positions -1087 and -524 were typed by polymerase chain reaction, automated fragment length analysis, and restriction fragment digestion. There was a significant decrease in the promoter alleles G12and G10 in the ReA group compared with the HLA-B27-positive controls indicating that these alleles might have a protective effect for the occurrence of ReA.[15] Although it is not clear at the moment whether these alleles are associated with a higher production of IL-10 this data raises the possibility that the relative increase of IL-10 found in ReA might, at least partially, be genetically determined.

IL-10 in Ankylosing Spondylitis

There are much less data on IL-10 in AS compared to ReA. In AS peripheral arthritis is less common than in ReA resulting in less access to synovial fluid and/or synovial membrane. As a consequence, all available data on IL-10 in AS are from peripheral blood. Before better data will be available from the site of inflammation no clear conclusions can yet be drawn.

In one study a correlation between IL-10 levels in plasma and disease activity was reported[16] while in another study no difference in median levels of IL10 between AS and controls could be observed.[19]

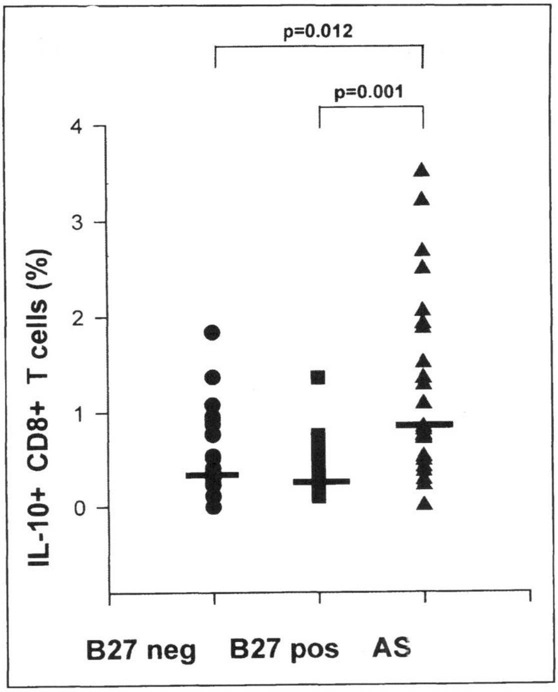

Figure 3. Percentage of IL-10 positive CD8+ T cells among patients with AS (all HLA-B27 positive), healthy HLA-B27 positive controls, and healthy HLA-B27 negative controls. The horizontal bars indicate the medians. Reproduced with permission from Rudwaleit et al, 2000. ©2000 BMJ Publishing Group.[17]

To address this question we measured cytokine-positive CD4+ and CD8+ T cells derived from peripheral blood after mitogenic in vitro stimulation by flow cytometry in 25 HLA-B27-positive patients with active AS in comparison to 18 healthy HLA-B27 positive controls and 22 healthy HLA-B27 negative controls.[17] AS patients had a significantly lower percentage of IFNγ- or TNFα-positive CD4+ T cells compared to HLA-B27-negative controls while the results for HLA-B27-positive healthy control group was in between these 2 groups. For IL-10-positive T cells, we found a significant increase in the CD8+ T cell subpopulation compared to the B27-positive and B27-negative controls (Fig. 3), but not in the CD4+ subpopulation. Thus, there is also a tendency in AS, similar to ReA, to find a higher amount of IL-10 produced by immunocompetent cells compared to controls. A more recent paper looked for IL-10 polymorphism in AS. No significant effect on susceptibility to AS but a possibly minor role in determining age of disease onset and disease activity was found.[18]

Summary and Conclusion

Our data indicate that an imbalance between IL-10 and TNFα/IFNγ might contribute to the persistence of bacteria in ReA, an imbalance which can be overcome by exogenous IL-12. We could clearly show that IL-10 is not only produced by monocytes/macrophages but also by T cells. These T cells might play a central role in the regulation of the immune response. The contribution of IL-10 for the ongoing immune response in AS is less clear. Some of the ReA patients move on to AS in the course of their disease, however other factors such as inflammatory bowel disease with interactions between immune cells and gut bacteria seems to be at least as important as an earlier bacterial infection. Thus, the chronic interaction between the host's immune system and bacteria are probably most relevant for the development and persistence of AS. For the persistence of bacteria an upregulation of IL-10 seems to be important but for the damage of gut mucosa a Th1-response is more relevant. Because different triggers might cause AS this might be an explanation why the picture regarding IL-10 and other cytokines is not so clear. The dramatic improvement of AS patients after treatment with TNFα-blockers suggest that a relative overproduciton of IL-10 can not be of major importance for the pathogenesis.

References

1. Sieper J, Braun J. Pathogenesis of spondylarthropathies. Persistent bacterial antigen, autoimmunity, or both? Arthritis Rheum 1995; 38(11):1547-54.
2. Sieper J, Braun J, Kingsley GH. Report on the fourth international workshop on reactive arthritis. Arthritis Rheum 2000; 43(4):720-34.
3. Autenrieth IB, Beer M, Bohn E et al. Immune responses to Yersinia enterocolitica in susceptible BALB/c and resistant C57BL/6 mice: An essential role for gamma interferon. Infect Immun 1994; 62(6):2590-9.
4. Yang X, HayGlass KT, Brunham RC. Genetically determined differences in IL-10 and IFN-gamma responses correlate with clearance of Chlamydia trachomatis mouse pneumonitis infection. J Immunol 1996; 156(11):4338-44.
5. Simon AK, Seipelt E, Sieper J. Divergent T-cell cytokine patterns in inflammatory arthritis. Proc Natl Acad Sci USA 1994; 91(18):8562-6.
6. Yin Z, Braun J, Neure L et al. Crucial role of interleukin-10/interleukin-12 balance in the regulation of the type 2 T helper cytokine response in reactive arthritis. Arthritis Rheum 1997; 40(10):1788-97.
7. Smeets TJ, Dolhain RJ, Breedveld FC et al. Analysis of the cellular infiltrates and expression of cytokines in synovial tissue from patients with rheumatoid arthritis and reactive arthritis. J Pathol 1998; 186(1):75-81.
8. Braun J, Yin Z, Spiller I et al. Low secretion of tumor necrosis factor alpha, but no other Th1 or Th2 cytokines, by peripheral blood mononuclear cells correlates with chronicity in reactive arthritis. Arthritis Rheum 1999; 42(10):2039-44.
9. Simon AK, Seipelt E, Wu P et al. Analysis of cytokine profiles in synovial T cell clones from chlamydial reactive arthritis patients: Predominance of the Th1 subset. Clin Exp Immunol 1993; 94(1):122-6.
10. Kotake S, Schumacher Jr HR, Arayssi TK et al. Gamma interferon and interleukin-10 gene expression in synovial tissues from patients with early stages of Chlamydia-associated arthritis and undifferentiated oligoarthritis and from healthy volunteers. Infect Immun 1999; 67(5):2682-6.
11. Keane-Myers A, Nickell SP. Role of IL-4 and IFN-gamma in modulation of immunity to Borrelia burgdorferi in mice. J Immunol 1995; 155(4):2020-8.
12. Uyemura K, Demer LL, Castle SC et al. Cross-regulatory roles of interleukin (IL)-12 and IL-10 in atherosclerosis. J Clin Invest 1996; 97(9):2130-8.
13. Mertz AK, Wu P, Sturniolo T et al. Multispecific CD4+ T cell response to a single 12-mer epitope of the immunodominant heat-shock protein 60 of Yersinia enterocolitica in Yersinia-triggered reactive arthritis: Overlap with the B27-restricted CD8 epitope, functional properties, and epitope presentation by multiple DR alleles. J Immunol 2000; 164(3):1529-37.
14. Thiel A, Wu P, Lauster R et al. Analysis of the antigen-specific T cell response in reactive arthritis by flow cytometry. Arthritis Rheum 2000; 43(12):2834-42.
15. Kaluza W, Leirisalo-Repo M, Marker-Hermann E et al. IL10.G microsatellites mark promoter haplotypes associated with protection against the development of reactive arthritis in Finnish patients. Arthritis Rheum 2001; 44(5):1209-14.
16. Claudepierre P, Rymer JC, Chevalier X. IL-10 plasma levels correlate with disease activity in spondyloarthropathy. J Rheumatol 1997; 24(8):1659-61.
17. Rudwaleit M, Siegert S, Yin Z et al. Low T cell production of TNFalpha and IFNgamma in ankylosing spondylitis: Its relation to HLA-B27 and influence of the TNF-308 gene polymorphism. Ann Rheum Dis 2001; 60(1):36-42.
18. Goedecke V, Crane AM, Jaakkola E et al. Interleukin 10 polymorphisms in ankylosing spondylitis. Genes Immun 2003; 4(1):74-6.
19. Vazquez-Del MM, Garcia-Gonzalez A, Munoz-Valle JF et al. Interleukin 1beta (IL-1beta), IL-10, tumor necrosis factor-alpha, and cellular proliferation index in peripheral blood mononuclear cells in patients with ankylosing spondylitis. J Rheumatol 2002; 29(3):522-6.
20. Zou J, Rudwaleit M, Brandt J et al. Up regulation of the production of tumour necrosis factor alpha and interferon gamma by T cells in ankylosing spondylitis during treatment with etanercept. Ann Rheum Dis 2003; 62(6):561-4.
21. Zou J, Rudwaleit M, Brandt J et al. Down-regulation of the nonspecific and antigen-specific T cell cytokine response in ankylosing spondylitis during treatment with infliximab. Arthritis Rheum 2003; 48(3):780-90.

CHAPTER 8B

Interleukin-10 and Psoriasis

Khusru Asadullah, W. Sterry and Hans-Dieter Volk

Abstract

Interleukin (IL)-10 is an important immunoregulatory cytokine. One of its main biological function seems to be the limitation and termination of inflammatory responses. Remarkably, a relative deficiency in IL-10 expression is found in psoriasis, a frequent inflammatory skin disease, characterized by a type 1 cytokine pattern. Induction of IL-10 expression was found by conventional antipsoriatic therapies, suggesting that IL-10 may be a key cytokine in psoriasis and that application of this cytokine may have therapeutic effects. In first clinical trials over 3-7 weeks in patients with established psoriasis IL-10 was well tolerated and clinical efficient. In a long term trial in patients with psoriasis in remission, IL-10 therapy decreased the incidence of relapse and prolonged the disease free interval. Laboratory investigations suggest that IL-10 exerts its antipsoriatic activity by effects on different cell populations including antigen presenting cells and T-cells. IL-10 led to a lasting type 1/ type 2 cytokine balance shift. Direct effects of IL-10 on keratinocytes, however, are unlikely to have contributed to the clinical response, since IL-10 unresponsiveness of keratinocytes was found in vitro. IL-10 seems to have major importance in psoriasis. Further investigations are necessary to determine whether its application may represent a promising new therapeutic approach.

Pathophysiology of Psoriasis

Psoriasis is a multigenic, cutaneous disorder characterized by inflammation and abnormal epidermal proliferation with a prevalence of 2-3% in the general population. It is defined as a clinical entity affecting skin, nails, and in approx. 5% of the patients joints. Typical findings are keratinocyte hyperproliferation with hyper- and parakeratotic differentiation, epidermal influx of polymorphonuclear leukocytes, and the presence of a mononuclear infiltrate in the papillary dermis and in the epidermis. In contrast to type 2, psoriasis type 1 psoriasis is characterized by early onset (before age 40) and a positive family history.

Several observations indicate that T lymphocytes and cytokines are of major importance in the pathogenesis of this chronic skin disease. These observations are supported by the beneficial effects of systemic administration of immunosuppressive drugs like cyclosporine A, FK506, DAB389IL-2, and anti-T-cell antibodies, known to act on T-cells and to influence the cytokine pattern. Finally, recent investigations using SCID mice demonstrated a central role of immunocytes, in particular T cells, in the pathophysiology of psoriasis. Table 1 summerizes the arguments for considering psoriasis a T-cell mediated dermatosis.[1]

All in all, based on the current knowledge, psoriasis is considered to be probably initiated by presentation of so far unknown "psoriasis-related antigens" by specialized cutaneous APC. Presentation of these antigens and/or superantigens activates T-effector lymphocytes leading to enhanced cytokine formation. These activated T-cells are considered to change epidermal growth homeostasis, leading to increased keratinocyte proliferation and abnormal differentiation. The cutaneous and systemic over-expression of several proinflammatory cytokines, in particular type 1 cytokines, has been demonstrated in psoriasis. This includes interleukin (IL-)2, IL-6,

Interleukin-10, edited by Francesco M. Marincola.

Table 1. Arguments for considering psoriasis a T cell mediated dermatosis[26]

- Presence of activated T-cells in the skin lesions
- Cure of the disease by bone marrow transplantation from healthy persons and transfer of the disease by transplantation of bone marrow from psoriatic patients
- Demonstration of the impact of immunocytes by SCID mice experiments
- Therapeutic effects of immunosuppressants targeting T lymphocytes (e.g., Cyclosporin-A, anti-T cell antibodies)

IL-8, IL-12, interferon (IFN-)γ, tumor necrosis factor (TNF)-α. Considering all data and the evidences that "classical" type-1 and type-2 patterns only mark extremes in a wide variety of motifs, the cytokine profile in psoriasis can be considered as "proinflammatory, type 1-like". The resulting immunological dysbalance is shown in Figure 1. Beside the regulatory effects of cytokines on inflammation which seems to be crucial, several out of them might directly enhance the keratinocyte proliferation. So, it has been shown that IL-1 and IL-6 are mitogenic for keratinocytes. However, several other factors are involved into the control of keratinocyte growth as well. It is important to notice that there might be a special mix of factors responsible for the keratinocyte hyperproliferation and that other factors not identified so far might be involved, too.

IL-10 Expression in Psoriasis

Its considerable anti-inflammatory effects and its ability to act as a main suppresser of cellular immunity[2] raises the question of the IL-10 expression under pathophysiological conditions, including in psoriasis.

Immunohistochemical investigations suggested a low cutaneous IL-10 protein expression.[3] Similar results were found by quantification of IL-10 protein in blister fluids.[4] We found that the cutaneous IL-10 mRNA expression in psoriasis was significantly lower than in atopic dermatitis or cutaneous T-cell lymphoma. The level of IL-10 mRNA expression did not differ from healthy skin, even though numerous pro-inflammatory cytokines are overexpressed. These results are in particular remarkable since TNF, a major inducer of IL-10 was highly expressed. All in all this indicates a relative IL-10 deficiency in psoriasis[5] which might have a genetic background.

Remarkably, well established antipsoriatic therapies such as UV radiation or vitamin D analogues are leading to an increase of IL-10 expression in vitro and in vivo.[5]

IL-10 and IL-10 Receptor Polymorphisms in Psoriasis

Genetic markers in cytokine genes are becoming widely used in studies of immune-mediated disease and it is becoming apparent that they can be markers of disease susceptibility as well as of disease severity.[6] When the two known polymorphic microsatellite markers in the human IL-10 promoter IL10.G and IL10.R. were investigated in psoriasis patients, no difference was noted in comparison to the control group. Also, no difference in allele distribution was observed when the psoriasis patients were stratified according to age of onset. However, a clear differential distribution was revealed when patients were grouped according to whether they had a positive family history of psoriasis. In this case, allele IL10.G13 was positively associated with familial psoriasis and this was also true when younger patients were considered - patients with age-of-onset of less than forty were three-fold more likely to have a psoriatic family background if they carried this allele. Thus, it appears that the IL-10 locus contributes to the genetic background in familial psoriasis.[7] More recent investigations in a larger population, however, do not support this.[8] Further investigations are necessary to determine whether the relative IL-10 deficiency in psoriasis has a genetic origin.

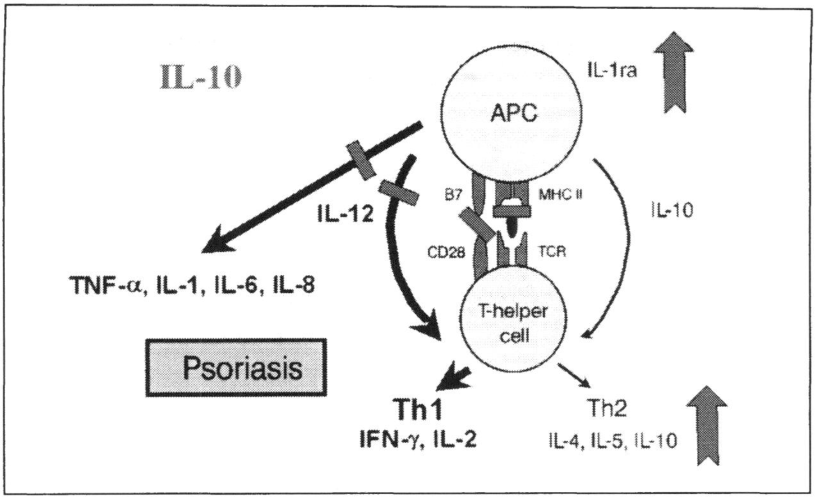

Figure 1. Effects of IL-10 on the T-helper (Th)1/Th2 dysbalance in psoriasis.[18] An immune deviation towards a type 1 cytokine aptetrn is a typical finding in several indications such as psoriasis, rheumatoid arthrism, inflammatory bowel disease, transplant rejection, multiple sclerosis. IL-10 reverses the Th1 cytokine pattern present. It promotes the development of a type 2 cytokine pattern by inhibiting the IFN-γ production ofT lymphocytes particularly via the suppression of IL-12 synthesis in accessory cells. Moreover it inhibits MHC class II and costimulatory molecule expression.

Clinical Effects of IL-10 Therapy in Psoriasis

The powerful immunomodulatory properties of IL-10 and the promising results from IL-10 delivery on the course of several inflammatory diseases in experimental models induced the interest on clinical application of recombinant human (rh) IL-10 (ilodecakin / Tenovil; Schering-Plough Corp., USA) After promising data from phase 1 trials in was used in several indications including in psoriasis. The first administration of human recombined IL-10 in human was performed in 1995.[9] Overall, these studies showed that IL-10 is well tolerated without serious side effects at doses up to 25 μg/kg; mild to moderate flu-like symptoms were observed in a fraction of recipients at doses up to 100 μg/kg.[10] Therapeutic effects of IL-10 in psoriatic patients has been studied in 7 trials so far (Table 2).

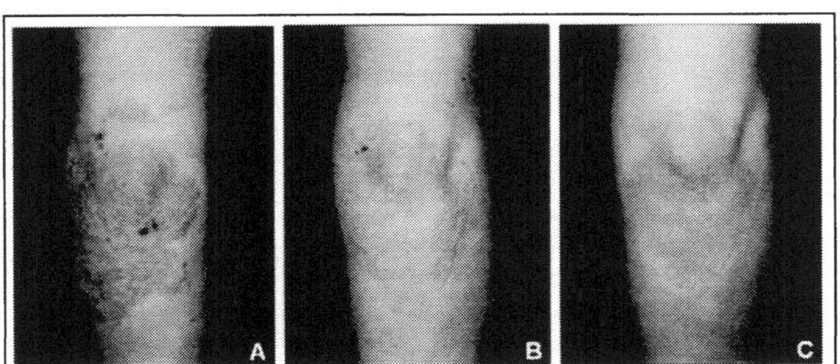

Figure 2. Clinical effects of short course IL-10 therapy in established psoriasis.[11] Example of a patient showing good response towards IL-10 therapy. Typical skin lesions before (A), during (B - day 15) and at the end of therapy (C- day 50) are shown. The patient received IL-10 injections in a dosage of 20 μg/kg 3 times per week. © American Medical Association.

Table 2. Clinical effects of IL-10 therapy in psoriasis[27]

Study	Design, Patients and Interventions	Outcome and Remarks	Ref.
Pilot study in exacerbated psoriasis vulgaris	Open label, not placebo controlled; 3 patients with moderate to severe psoriasis; 8 µg/kg/d s.c. over a 24 day period	Safe, clinical and histological confirmed improvement	5
Pilot study in exacerbated psoriasis vulgaris	Open-label; not placebo controlled; 10 patients with moderate to severe psoriasis: 4 µg/kg/d s.c. over 42 days	Safe, good to moderate response, histological confirmed; significant mean decrease in PASI by 40% after 3 weeks and 68% after 6 weeks. These patients are a subgroup of the study Reich et al, 2001 (see below).	12
Phase II in exacerbated psoriasis vulgaris	Open-label; not placebo controlled; 10 patients with moderate to severe psoriasis; 8 µg/kg daily (n=5) or 20 µg/kg three times per week (n=5) s.c. over 49 days	Safe, good to moderate response, histological confirmed; significant mean decrease in PASI by 55% after 7 weeks	11
Phase II in exacerbated psoriasis vulgaris	Open-label, not placebo-controlled; 15 patients with moderate to severe disease; 4 µg/kg/d s.c. over 42 days	Safe, good—moderate response histological confirmed; 14 patients evaluated, mean decrease in PASI by 50% after 4 weeks and 59% after 6 weeks.	13
Phase II in psoriatic arthritis	Placebo-controlled, double-blind; 29 patients with psoriatic arthritis, dose-escalating with 1, 5, or 10 µg/kg s.c. daily for 28 days	Well tolerated, significant clinical improvement in skin (> 30% PASI reduction: 50% in IL10 group (10 µg/kg) vs 10% in placebo after 4 weeks. No decrease in articular disease activity.	15
Phase II in exacerbated psoriasis vulgaris	Randomized, double-blind, placebo controlled; 28 patients with moderate to severe psoriasis; 20 µg/kg three times per week for 12 weeks	Modest trend towards improvement after 6 and 8 weeks (mean decrease in PASI by 31% and 35% in IL10 group versus 7% and 13% in placebo group, respectively), but no significant difference at week 12 (17% vs 13% in placebo group).	14
Phase II in psoriasis vulgaris in remission	Placebo-controlled, double-blind; 17 patients with moderate to severe psoriasis in remission; 10 µg/kg; IL-10 (n=7) or placebo (n=10) 3 x week for 4 month	Well tolerated; 90% relapse in the placebo group versus 28% in the IL-10 treated group; Significant prolongation of relapse-free interval (101 days vs 66 in placebo).	16

In our pilot trial starting in 1997, daily injections of 8 µg rhIL-10 /kg body weight directly under a psoriatic plaque over a 24 day period led to complete clearance of the plaque in one of two patients.[5] Moreover, some systemic antipsoriatic effects were observed in all 3 patients treated in this pilot trial (subcutaneous injections under nonlesional skin in the third patient).

In a second trial (open-label phase II), ten psoriatic patients received subcutaneously rhIL-10 over a 7 week period in a dosage of 8 µg/kg daily (n=5) or 20 µg/kg three times per week (n=5), respectively.[11] Patients were followed up for an additional 7 weeks. The treatment was well tolerated. We found antipsoriatic effects in 9 out of 10 patients resulting in a significant decrease of the psoriasis area and severity index (PASI) by 55.3 ± 11.5% (mean ± SEM, p<0.02). The antipsoriatic effect was confirmed by histological examination. Heterogeneity in the effectiveness was found among the patients, but seemed to be independent of the dosage regime[11] (Fig. 2).

Similar clinical effectiveness of IL-10 application has been reported by Reich et al.[12] In this open-label phase II trial ten patients were treated subcutaneously with 4 µg/kg rhIL-10 daily. The mean of the disease activity score PASI decreased by 67.9% after 6 weeks of treatment and was associated with improvement of histological parameters.[13] The clinical response was associated with a significant decrease of cutaneous cell infiltration and the lesional expression of type1 cytokines (IFNγ, TNF), IL-17 IL-8, and IL-8 receptor CXCR2. There was some evidence that genetic factors are involved in the response to IL-10.[13]

In a more recent study[14] 28 patients with moderate-to-severe psoriasis received rhIL-10 (20 µg/kg) or placebo subcutaneously 3 times weekly for 12 weeks in a randomized, double-blind manner. Remarkably, treatment with rhIL-10 resulted in only temporary clinical improvement after 6 and 8 weeks, despite sustained systemic decreases in proinflammatory and type 1 cytokine production.

The effect of IL-10 in psoriatic arthritis patients has been investigated by McInnnes and coworkers. IL-10 was given sc for 28 consecutive days in a double blind, placebo-controlled study including 29 patients (0, 1, 5, or 10 µg/kg). Modest, but significant clinical improvement in skin, but not articular disease activity scores with only minor adverse effects was observed.[15]

More recently, we investigated the effects of long-term IL-10 application on the immune system and duration of psoriasis remission.[16] We performed a placebo-controlled, double-blind, phase II trial using IL-10 in patients with chronic plaque psoriasis in remission. Patients received subcutaneous injections with either IL-10 (10 µg/kg body weight; n=7) or placebo (n=10) 3 times per week until relapse or study termination after 4 months. The treatment was well tolerated. In the placebo group almost all patients (90%) showed a relapse during the observation period. In contrast to this, only 2 out of 7 patients (28.6%) relapsed in the IL-10-treated group. Kaplan Meier analysis revealed a significantly lower relapse incidence in the IL-10 than in the placebo group (p=0.02). The mean relapse-free interval time was 101.6 ± 12.6 days in the IL-10 group in comparison to 66.4 ± 10.4 days in the placebo group (Fig. 3).

Immunological Effects of IL-10 Therapy in Psoriasis

Analysing the immunological effects of IL-10 therapy in psoriasis should lead to an better understanding of its antipsoriatic mode of action. It seems that IL-10 exerts its antipsoriatic activity by effects on different cell populations including T-cells and APCs as well as their mutual interaction. IL-10 is able to suppress the APC activity of monocytes/macrophages and the development of dendritic cells. In fact, depressed monocytic HLA-DR and CD86 expression as well as TNF-α and IL-12 secretion capacities were observed by us during IL-10 therapy. Moreover, IL-10 led to a lasting type 1/ type 2 shift (increasing proportion of IL-4, IL-5, and IL-10 producing T-cells, selective increase in IgE serum levels).[17] A significant negative correlation was demonstrated between the IL-4 secretion capacity and Psoriasis Area and Severity Index score was found in our long term trial. The physiological significance of these findings was reflected by the depressed DTH reaction to recall antigens during IL-10 therapy.[11,18] Interestingly, IL-10 therapy led to a decrease in cutaneous IL-8 and an increase in IL-4 expression, both of which might contribute to the antipsoriatic effect.[13,19] This is supported by the recently reported antipsoriatic effects of IL-4 itself.[20] Direct effects of IL-10 on keratinocytes, however, are unlikely to have contributed to the clinical response, since the IL-10 unresponsiveness of keratinocytes has been demonstrated by us recently.[21,22]

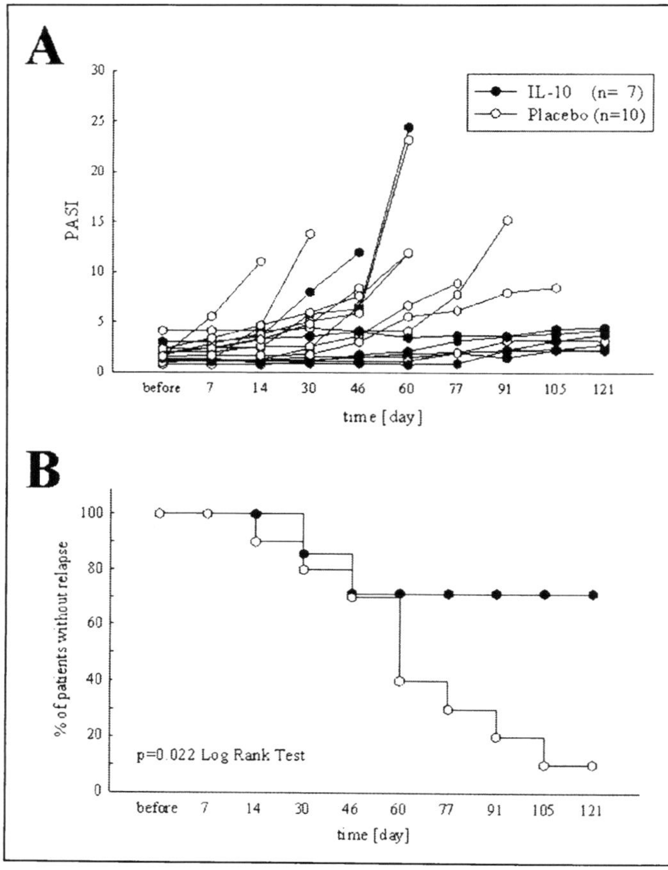

Figure 3. Clinical effects of long term IL-10 therapy in psoriasis.[16] A) Individual courses of psoriatic disease under long term IL-10 application to prevent the reoccurenace of disease. The patients dropped out of the analysis when fulfilling the criteria for psoriasis relapse. B) Kaplan Meier analysis of relapse-risk.

Perspectives

Some effectiveness of IL-10 application for the therapy of established, exacerbated psoriasis has been demonstrated by the majority of early phase II trials. Further investigations, in particular multicenter, placebo-controlled, double blind trials, however, are required to better determine the therapeutic potential of this cytokine. This will lead to the answer whether IL-10 could finally become an approved, successful antipsoriatic drug. Independent from this, the effects of IL-10 in psoriasis are contributing to an better pathopysiological understanding of this immune skin disease.[1]

Regarding the clinical response the picture is already emerging that the potency of this treatment seems to be overall below that of other promising approaches such as the anti-TNF-α strategies. Thus, a complete disease clearance is usually a rare event after IL-10 application, suggesting a relatively low anti-inflammatory potential for this therapy. Moreover, a recent report[23] and our recent unpublished observations suggest even pro-inflammatory properties of IL-10. In agreement with this we observed an increase of the sIL-2R levels in IL-10 long-term-treated patients. Therefore, we speculate that IL-10 might not be the major "break through" in anti-inflammatory therapy of exacerbated disease. In contrast, impressive clinical effects with regard to prevention of psoriasis relapse, were found in a single study, rather reflecting long-term immunomodulatory than anti-inflammatory properties of IL-10. It might be that not anti-inflammatory but complex immunomodulatory effects might be primarily responsible for the prolongation of the disease-free interval we observed in psoriasis. Such a hypothesis is supported by recent data from animal and in vitro experiments, demonstrating that IL-10 can induce the formation of regulatory T-cells with major impact on the immunoregulation.[24,25] Taken together, both first clinical and immunological data suggest that IL-10 application seems to be even more attractive in a secondary prophylactic (prevention of relapse) than in a therapeutic approach in psoriasis. It might be speculated, that IL-10 long term therapy could be equally effective in other entities as inflammatory bowel diseases or rheumatoid arthritis and transplantation.

If it could be determined why certain patients show a good response whereas other do not respond at all, this may lead to the identification of patients suitable for IL-10 therapy. Determining the cytokine and cytokine receptor pattern might be an approach. Currently, however, no real predictive test parameter is available.

So far, IL-10 as a cytokine is therapeutically has been used as a recombinant protein i.e., a large molecule. Therefore its quite expensive to produce, must be administered by injection, which is quite inconvenient for the patient, and the induction of neutralizing antibodies, which might limit their effect, has to be excluded for long term application. Identification of molecules mediating the effects of this cytokine which are suitable for pharmacological intervention with small molecules will, therefore, of increasing interest. Interference of protein–protein interaction with small molecules, however, is almost impossible. Therefore, it is unlikely that IL-10 receptor agonists suitable for oral application will be discovered. Molecules acting downstream of the cytokine receptors targeting for example certain kinases or phosphatases or signal transduction molecules might represent much better "drugable" approaches. The problem here, however, is the specificity, since the known IL-10 signal transduction molecules are shared by different cytokines. Induction of cytokine production with low molecular weight compounds may represent novel approaches. In any case, further investigations on the regulation of IL-10 expression and signalling are of major importance and may lead to better therapeutic approaches.

References

1. Asadullah K, Volk HD, Sterry W. Novel immunotherapies for psoriasis. Trends in Immunology 2002; 23:47-53.
2. Spits H, De Waal Malefyt R. Functional characterization of human IL-10. Int Arch Allergy Immunol 1992; 99:8-15.
3. Nickoloff BJ, Fivenson DP, Kunkel SL et al. Keratinocyte interleukin-10 expression is upregulated in tape-stripped skin, poison ivy dermatitis, and Sezary syndrome, but not in psoriatic plaques. Clin Immunol Immunopathol 1994; 73:63-68.
4. Mussi A, Bonifati C, Carducci M et al. IL-10 levels are decreased in psoriatic lesional skin as compared to the psoriatic lesion-free and normal skin suction blister fluids. J Biol Regul Homeost Agents 1994; 8:117-120.
5. Asadullah K, Sterry W, Stephanek K et al. IL-10 is a key cytokine in psoriasis: Proof of principle by IL-10 therapy - a new therapeutic approach. J Clin Invest 1998; 101:783-794.
6. Bidwell J, Keen L, Gallagher G et al. Cytokine gene polymorphism in human disease: On-line databases. Genes Immun 1999; 1:3-19.
7. Asadullah K, Eskdale J, Wiese A et al. Interleukin-10 promotor polymorphism in psoriasis. J Invest Dermatol 2001a; 116:975-978.
8. Hensen P, Asadullah K, Windemuth C et al. Interleukin-10 promoter polymorphism IL10. G and familial early onset psoriasis. Br J Dermatol 2003; 149(2):381-385.
9. Chernoff AE, Granowitz EV, Shapiro L et al. A randomized, controlled trial of IL-10 in humans. Inhibition of inflammatory cytokine production and immune responses. J Immunol 1995; 154:5492-5499.
10. Moore KW, de Waal Malefyt R, Coffman RL et al. Interleukin-10 and the interleukin-10 receptor. Annu Rev Immunol 2001; 19:683-765.
11. Asadullah K, Döcke WD, Ebeling M et al. Interleukin-10 treatment of psoriasis - clinical results of a Phase II trial. Arch Dermatol 1999a; 135:187-192.

12. Reich K, Bruck M, Grafe A et al. Treatment of psoriasis with interleukin-10. J Invest Dermatol 1998; 111:1235-1236.
13. Reich K, Garbe C, Blaschke V et al. Response of psoriasis to Interleukin-10 is associated with suppression of cutaneous type 1 inflammation, downregulation of the epidermal Interleukin-8/CXCR2 pathway and normalization of keratinocyte maturation. J Invest Dermatol 2001; 116:319-329.
14. Kimball AB, Kawamura T, Tejura K et al. Clinical and immunologic assessment of patients with psoriasis in a randomized, double-blind, placebo-controlled trial using recombinant human interleukin 10. Arch Dermatol 2002; 138:1341-1346.
15. McInnes IB, Illei GG, Danning CL et al. IL-10 improves skin disease and modulates endothelial activation and leukocyte effector function in patients with psoriatic arthritis. J Immunol 2001; 167:4075-4082.
16. Friedrich M, Döcke WD, Klein A et al. Immunomodulation by IL-10 therapy decreases the incidence of relapse and prolongs the relapse-free interval in psoriasis. J Invest Dermatol 2002; 118:672-677.
17. Asadullah K, Döcke WD, Sabat R et al. The treatment of psoriasis with interleukin-10: Rationale and review of the first clinical trials. Expert Opin Investig Drugs 2000a; 9:95-102.
18. Asadullah K, Sabat R, Wiese A et al. Interleukin-10 in cutaneous disorders: Implications for its pathophysiological importance and therapeutic use. Arch Dermatol Res 1999b; 291:628-636.
19. Asadullah K, Friedrich M, Hanneken S et al. Effects of systemic interleukin-10 therapy on psoriatic skin lesions: Histological, immunohistological and molecularbiological findings. J Invest Dermatol 2001b; 116:721-727.
20. Ghoreschi K, Thomas P, Breit S et al. Interleukin-4 therapy of psoriasis induces Th2 responses and improves human autoimmune disease. Nat Med 2003; 9:40-46.
21. Seifert M, Sterry W, Effenberger E et al. The antipsoriatic activity of IL-10 is rather caused by effects on peripheral blood cells than by direct effect on human keratinocytes. Arch Derm Res 2000; 292:164-172.
22. Seifert M, Gruenberg BH, Sabat R et al. Keratinocyte unresponsiveness towards interleukin-10: Lack of specific binding due to deficient IL-10 receptor 1 expression. Exp Dermatol 2003; 12:137-44.
23. Lauw FN, Pajkrt D, Hack CE et al. Proinflammatory effects of IL-10 during human endotoxemia. J Immunol 2000; 165:2783-2789.
24. Cottrez F, Hurst SD, Coffman RL et al. T regulatory cells 1 inhibit a Th2-specific response in vivo. J Immunol 2000; 165:4848-4853.
25. Levings MK, Sangregorio R, Roncarolo MG. Human CD25(+)CD4(+) T regulatory cells suppress naive and memory T cell proliferation and can be expanded in vitro without loss of function. J Exp Med 2001; 193:1295-1302.
26. Asadullah K, Sterry W (to be published). Psoriasis, molecular basis. In: Ganten D, Ruckpaul K, eds. Encyclopedic Reference of Genomics and Proteomic in Molecular Medicine. Berlin Heidelberg, New York: Springer Verlag.
27. Asadullah K, Sterry W, Volk HD. Interleukin-10 therapy - review of a new approach. Pharmacol Rev 2003; 55:241-269.

Chapter 8C

IL-10:
The Role in Crohn's Disease

James Lindsay

Abstract

The gastrointestinal tract serves as a barrier between the host and the vast array of foreign antigen that is contained within its lumen. The mucosal immune system must balance two opposing functions; to mount an immune response to pathogens, whilst maintaining tolerance to antigens derived from commensal bacteria and food. This balance is regulated by both cellular interactions and the release of soluble mediators such as cytokines. Diseases such as Crohn's disease are characterized by alterations in the balance of pro-inflammatory and regulatory cytokines. There is considerable interest in therapeutic strategies that manipulate the cytokine balance within the mucosa. The cytokine interleukin-10 (IL-10) plays a pivotal role within the mucosal immune system impacting on both cell mediated and innate immune responses whilst promoting regulatory lymphocyte function. This role is highlighted both by the chronic ileocolitis that develops in gene targeted IL-10 knockout mice, and by the therapeutic efficacy of IL-10 injection in several animal models of colitis. However, trials of daily systemic IL-10 administration in patients with Crohn's disease have reported only a modest clinical response. There is increasing evidence that focused IL-10 delivery using gene therapy strategies may enhance the therapeutic effect of IL-10 in intestinal inflammation.

Introduction

Crohn's disease is a chronic relapsing inflammatory condition of the gastrointestinal tract, leading to transmural ulceration of any part of the bowel with associated stricturing and fistula formation (Fig. 1). The spectrum of disease is vast, thus patients may present with abdominal pain, bloody diarrhoea, weight loss, peri-anal sepsis, malabsorption or one of many systemic manifestations such as liver, skin or joint disease.[1] In all cases there is evidence of systemic immune activation such as raised inflammatory markers, pyrexia and a peripheral leucocytosis. Intestinal inflammation in Crohn's disease is associated with a vast increase in the activity of the mucosal immune system.[2] Thus, increased numbers of monocytes, macrophages, lymphocytes and neutrophils secrete lipid mediators (eicosonoids, and platelet activating factor) and both reactive oxygen and nitrogen metabolites, which lead to tissue destruction. The influx of leukocytes from the peripheral circulation is orchestrated by increased expression of adhesion molecules on the vascular endothelium and the production of cytokines and chemokines within the mucosa (Fig. 2).[3] No single etiological factor has been identified; thus it is postulated that the intestinal inflammation associated with Crohn's disease results from a combination of genetic and environmental factors leading to a persistent Th1 lymphocyte mediated immune response driven by a loss of tolerance to commensal enteric bacteria.[1,4,5] In addition, recent evidence has suggested that genetic defects within the innate immune system's response to luminal bacteria may play a role in the initiation of the disease.[6] Animal models in which IL-10 signalling is disrupted highlight the crucial role that IL-10 plays in the regulation of mucosal

Interleukin-10, edited by Francesco M. Marincola. ©2006 Eurekah.com.

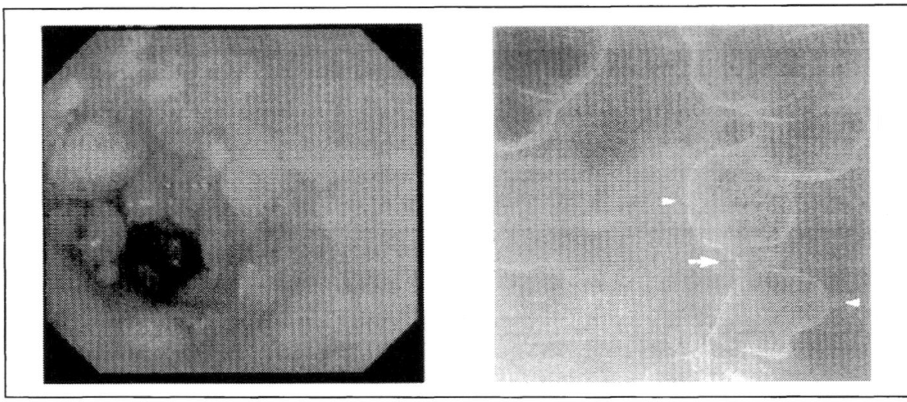

Figure 1. The features of Crohn's disease. A) A colonoscopic view of the mucosa of a patient with active Crohn's disease illustrating snail track ulceration, mucosal erythema and stricture formation. B) A double contrast barium enema demonstrates pseudosaculations of the colon (arrow heads) and strictured segments (arrow) in another patient with active disease.

immune responses.[7,8] This chapter focuses on the role of IL-10 in intestinal inflammation and explores its potential as a therapy for Crohn's disease.

Animal Models and IL-10

The study of animal models of colitis has demonstrated that disruption of regulatory pathways within the mucosal immune system are involved in the pathogenesis of intestinal inflammation. Of particular relevance to this chapter are models that target IL-10 signalling and the CD4$^+$ CD45RB lymphocyte transfer model of colitis.

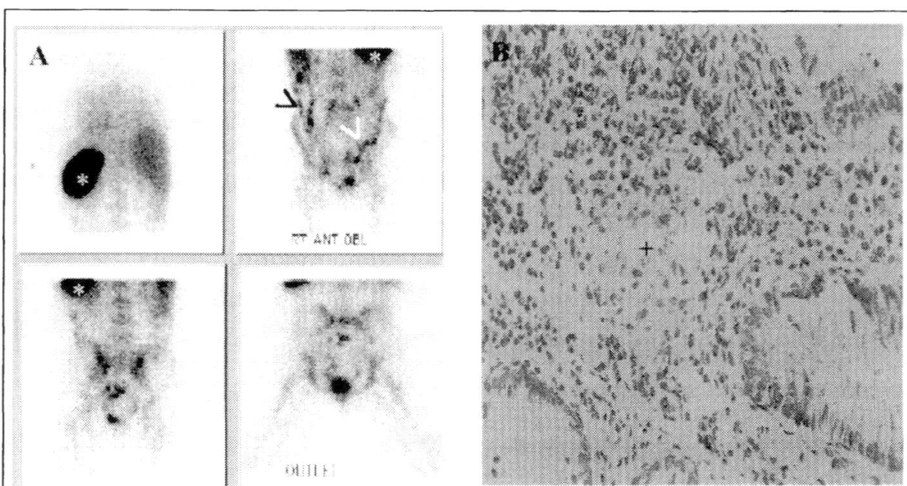

Figure 2. Leukocytic infiltration in active Crohn's disease. Intestinal inflammation in Crohn's disease is associated with an influx of leukocytes within the mucosa. A) A ^{99}tehnetium labelled leukocyte scan from a patient with active disease. In this study, leukocytes are labelled with 99Technetium ex vivo prior to reinjection into the patient, where they are taken up by areas of inflammation (arrows) and the spleen (*). B) A histological section of the mucosa in active Crohn's disease demonstrating the classical features of partial crypt destruction, a mononuclear cell infiltrate and granuloma formation (+).

IL-10 Knock-Out Mice

Both gene targeted IL-10 knockout (IL-10 -/-) mice and mice with a targeted defect in the IL-10 receptor (CRF2-4) develop a spontaneous enterocolitis.[7,8] Furthermore, mice with cell-type specific STAT-3 deficiency, in which neither neutrophils nor macrophages are able to respond to IL-10, spontaneously develop colitis, whereas no intestinal inflammation develops in mice in which only T cells are STAT-3 deficient.[9] This demonstrates that IL-10 is required for normal intestinal homeostasis, but that an intact IL-10 response in T lymphocytes is not essential.

Further characterisation of the disease in IL-10 -/- mice reveals marked similarities with Crohn's disease such as transmural multi-focal lesions throughout the colon and the small bowel. Associated pathological changes include an inflammatory cell infiltrate in the mucosa and submucosa, crypt abscess formation, and epithelial cell hyperplasia (reviewed in ref. 10). Immunohistochemistry shows that the inflammatory infiltrate is mainly composed of mononuclear cells with some neutrophils and eosinophils. The mononuclear cells include activated macrophages, B cells, and $CD4^{+}/CD8^{+}$ T cells expressing the activation markers CD62L, CD44 and CD69. The majority of colonic CD4+ lymphocytes express a memory phenotype as reflected by low levels of CD45RB. Affected tissue spontaneously secretes increased levels of TNF-α, IL-1, IL6 and IFN-γ, but little IL-4.[11]

Further similarities with Crohn's disease are highlighted by the transfer of specific colonic T cell subsets from the IL-10 -/- mice into the RAG-2 -/- (lymphocyte deplete) mouse that demonstrates that the colitis is mediated by $CD4^{+}$ Th1 lymphocytes.[12] Furthermore, the observation that IL-10 -/- mice reared in germ free conditions do not develop colitis highlights the crucial role of intestinal flora in this disease.[13] Recolonization of the intestinal tract with certain pathogens such as Helicobacter hepaticus promotes disease via a Th1 response in IL-10 -/- mice, but not in wild type mice.[14] In addition IL-10 -/- mice have an increase in adherent and translocated mucosal bacteria associated with a reduced intestinal colonization with Lactobacillus species (another feature of human Crohn's disease). Repopulation of the colon with Lactobacillus species reduces the level of adherent and translocated bacteria, and attenuates the development of colitis.[15]

Thus, disease in the IL-10 -/- mouse is caused by a failure to maintain tolerance to components of normal intestinal flora. One hypothesis for the pathogenesis of inflammation in this model is that in the IL-10 deficient GI tract, persistent antigenic stimulation and the uncontrolled release of IL-12, IL-18 and IFN-γ leads to excessive generation and uncontrolled activation of Th1 T cells, converting protective immunity into a pathological response (Fig. 3). This is supported by the fact that weaning IL-10 -/- mice treated with either anti-IL-12 or anti-IFN-γ antibodies remain free from disease. Anti-IL-12, but not anti IFN-γ, treatment is also able to attenuate established disease in adult IL-10 -/- mice, suggesting that IL-12 supports the expansion of a population of differentiated Th1 cells that mediate disease independently of their IFN-γ production.[16]

Whilst daily treatment of weaning IL-10 -/- mice with intra peritoneal IL-10 prevents disease, similar treatment of adult mice is only partially able to suppress disease activity.[11] This may reflect the pharmacokinetics of IL-10 delivery, or suggest that once initiated, Th1 T cells are committed to perpetuate inflammation, and that the partial effect of IL-10 is to suppress the production other pro-inflammatory mediators such as TNF-α.

Transfer Models of Colitis

Transfer of specific subsets of T lymphocytes into immunocompromised (SCID) mice yields further information about the role of both IL-10 and T cell subsets in the regulation of the intestinal immune response. Transfer of the $CD45RB^{high}$ (naïve) subset of $CD4^{+}$ cells into SCID mice induces colitis that does not occur with transfer of the $CD45RB^{low}$ (memory) subset or coadministration of unfractionated $CD4^{+}$ T cells.[17] This suggests that there are regulatory factors within the $CD45RB^{low}$ subset that are able to suppress the response of the $CD45RB^{high}$ subset to enteric antigens. The protective effects of the $CD45RB^{low}$ population are independent of IL-4 suggesting that they are functionally distinct from Th2 cells.[18] IL-10

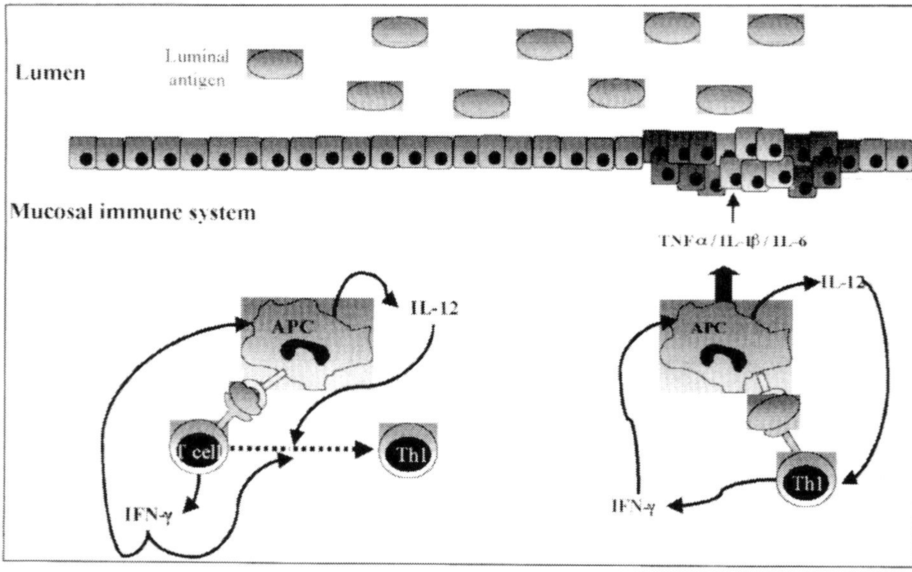

Figure 3. The immunopathology of IL-10 -/- mice. In the absence of IL-10, the presentation of luminal antigen to mucosal T-cells leads to uninhibited IFN-γ release that stimulates antigen presenting cell IL-12 release; this, in conjunction with the IFN-γ leads to Th1 T cell differentiation. When presented with antigen, the Th1 cells drive macrophage pro-inflammatory cytokine release, which results in soluble mediator release and tissue damage.

therapy attenuates the colitis induced by CD45RB^{high} CD4$^+$ cells, conversely CD45RB^{low} CD4$^+$ cells from an IL-10 -/- mouse are unable to prevent the colitis induced by CD45RB ^{high} CD4$^+$ cells.[19] Indeed, both CD45RB ^{low} CD4$^+$ and CD45RB^{high} CD4$^+$ subsets from IL-10 -/- mice induce colitis when transferred to RAG-2 -/- mice. These results cannot be attributed to developmental abnormalities in the IL-10 -/- mouse as blockade of IL-10 function using a monoclonal antibody against the IL-10 receptor abrogated the protective effect of CD45RB^{low} CD4$^+$ cells from wild type mice.[20] These results implicate IL-10 in the development and function of a subset of regulatory cells that protect against pathogenic responses to intestinal antigen.

Crohn's Disease – A Failure of Mucosal Immunoregulation?

Implicit in the function of the mucosal immune system is the ability to mount an immune response to pathogens whilst maintaining tolerance to the vast array of luminal antigens derived from food and commensal bacteria. The fact that many animal models of Crohn's disease remain disease-free if reared in germ free conditions suggest that luminal bacteria are involved in either the initiation or prolongation of intestinal inflammation. Further evidence for this hypothesis is provided by clinical studies showing that diversion of the faecal stream from an area of active inflammation via a defunctioning split ileostomy induces remission in 65-70% of patients with Crohn's disease, the majority of whom will relapse after the reintroduction of ileal effluent.[21,22] A lack of tolerance to intestinal bacteria is probable, as the activity of the ileal effluent is lost after sterilization through a 0.22 μm filter. Furthermore, mucosal mononuclear cells isolated from areas of active Crohn's disease proliferate in response to autologous intestinal bacteria, whereas mononuclear cells harvested from noninflamed intestine show no response to autologous bacteria, but will proliferate in response to heterologous intestinal flora.[5] It is possible that this lack of tolerance is mediated by a failure in the generation or function of similar regulatory T lymphocytes that have been demonstrated to prevent disease in the CD4$^+$CDRB^{high} transfer model of colitis discussed above.

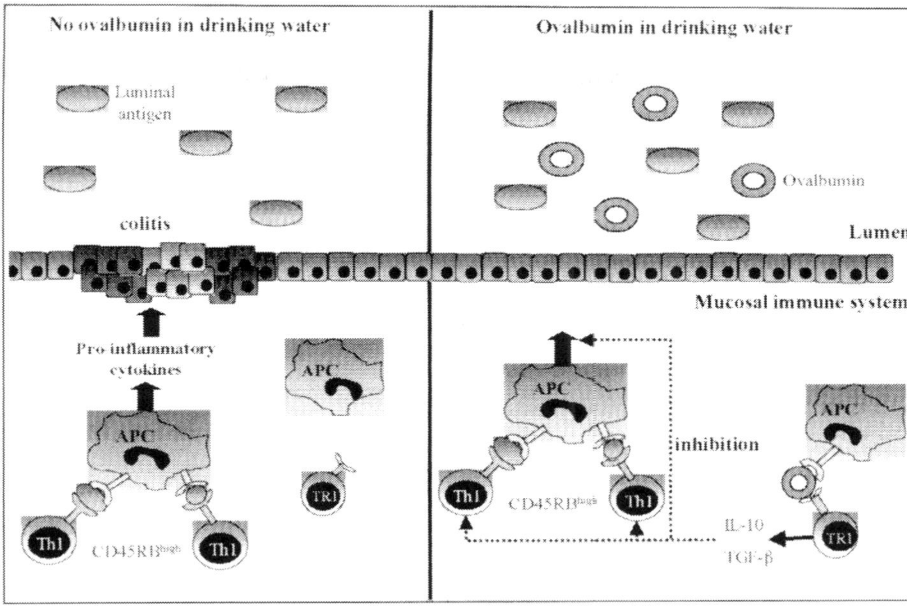

Figure 4. Antigen specific regulatory T cells. The transfer of CD45RB^{high} T cells into a C.B17 SCID mouse leads to a Th1 T-cell driven colitis that is dependent on the presence of luminal antigen. Cotransfer of regulatory T cell clones (Tr1) expressing a transgenic T-cell receptor specific to ovalbumin are able to suppress the CD45RB^{high} colitis by secreting IL-10 and TGF-β. However, this regulatory effect will only occur in mice that are fed ovalbumin in their drinking water.

Tolerance in the Mucosal Immune System

The immune system requires a mechanism for generating antigen specific tolerance to maintain immune homeostasis and control the harmful effects of auto-antigen reactive T lymphocytes. Several mechanisms have been proposed to allow the immune system to suppress the activity of certain T cell clones including clonal deletion, the induction of anergy and antigen specific suppression (reviewed in ref. 23). Recent experiments both in vitro and in animal models of autoimmune disease have characterised different populations of regulatory T lymphocytes that suppress antigen specific lymphocyte responses by the release of cytokines and also by cell:cell interactions.

Groux et al have demonstrated that in vitro repetitive antigenic stimulation of human and murine T cells in the presence of IL-10 leads to the formation of a subset of regulatory T cells termed T regulatory 1 (Tr1) cells.[24] These cells have low proliferative capacity and secrete high levels of IL-10, no IL-4 and low levels of IL-2. They are immune suppressive in vitro and will inhibit the antigen-induced proliferative response of naïve elutriated T cells in a contact independent manner. More importantly, Tr1 cells can be shown to be immunosuppressive in vivo. Transfer of Tr1 clones derived from CD4^{+} T cells expressing a transgenic T cell receptor specific for ovalbumin are able to suppress the CD45RB^{high} CD4^{+} T cells induced colitis in SCID mice. This effect is dependent on the in vivo activation of the Tr1 clones, as they only inhibit colitis in recipients receiving ovalbumin in their drinking water (Fig. 4).[24] Although the initiation of this regulation is antigen specific, the effector response is not, as the Tr1 clones are able to suppress the function of T cells responding to many differing intestinal antigens (a phenomenon termed antigen-driven bystander suppression).

Powrie et al have suggested that these Tr1 clones are the 'in vitro' equivalent of the regulatory cells that mediate the protective functions of the CD45RB^{low} subset of CD4^{+} T cells in the transfer model of colitis.[25] Further characterisation of the regulatory subset of CD45RB^{low} CD4^{+} cells in this model has demonstrated that they express the IL-2 receptor α-chain (CD25) and comprise 5-10% of circulating CD4^{+} cells.[25] This finding is supported by reports that the elimination of the CD4^{+}CD25^{+} cells leads to a variety of autoimmune conditions in genetically susceptible mice.[26] It is not clear whether these cells mediate their effects via cytokine release or cell:cell contact, although several experiments have demonstrated that the release of IL-10 and TGF-β is important.[24,27]

The ability of regulatory T cells to inhibit inflammatory responses may involve impairment of antigen presentation by dendritic cells via a reduction in costimulatory molecule expression. Alternatively regulatory T cells may interrupt the proliferative capacity of activated T cells, preventing their accumulation within the intestine. Finally, the release of immunoregulatory cytokines (such as IL-10 and TGF-β) by regulatory T cells may reduce the release of pro-inflammatory cytokines and chemokines by activated macrophages.

IL-10 in the Normal and Inflamed Gastrointestinal Mucsosa

Thus, it is clear that IL-10 plays a pivotal role in mucosal homeostasis, and that disruption of IL-10 signaling may be involved in the initiation and progression of intestinal inflammation. However, this does not automatically imply a role for IL-10 in the pathogenesis of Crohn's disease. Important questions to be addressed include the levels of IL-10 in inflamed and healthy mucosa, the responsiveness of mucosal mononuclear cells to IL-10, and the presence of functional polymorphisms in genes involved in IL-10 signaling in patients with Crohn's disease.

Several groups have studied serum and tissue levels of IL-10 in patients with active Crohn's disease and healthy controls. Studies of serum IL-10 levels in patients with Crohn's disease and healthy controls have yielded conflicting results.[28-30] Furthermore, immunohistochemical studies demonstrate abundant IL-10 protein and mRNA within intestinal epithelial cells of both healthy and inflamed colon.[31] However, in patients with active inflammation, IL-10 expression is increased in sub mucosal antigen presenting cells and lamina propria lymphocytes.[31] This paradoxical rise in immunoregulatory IL-10 levels in inflamed mucosa should not be surprising given the marked Th1 immune response that is demonstrated. Despite this elevation in endogenous IL-10, lamina propria monocytes isolated from areas of active inflammation still respond to the addition of exogenous IL-10 by reducing the release of pro-inflammatory cytokines such as TNF-α and IL-1β.[28] In addition, a recent report has suggested that low levels of IL-10 mRNA in a non inflamed section of ileum at the time of surgery for active Crohn's disease predict subsequent relapse.[32] This suggests that exogenous IL-10 may have a role in suppressing active disease and preventing relapse.

Preliminary data indicate a potential association between disease extent and a functional polymorphism in the promoter region of the IL-10 gene; however, this has not been confirmed in more detailed studies.[33-35] Two novel polymorphisms have been reported in the IL-10 receptor 1 gene, a serine 138-to-glycine (S138G) and a glycine 330-to-arginine (G330R) substitution. Despite evidence that the G330R substitution is a loss of function allele, no association has been demonstrated with IBD.[36] Thus, at present there is no convincing evidence that genetic polymorphisms in the IL-10 signaling pathway predispose to Crohn's disease.

Systemic IL-10 as a Therapy for Crohn's Disease

Both the role of IL-10 in the maintenance of mucosal homeostasis and the therapeutic efficacy of systemic IL-10 injections in animal models of colitis indicated its potential as a novel therapy in Crohn's disease.[37-39] In vitro studies showed that exogenous IL-10 down-regulates the enhanced pro-inflammatory cytokine release from lamina propria mononuclear cells isolated from patients with Crohn's disease.[28] Furthermore, in an initial trial of 46 patients with steroid-resistant Crohn's disease, seven daily IL-10 injections induced clinical remission in 50% of patients over the three week follow up compared to only 23% given placebo.[40]

In contrast, there was no beneficial effect of IL-10 over placebo in a study of the effect IL-10 on post-operative endoscopic recurrence in patients with ileocolonic Crohn's disease.[41] Furthermore, two recent studies have investigated the efficacy and safety of 28 daily subcutaneous injections of recombinant IL-10 in patients with moderately active and steroid refractory Crohn's disease respectively.[42,43]

Fedorak et al report a 24 week double blind placebo controlled study of 95 patients with moderately active Crohn's disease who were not concurrently receiving steroid, mesalasine or other immunosuppressive therapy. Patients were randomized to receive subcutaneous rhuIL-10 at one of four doses (1, 5, 10, and 20 μg/kg) or placebo daily for 28 days. During the 20-week follow period a dose dependent, but fully reversible, anemia and thrombocytopenia were noticed. At the end of the treatment period 23.5% of patients receiving 5 μg/kg rhuIL-10 (n=17) experienced clinical remission and endoscopic improvement compared to 0% of patients receiving placebo (n=23). Interestingly higher doses were less effective, although there was no difference in reported adverse events between doses. During the course of the 20-week follow period disease recurrence requiring therapeutic intervention occurred in 70% of placebo treated patients compared to 55% of the rhuIL-10 treated group.[43]

In contrast, Schreiber et al report their experience of rhuIL-10 (1, 4, 8, and 20 μg/kg) in 329 therapy refractory patients with active Crohn's disease. No differences in the induction of clinical remission (Crohn's disease activity index (CDAI) <150) were observed between placebo and rhuIL-10 at any dosing schedule, although clinical improvement (a fall in CDAI of at least 100 points) was seen in 46% of patients in the 8 μg/kg group compared to 27% in patients receiving placebo. The authors also determined nuclear levels of the transcription factor NF-κB p65 and cytoplasmic concentrations of its inhibitor IκBα in ileocolonic biopsy specimens. They noticed significant differences between patients who responded to rhuIL-10 treatment and those who did not, implicating the inhibition of NF-kB nuclear translocation as one possible mechanism of action of this therapy.[42]

Thus, the two recent trials have demonstrated that daily subcutaneous rhuIL-10 is safe and well tolerated. However, in the patient groups studied, the therapeutic effects are very modest when compared to a clinical response in 65% and remission in 30% of patients receiving a single *Infliximab* infusion.[44] There are several explanations for these results; firstly, it is possible that response to IL-10 is limited to a subgroup of patients, as has been shown with other biological therapies such as anti-TNF-α antibody therapy.[44] Crohn's disease is a heterogeneous condition with polygenic influences on disease phenotype. Thus, therapies that target an individual step within a complicated immunological cascade may only be applicable to a proportion of patients.[45] This theory is supported by the concordance between a clinical response to IL-10 and a decrease in the ratio of nuclear NF-κB to cytoplasmic IκBα.[42] As discussed previously, both the IL-10 and the IL-10R genes are polymorphic. Whether response to IL-10 therapy is limited to patients with specific alleles is a subject of on-going research.

Secondly, it is possible that changes in the systemic dose of rhuIL-10 alter the balance between its immunoregulatory and immunostimulatory effects. This may explain the bell shaped curve in therapeutic efficacy reported in the clinical trials.[42-43] In fact, the higher doses of IL-10 were associated with an increase in the release of T-helper cell pro-inflammatory cytokines from in vitro stimulated peripheral blood mononuclear cells.[46] A comparison between rhuIL-10 and vIL-10, which has no immunostimulatory properties, would make this clearer.

Finally, it is possible the pharmacodynamics of daily systemic IL-10 delivery do not lead to sufficient mucosal penetration to achieve a clinical response. Although there was no antibody response to the systemic administration of IL-10 in the clinical trials, the serum half-life was between 1.1 and 2.6 hours.[40] Local IL-10 delivery by enema has been shown to have immunoregulatory effects in patients with ulcerative proctitis, although the histological response was not reported.[28] Thus alternative delivery systems may enhance the response to IL-10 therapy.

Novel Strategies for IL-10 Delivery

Gene therapy strategies may permit prolonged IL-10 delivery to the inflamed mucosa, resulting in enhanced immunoregulation and perhaps increased regulatory cell activity. This concept is supported by the study of mice that over express IL-10. Transgenic mice in which IL-10 transcription is under the control of the *fabpi* promoter exclusively over-express IL-10 in mature colonic epithelial cells and are resistant to the development of experimental colitis.[47] Furthermore the population of $CD4^+$ $CD25^+$ cells within the mucosa is expanded, and isolated lamina propria mononuclear cells secrete increased levels of TGF-β suggesting enhanced regulatory T cell function. In addition, $CD45RB^{high}$ $CD4^+$ cells from transgenic mice that express IL-10 under the control of the IL-2 promoter are nonpathogenic when transferred to SCID mice.[48] Thus the ideal vector for IL-10 gene therapy would deliver long term IL-10 to the intestine targeted to areas of inflammation with minimal systemic exposure and immunogenicity.

Initial studies of IL-10 gene therapy involved the systemic delivery of adenoviral vectors encoding murine IL-10 (AdvmuIL-10) under the control of a constitutive promoter to murine and rat models of colitis.[49,50] A single systemic injection predominantly targets hepatocytes, but low-level bioactive IL-10 release can also be demonstrated for up to 10 weeks in the spleen and colon.[49] The predominant sites of transgene expression within the colon are colonocytes, but some LPMN cell expression can be demonstrated, particularly in the setting of intestinal inflammation.[51,52] IL-10 could also be demonstrated in the serum for approximately one week.[49]

A single systemic injection of AdvmuIL-10 administered at weaning was sufficient to prevent the onset of disease in IL-10 -/- mice for at least 10 weeks. In addition, unlike daily systemic injections of recombinant protein, AdvmuIL-10 therapy induced clinical, immunological and histological remission in mice with established disease.[49] Systemic AdvmuIL-10 also prevents the onset of TNBS induced colitis in both mice and rats.[50,53] However, the systemic delivery of adenoviral vectors induces both humoral and cell mediated anti-adenoviral responses that may limit the duration of transgene expression and prevent retreatment. Although the delivery of the IL-10 transgene diminished the host anti-adenoviral antibody response in IL-10 -/- mice, this effect was not noted in IL-10 competent controls.[49] Furthermore, the systemic expression of IL-10 resulted in diminished pro-inflammatory cytokine release from LPS stimulated splenocytes, which may limit its potential in human trials. Finally there has been one death after systemic adenoviral vector injection and two recently publicized episodes of insertional mutagenesis after gene therapy for SCID (reviewed in ref. 54). Thus despite its therapeutic efficacy, it is unlikely that the systemic delivery of adenoviral vectors encoding IL-10 would be appropriate in patients with Crohn's disease.

Targeting IL-10 delivery to the colon may bypass some of the safety issues related to systemic delivery whilst maintaining therapeutic efficacy. Local delivery of AdvmuIL-10 as an enema induces colonic IL-10 release with minimal systemic expression and low immunogenicity.[55] This strategy induced clinical and histological remission in IL-10 -/- mice with established disease. It may even be possible to target the therapy to sites of active disease using an inflammation inducible C3-*tat*/HIV promoter that induces IL-10 release in response to inflammatory stimuli such as TNF-α.[56] Finally the use of alternative vectors such as adeno-associated viruses, which integrate into the host genome, may allow longer transgene expression.

An alternative means of achieving local IL-10 delivery is the oral administration of nonpathogenic bacteria that have been genetically modified to release IL-10. Thus, a genetically modified IL-10 secreting *Lactococcus lactis* strain was able to prevent the onset of disease in IL-10 -/- mice and was therapeutic in chronic dextran sodium sulfate colitis.[57] This exciting development is currently undergoing initial trials in patients with Crohn's disease. The environmental concern about the release of genetically modified bacteria into the environment has been addressed by inserting the therapeutic human IL-10 cassette into the bacterial genome in place of the bacterial thiamine synthase gene.[58] Thus, the bacteria are only viable when in contact with a thiamine source. Preliminary studies have demonstrated that the thiamine content of the gastrointestinal tract is sufficient to support bacterial survival, but on excretion into the environment the viability falls by 8 log.

Finally, targeted local IL-10 delivery may also be achieved by the infusion of autologous $CD4^+$ leucocytes that have been modified ex vivo to express a human IL-10 gene under a constitutive promoter.[59,60] Activated human $CD4^+$ leucocytes infected with an IL-10 encoding retrovirus release bioactive IL-10 for at least 17 weeks and are potent inhibitors of both T lymphocyte proliferation and dendritic cell IL-12 release.[59] These cells have gut homing properties as they express the mucosal integrin $\alpha4\beta7$ and display efficient binding to MAdCAM-1 expressing cells in vitro. The therapeutic efficacy of this strategy has been tested in the CD45RB transfer model of colitis. Murine splenic $CD4^+$ cells that were transduced by a retrovirus encoding murine IL-10 were able to inhibit $CD45RB^{high}$ cell induced colitis, even if infused 14 days after the pathogenic $CD4^+$ cells. There was convincing evidence that the transduced $CD4^+$ splenocytes migrated to the intestine and released IL-10 for several weeks.[60]

Conclusion

It remains to be seen whether these novel IL-10 delivery strategies will prove effective in human disease. Although it is possible to safely target IL-10 delivery to the intestinal mucosa, several questions remain. Despite its obvious immunoregulatory properties and its role in the pathogenesis of many animal models of colitis, IL-10 levels are already elevated in the intestine of patients with active Crohn's disease; will the addition of supra-physiological doses yield further therapeutic effect? Should IL-10 be considered a therapy for acute flares or maintenance of remission? Finally, are there genetically determined subgroups of patients more likely to respond?

In summary, IL-10 plays a pivotal role in mucosal homeostasis and the control of intestinal inflammation. Systemic IL-10 delivery is safe and well tolerated, but ineffective as a therapy for Crohn's disease. Targeted intestinal IL-10 delivery using gene therapy strategies is an exciting new concept in the therapy of intestinal inflammation and is currently being evaluated in patients with Crohn's disease.

References

1. Fiocchi C. Inflammatory bowel disease: Etiology and pathogenesis. Gastroenterology 1998; 115:182-205.
2. Pallone F, Fais S, Squarcia O et al. Activation of peripheral blood and intestinal lamina propria lymphocytes in Crohn's disease. In vivo state of activation and in vitro response to stimulation as defined by the expression of early activation antigens. Gut 1987; 28(6):745-753.
3. Carlos TM, Harlan JM. Leukocyte-endothelial adhesion molecules. Blood 1994; 84:2068-2101.
4. Satsangi J, Jewell DP, Bell JI. The genetics of inflammatory bowel. Gut 1997; 40:572-574.
5. Duchmann R, Kaiser I, Hermann E et al. Tolerance exists towards resident intestinal flora but is broken in active inflammatory bowel disease (IBD). Clin Exp Immunol 1995; 102(3):448-455.
6. Hisamatsu T, Suzuki M, Reinecker H-C et al. CARD15/NOD2 functions as an antibacterial factor in human intestinal epithelial cells. Gastroenterology 2003; 124:993-1000.
7. Kuhn R, Lohler J, Rennick D et al. Interleukin-10-deficient mice develop chronic enterocolitis. Cell 1993; 75(2):263-74.
8. Spencer SD, Di Marco F, Hooley J et al. The orphan receptor CRF2-4 is an essential subunit of the interleukin-10 receptor. J Exp Med 1998; 187(4):571-8.
9. Takeda K, Clausen BE, Kaisho T et al. Enhanced Th1 activity and development of chronic enterocolitis in mice devoid of Stat3 in macrophages and neutrophils. Immunity 1999; 10:39-49.
10. Rennick DM, Fort MM, Davidson NJ. Studies with IL-10-/- mice: An overview. Journal of Leukocyte Biology 1997; 61:389-396.
11. Berg DJ, Davidson N, Kuhn R et al. Enterocolitis and colon cancer in interleukin-10-deficient mice are associated with aberrant cytokine production and CD4(+) TH1- like responses. J Clin Invest 1996; 98(4):1010-1020.
12. Davidson NJ, Leach MW, Fort MM et al. T helper cell 1-type CD4+ T cells, but not B cells, mediate colitis in interleukin 10-deficient mice. J Exp Med 1996; 184:241-251.

13. Sellon RK, Tonkonogy S, Schultz M et al. Resident enteric bacteria are necessary for development of spontaneous colitis and immune system activation in interleukin-10-deficient mice. Infection and Immunity 1998; 66:5224-5231.
14. Kullberg MC, Ward JM, Gorelick PL et al. Helicobacter hepaticus triggers colitis in specific-pathogen-free interleukin-10 (IL-10)-deficient mice through an IL-12- and gamma interferon-dependent mechanism. Infection and Immunity 1998; 66:5157-5166.
15. Madsen KL, Doyle JS, Jewell LD et al. Lactobacillus species prevents colitis in interleukin 10 gene-deficient mice. Gastroenterology 1999; 116:1107-1114.
16. Davidson NJ, Hudak SA, Lesley RE et al. IL-12, but not IFN-gamma, plays a major role in sustaining the chronic phase of colitis in IL-10-deficient mice. J Immunol 1998; 161:3143-3149.
17. Leach MW, Bean AG, Mauze S et al. Inflammatory bowel disease in C.B-17 scid mice reconstituted with the CD45RBhigh subset of CD4+ T cells. Am J Path 1996; 148:1503-1515.
18. Powrie F, Carlino J, Leach MW et al. A critical role for transforming growth factor-beta but not interleukin 4 in the suppression of T helper type 1-mediated colitis by CD45RB(low) CD4+ T cells. J Exp Med 1996; 183:2669-2674.
19. Powrie F, Leach MW, Mauze S et al. Inhibition of Th1 responses prevents inflammatory bowel disease in scid mice reconstituted with CD45RBhi CD4+ T cells. Immunity 1994; 1(7):553-562.
20. Asseman C, Mauze S, Leach MW et al. An essential role for interleukin 10 in the function of regulatory T cells that inhibit intestinal inflammation. J Exp Med 1999; 190:995-1004.
21. Harper PG, Lee ECG, Kettlewell MGW et al. Role of faecal stream in the maintenance of Crohn's colitis. Gut 1985; 26:279-284.
22. Fasoli R, Kettlewell MGW, Mortensen NJ et al. Response to fecal challange in defunctioned colonic Crohn's disease: Prediction of long term response. British Journal of Surgery 1990; 77:616-617.
23. Groux H. An overview of regulatory T cells. Microbes and Infection 2001; 3:883-9.
24. Groux H, O'Garra A, Bigler M et al. A CD4+ T-cell subset inhibits antigen-specific T-cell responses and prevents colitis. Nature 1997; 389:737-742.
25. Toms C, Powrie F. Control of intestinal inflammation by regulatory T cells. Microbes and Infection 2001; 3:929-35.
26. Sakaguchi S, Sakaguchi N, Asano M et al. Immunologic self-tolerance maintained by activated T cells expressing IL-2 receptor alpha-chains (CD25). Breakdown of a single mechanism of self-tolerance causes various autoimmune diseases. J Immunol 1995; 155:1151-64.
27. Fuss IJ, Boirivant M, Lacy B et al. The interrelated roles of TGF-beta and IL-10 in the regulation of experimental colitis. J Immunol 2002; 168:900-908.
28. Schreiber S, Heinig T, Thiele HG et al. Immunoregulatory role of interleukin 10 in patients with inflammatory bowel disease. Gastroenterology 1995; 108(5):1434-1444.
29. Gasche C, Bakos S, Dejaco C et al. IL-10 secretion and sensitivity in normal human intestine and inflammatory bowel disease. J Clin Immunol 2000; 20:362-70.
30. Kucharzik T, Stoll R, Lugering N et al. Circulating antiinflammatory cytokine IL-10 in patients with inflammatory bowel disease (IBD). Clin Exp Immunol 1995; 100:452-456.
31. Autschbach F, Braunstein J, Helmke B et al. In situ expression of interleukin-10 in noninflamed human gut and in inflammatory bowel disease. Am J Path 1998; 153:121-130.
32. Meresse B, Rutgeerts P, Malchow H et al. Low ileal interleukin 10 concentrations are predictive of endoscopic recurrence in patients with Crohn's disease. Gut 2002; 50:25-28.
33. Parkes M, Satsangi J, Jewell D. Contribution of the IL-2 and IL-10 genes to inflammatory bowel disease (IBD) susceptibility. Clin Exp Immunol 1998; 113:28-32.
34. Tagore A, Gonsalkorale WM, Pravica V et al. Interleukin-10 (IL-10) genotypes in inflammatory bowel disease. Tissue Antigens 1999; 54:386-390.
35. Klein W, Tromm A, Griga T et al. The IL-10 gene is not involved in the predisposition to inflammatory bowel disease. Electrophoresis 2000; 21:3578-3582.
36. Gasche C, Grundtner P, Zwirn P et al. Novel variants of the IL-10 receptor 1 affect inhibition of monocyte TNF-alpha production. J Immunol 2003; 170(11):5578-82.
37. Duchmann R, Schmitt E, Knolle P et al. Tolerance towards resident intestinal flora in mice is abrogated in experimental colitis and restored by treatment with interleukin-10 or antibodies to interleukin-12. Eur J Immunol 1996; 26:934-938.
38. Tomoyose M, Mitsuyama K, Ishida H et al. Role of interleukin-10 in a murine model of dextran sulfate sodium-induced colitis. Scand J Gastroenterol 1998; 33(4):435-440.
39. Ribbons KA, Thompson JH, Liu X et al. Anti-inflammatory properties of interleukin-10 administration in hapten-induced colitis. Eur J Pharmacol 1997; 323:245-254.
40. van Deventer SJ, Elson CO, Fedorak RN. Multiple doses of intravenous interleukin 10 in steroid-refractory Crohn's disease. Crohn's Disease Study Group. Gastroenterology 1997; 113:383-389.

41. Colombel JF, Rutgeerts P, Jacyna M et al. Safety and tolerance Of subcutaneous recombinant human interleukin-10 (IL-10) in subjects with Crohn's disease after first ileal Or ileocolonic resection. Gastroenterology 1999:Abstract G 3000.
42. Schreiber S, Fedorak RN, Nielsen OH et al. Safety and efficacy of recombinant human interleukin 10 in chronic active Crohn's disease. Gastroenterology 2000; 119(6):1461-1472.
43. Fedorak RN, Gangl A, Elson CO et al. Recombinant human interleukin 10 in the treatment of patients with mild to moderately active Crohn's disease. Gastroenterology 2000; 119(6):1473-1482.
44. Targan SR, Hanauer SB, van Deventer SJ et al. A short-term study of chimeric monoclonal antibody cA2 to tumor necrosis factor alpha for Crohn's disease. Crohn's Disease cA2 Study Group. New England Journal of Medicine 1997; 337:1029-1035.
45. Lindsay JO, Hodgson HJF. Review article: The immunoregulatory cytokine IL-10 – a therapy for Crohn's disease? Alimentary Pharmacology and Therapeutics 2001; 15:1709-1716.
46. Tilg H, van Montfrans C, van den Ende A et al. Treatment of Crohn's disease with recombinant human interleukin 10 induces the pro-inflammatory cytokine interferon gamma. Gut 2002; 50:191-5.
47. De Winter H, Elewaut D, Turovskaya O et al. Regulation of mucosal immune responses by recombinant interleukin 10 produced by intestinal epithelial cells in mice. Gastroenterology 2002; 122:1829-1841.
48. Hagenbaugh A, Sharma S, Dubinett SM et al. Altered immune responses in IL-10 transgenic mice. J Exp Med 1997; 185:2101-10.
49. Lindsay JO, Ciesielski CJ, Scheinin T et al. The prevention and treatment of murine colitis using gene therapy with adenoviral vectors encoding IL-10. J Immunol 2001; 166(12):7625-33
50. Barbara G, Xing Z, Hogaboam CM et al. Interleukin-10 gene transfer prevents experimental colitis in rats. Gut 2000; 46:344-349
51. Foreman PK et al. Adenovirus-mediated transduction of intestinal cells in vivo. Hum Gene Ther 1998; 9:1313-1321.
52. Wirtz S, Galle PR, Neurath MF. Efficient gene delivery to the inflamed colon by local administration of recombinant adenoviruses with normal or modified fibre structure. Gut 1999; 44:800-807.
53. Lindsay JO, van Montfrans C, Brennan FM et al. IL-10 gene therapy prevents TNBS-induced colitis. Gene Therapy 2002; 9:1715-1721.
54. Check E. Gene therapy – shining hopes dented, but not dashed. Nature 2002; 420:735.
55. Lindsay JO, Ciesielski CJ, Scheinin T et al. Local delivery of adenoviral vectors encoding murine IL-10 induces colonic IL-10 production and is therapeutic for murine colitis. Gut 2003; 52:981-987.
56. Miagkov AV, Varley AW, Munford RS et al. Endogenous regulation of a therapeutic transgene restores homeostasis in arthritic joints. J Clin Invest 2002; 109:1223.
57. Steidler L, Hans W, Schotte L et al. Treatment of murine colitis by Lactococcus lactis secreting interleukin-10. Science 2000; 289:1352-5.
58. Steidler L, Neirynck S, Huyghebaert N et al. Biological containment of genetically modified Lactococcus lactis for intestinal delivery of human interleukin 10. Nat Biotechnol. 2003; 21(7):785-9.
59. Van Montfrans C, Hooijberg E, Pena MS et al. Generation of regulatory gut-homing human T-lymphocytes using ex vivo IL-10 gene transfer. Gastroenterology 2002; 123:1877-1888.
60. Van Montfrans C, Pena MS, Pronk I et al. Prevention of colitis by IL-10 transduced T lymphocytes in the SCID mice transfer model. Gastroenterology 2002; 123:1865-1876.

Chapter 8D

IL-10 and Systemic Lupus Erythematosus

Simone Mocellin

Systemic lupus erythematosus (SLE) is a multi-organ autoimmune disease characterized by B cell hyperactivity leading to excessive auto-antibody production.[1,2] SLE patients have high levels of circulating auto-antibodies against a number of nuclear antigens including double stranded DNA (dsDNA) and ribonucleoproteins. Tissue damage is triggered by an inflammatory response to immune complex deposition and is widespread (Fig. 1). The primary cause of morbidity and mortality is glomerulonephritis which develops in about 60% of patients. Upon discovery that IL-10 both inhibits Th1 responses and is a potent cofactor for B cell survival, proliferation, differentiation, and immunoglobulin secretion, a potential role in the pathogenesis of SLE has been hypothesized.[3-5] Overal lines of evidence support this hypothesis and prompt for the clinical implementation of the anti-IL-10 therapeutic approach for the treatment of patients affected with this disease.[4]

IL-10 and Murine SLE

The suspicion that IL-10 may play a role in SLE was boosted early on by a study in SLE prone NZB/W F1 mice.[6] ZB/W F1 mice spontaneously develop a SLE-like disease characterized by anti-dsDNA auto-antibodies and glomerulonephritis.[7] Ishida and coworkers found that treatment of NZB/W F1 mice from birth with anti-IL-10 mAb improved 34-week survival from about 10 to 80%.[6] his improved survival correlated with decreased anti-dsDNA auto-antibodies, decreased proteinuria, and decreased incidence and severity of glomerular lesions in anti-IL-10 treated mice.

Furthermore, continuous administration of IL-10 to NZB/W F1 mice from 4 weeks of age accelerated development of autoimmunity. The protection mediated by anti-IL-10 treatment was shown to correlate with increased TNFα production and could be abrogated by treatment with anti-TNFα antibodies. Thus, the efficacy of anti-IL-10 in this mouse model of lupus nephritis is likely due to the blockage of the effects of IL-10 on TNFα-producing cells such as macrophages and T lymphocytes. The role of IL-10 in mouse models of SLE may not reflect the full extent of IL-10 involvement in the pathogenesis of SLE in humans because mouse B cells do not respond to IL-10 as human B cells do. While IL-10 dramatically increases proliferation, differentiation, and antibody production by anti-CD40 stimulated human B cells in culture, no such activity is seen on mouse splenic B cells.

IL-10 and Human SLE: The Cytokine Effect on PBMC

Llorente et al published the first paper describing IL-10 production by peripheral blood mononuclear cells (PBMC) from SLE patients.[8] This study, conducted on 47 patients with clinical conditions ranging from inactive to severe, stands out in that none of the patients were on corticosteroids, nonsteroidal anti-inflammatory drugs, or other immunosuppressive drugs. By RT-PCR, PBMC from 34 of 47 SLE patients were positive for IL-10 transcripts, while only 1 of 34 healthy controls was positive. When PBMC from 20 of these patients (19 RT-PCR positive, 1 RT-PCR negative) were cultured unstimulated for 24 h, they produced on average 33-fold more IL-10 than did healthy control PBMC. Separation of PBMC before culture

Interleukin-10, edited by Francesco M. Marincola. ©2006 Eurekah.com.

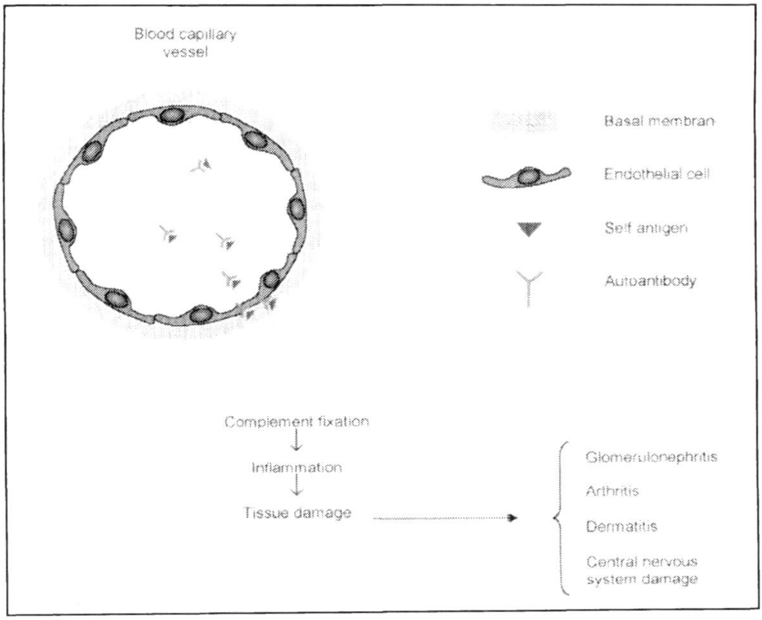

Figure 1. Mechanism of tissue damage in systemic lupus erythematosus. After generation of autoantibodies to self antigens (e.g., DNA, laminin, heparan sulphate), immune complexes deposit in the basal membrane of endothelium where they fix the complement, which starts the inflammation cascade. Consequently, tissue damage occurs in differnet organs.

revealed that most of the IL-10 was made by monocytes and B lymphocytes, with a small contribution from T lymphocytes. Though IL-10 production was clearly higher in SLE patients, no correlation of RT-PCR or culture supernatant IL-10 could be made with severity of disease as assessed by clinical index or anti-dsDNA level. Following the study of Llorente and coworkers, production of IL-10 by SLE patient PBMC was examined under a variety of conditions. PBMC have been tested ex vivo and after culture, either unstimulated or stimulated by PHA, SAC, LPS, or anti-CD3.[9-18] Methods for measuring PBMC IL-10 have assessed either total IL-10 production (ELISA or RT-PCR), or number of IL-10 producing cells (ELISPOT, single cell staining, or in situ hybridization). Most, but not all, of these studies verified the findings of Llorente et al. The largest differences between PBMC from SLE patients versus healthy controls were seen by ELISA on supernatants of cultures that were not stimulated. Supernatants from unstimulated PBMC cultures had IL-10 levels that averaged from 6- to 50-fold (average 23-fold) higher in SLE patient PBMC compared to healthy control PBMC. Cultures of stimulated PBMC produced much higher levels of IL-10 than did unstimulated cultures, but the difference in IL-10 production between PBMC cultures of SLE patients and healthy controls was just two–three-fold. Measurements of the number of PBMC producing IL-10 in SLE patients versus healthy controls also showed a modest increase of just two-fold in most cases. These findings indicate that the higher IL-10 production is likely due to a higher baseline activation state of certain cells within SLE patient PBMC. This possibility is supported by studies of activation markers on PBMC from SLE patients which show an increase in B cell, T cell, and monocyte activation.[19-26] Consistent with the findings of Llorente and coworkers, subsequent studies verified that most of the IL-10 made by PBMC from SLE patients is derived from monocytes and B cells, with a minor contribution from T cells.[8-10,12,25,27-29] Despite the finding of increased IL-10 production by PBMC from SLE patients, of the eight studies that looked for correlation between the level of PBMC IL-10 and disease severity, only one found a weak correlation.[8,15-17,28,30-32]

IL-10 Serum Levels and Human SLE Activity

Actual measurements of circulating serum IL-10 levels might be expected to give a clearer picture of disease association for two reasons. First, serum IL-10 reflects IL-10 production from many sources, including PBMC, cells in tissues such as macrophages and certain epithelial cells, and viruses such as Epstein Barr virus (EBV). Second, measurement of serum IL-10 circumvents the variability inherent in studies involving ex vivo manipulation of cells. Indeed, levels of IL-10 in the serum of SLE patients have consistently been shown to be 3–12-fold higher than in healthy controls.[5,12,15,25,27,33-37] Furthermore, correlation of serum IL-10 levels with disease activity has been demonstrated in all studies that examined this.[15,27,34,36-40] In one particularly well designed study, Park et al demonstrated a correlation of serum IL-10 levels with disease state by measuring serum IL-10 levels in 41 patients, both at the time they presented with flares, and 4 weeks after initiation of treatment.[36] Circulating IL-10 levels were nearly nine-fold higher in SLE patients with active disease compared with normal controls, and correlated with a clinical score for disease severity. Four weeks after initiation of treatment, 36 patients improved, while 5 deteriorated. Using paired samples from the same patients at the time of presentation with flare and 4 weeks later, the level of serum IL-10 was shown to correlate with the change in the clinical score after treatment.

IL-10 Human Gene Polymorphisms and SLE

The above studies show a correlation of high IL-10 levels with SLE, but do not answer the question of whether high levels of IL-10 are simply a byproduct of SLE and flares, or whether IL-10 plays a role in the disease process. There are some clues, however, that IL-10 may predispose to SLE. Studies showing that healthy relatives of SLE patients have increased numbers of IL-10-producing PBMC[15] or increased levels of IL-10 production by PBMC[10,41] imply that part of the hereditary nature of SLE may stem from differences in tendency to make IL-10. Several studies have attempted to dissect the molecular basis of increased production of IL-10 in SLE patients. Two CA-repeat microsatellites, IL-10.R (-4 kb) and IL-10.G (-1.1 kb), and three single nucleotide polymorphisms (SNP) at -1082 (G/A), -819 (C/T), and -592 (C/A) have been identified and shown to affect IL-10 production.[41-44] Attempts to find associations between these IL-10 promoter polymorphisms and SLE have yielded varying results. IL-10.G has been shown to associate with SLE incidence in Scottish, Mexican–American, and Italian populations,[45-47] but not in Mexican, Swedish, or Taiwanese populations.[48-50] However, in the Taiwanese population, an association was noted between IL-10.G and rapid renal failure in the small subset of patients with this manifestation. Alleles of IL-10.G differ in the populations that have association with SLE incidence, implying that IL-10.G may only be a marker for the presence of another nearby sequence variation that actually causes disease association. No association of the three identified -1082/-819/-592 SNP haplotypes with SLE incidence has been found in five studies of Chinese, Dutch and British populations.[41,51-54] However, the GCC haplotype was associated with anti-Ro auto-antibodies in British Caucasians, while the ATA haplotype was associated with renal disease in Chinese[52] but not Dutch[53] or British[54] populations, and with neuropsychiatric SLE in Dutch Caucasians [53] but not Chinese patients.[52] These differences in findings may be due simply to genetic differences in the populations studied. However, a recent finding by Gibson et al of seven more SNP in the -1.3 to -4 kb region of the IL-10 promoter means that further evaluation of previously studied populations may be warranted.[55] These new SNP, along with the previously identified polymorphisms, were examined in an African–American cohort. Haplotypes associated with high IL-10 production were associated with SLE in this population.[55] These studies suggest that high levels of IL-10 may be one of several predisposing factors for SLE incidence and severity.

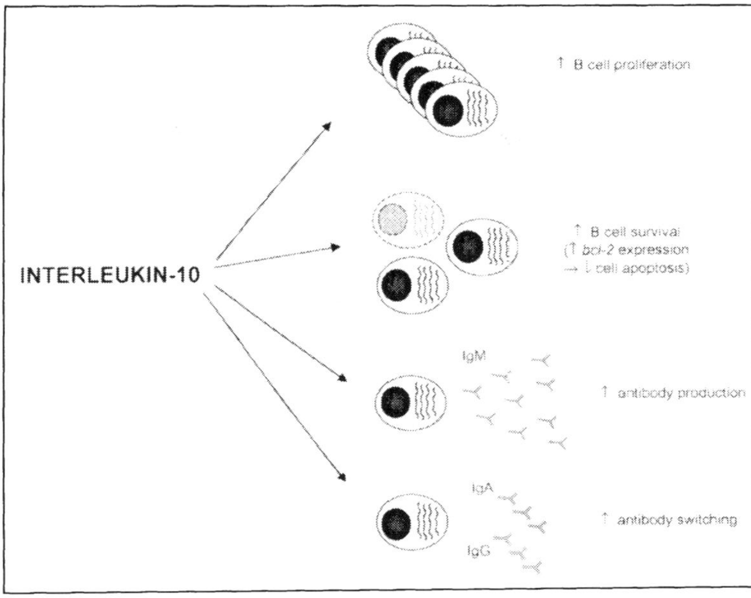

Figure 2. Interleukin-10 effects on B cells.

IL-10 and B Cell Activation in Humans

One predicted role of IL-10 in SLE is its ability to promote B cell differentiation and auto-antibody production (Fig. 2). IL-10 is a cofactor for these activities and requires B cell activation via the B cell receptor or costimulatory molecules such as CD40. Due to the constant presence of nuclear antigens such as dsDNA, and antibodies directed against them, B cells in SLE patients are continuously primed for costimulation by IL-10. One might then expect decreased auto-antibody production upon blocking of IL-10 in vivo. Llorente et al tested this hypothesis by transferring PBMC from SLE patients into SCID mice, resulting in auto-antibody production in these mice.[56] Treatment of the reconstituted mice with anti-human IL-10 mAb resulted in a drop of anti-dsDNA auto-antibodies in 10 of 12 mice to undetectable levels. Using the same SCID model, Kalechman et al treated reconstituted mice with the immunomodulator AS101 in order to decrease IL-10 levels, and observed also a decrease in auto-antibody production.[11] Another possible role for IL-10 in SLE could stem from its ability to promote survival of B cells. IL-10 has been shown to prevent spontaneous cell death of germinal center B cells in vitro by inducing BCL-2 expression.[57] Over expression of BCL-2 in transgenic mice results in an excess of B lymphocytes and immunoglobulin-producing cells, and development of an autoimmune disease similar to SLE.[58] The BCL-2 transgenic mice spontaneously produce auto-antibodies to nuclear antigens, and 60% develop terminal immune complex-mediated glomerulonephritis between 3 and 12 months of age. However, efforts to demonstrate a role for BCL-2 in SLE have met with mixed results.[38,46,47,49,59-63] In a recent study, Miret et al found that serum level of IL-10 and amount of BCL-2 protein in PBMC lysates did not correlate in SLE patients.[38] However, increased BCL-2 and increased IL-10 did individually correlate with higher SLEDAI disease index. Further study is necessary to understand the importance of a role for IL-10 mediated prevention of spontaneous B cell apoptosis in SLE. In a small, open-label clinical trial, anti-IL-10 treatment of lupus patients with cutaneous and joint manifestations resulted in improvement to clinically inactive disease in five of six patients within 6 months of the 3-week treatment regimen.[64] Though anti-dsDNA auto-antibody levels, C3 and C4 levels, and lymphocyte count did not change, other markers of biological activity were affected. By day 10 of treatment, serum markers of T lymphocyte and endothelial cell activation were decreased, and T lymphocyte function, as measured by responsiveness to PHA stimulation, was partially restored. Thus, IL-10 is likely to play a role in many facets of SLE development and progression.

Conclusions

The concept of cytokines acting as a two-edged sword is well established. The balance between beneficial and deleterious effects of cytokine therapy depends on the context of the challenge that the immune system faces. In the case of IL-10, administration of this cytokine can be beneficial in situations where excessive inflammatory responses of either Th1 or Th2 type lead to pathology. On the other hand, overproduction or inappropriate production of IL-10 can lead to immune suppression or, particularly in human, complications due to uncontrolled humoral responses. Administration of IL-10 at the right time and place can inhibit the pathology resulting from Th1 responses leading to immune-mediated tissue destruction in several autoimmune diseases, such as multiple sclerosis, experimental autoimmune encephalitis, experimental autoimmune uveitis, experimental autoimmune uveoretinitis[65,66] experimental autoimmune thyroiditis[67,68] Hashimoto's thyroiditis[69] insulitis,[70] autoimmune neuritis,[71] Sjogren's syndrome,[72] and experimental autoimmune myocarditis.[73] At the other end of the spectrum, pathology is caused by production of antibodies against autoantigens, resulting in immune-complex formation and deposition which eventually leads to tissue destruction as well. Administration of IL-10 antagonists can counteract the pathologic effects of IL-10 overproduction and interfere with the immune stimulatory arm of IL-10 function. In the case of SLE, administration of anti-IL-10 mAbs can reduce the expansion and activation of B cells producing anti-dsDNA antibodies. Inhibition of humoral autoimmune pathology by targeting the B cell stimulatory effects of IL-10 could potentially be extended to other auto-antibody mediated disease such as idiopathic autoimmune hemolytic anemia,[74] pemphigus vularis and pemphigoid,[75,76] myasthenia gravis,[77,78] autoimmune lymphoproliferative syndrome[79] and IgA nephropathy as well. In addition, reducing IL-10 levels or production will have an stimulatory effect on cellular immune responses due to the its impact on the Th1/Th2 cross-regulatory network. Clinical trials are warranted in order to test this hypothesis.

References

1. Mills JA. Systemic lupus erythematosus. N Engl J Med 1994; 330:1871-1879.
2. Ruiz-Irastorza G, Khamashta MA, Castellino G et al. Systemic lupus erythematosus. Lancet 2001; 357:1027-1032.
3. Yin Z, Bahtiyar G, Zhang N et al. IL-10 regulates murine lupus. J Immunol 2002; 169:2148-2155.
4. Emilie D, Mariette X. Interleukin 10: A new therapeutic target in systemic lupus erythematosus? Joint Bone Spine 2001; 68:4-5.
5. Lauwerys BR, Garot N, Renauld JC et al. Interleukin-10 blockade corrects impaired in vitro cellular immune responses of systemic lupus erythematosus patients. Arthritis Rheum 2000; 43:1976-1981.
6. Ishida H, Muchamuel T, Sakaguchi S et al. Continuous administration of anti-IL-10 antibodies delays onset of autoimmunity in NZB/W F1 mice. J Exp Med 1994; 179:305-310.
7. Howie JB, Helyer BJ. Autoimmune disease in mice. Ann N Y Acad Sci 1965; 124:167-177.
8. Llorente L, Richaud-Patin Y, Wijdenes J et al. Spontaneous production of interleukin-10 by B lymphocytes and monocytes in systemic lupus erythematosus. Eur Cytokine Netw 1993; 4:421-427.
9. Llorente L, Richaud-Patin Y, Fior R et al. In vivo production of interleukin-10 by nonT cells in rheumatoid arthritis, Sjogren's syndrome, and systemic lupus erythematosus. A potential mechanism of B lymphocyte hyperactivity and autoimmunity. Arthritis Rheum 1994; 37:1647-1655.
10. Llorente L, Richaud-Patin Y, Couderc J et al. Dysregulation of interleukin-10 production in relatives of patients with systemic lupus erythematosus. Arthritis Rheum 1997; 40:1429-1435.
11. Kalechman Y, Gafter U, Da JP et al. Delay in the onset of systemic lupus erythematosus following treatment with the immunomodulator AS101: Association with IL-10 inhibition and increase in TNF-alpha levels. J Immunol 1997; 159:2658-2667.

12. Horwitz DA, Gray JD, Behrendsen SC et al. Decreased production of interleukin-12 and other Th1-type cytokines in patients with recent-onset systemic lupus erythematosus. Arthritis Rheum 1998; 41:838-844.
13. Georgescu L, Vakkalanka RK, Elkon KB et al. Interleukin-10 promotes activation-induced cell death of SLE lymphocytes mediated by Fas ligand. J Clin Invest 1997; 100:2622-2633.
14. Andersen LS, Petersen J, Svenson M et al. Production of IL-1beta, IL-1 receptor antagonist and IL-10 by mononuclear cells from patients with SLE. Autoimmunity 1999; 30:235-242.
15. Grondal G, Kristjansdottir H, Gunnlaugsdottir B et al. Increased number of interleukin-10-producing cells in systemic lupus erythematosus patients and their first-degree relatives and spouses in Icelandic multicase families. Arthritis Rheum 1999; 42:1649-1654.
16. Funauchi M, Ikoma S, Enomoto H et al. Decreased Th1-like and increased Th2-like cells in systemic lupus erythematosus. Scand J Rheumatol 1998; 27:219-224.
17. Jones BM, Liu T, Wong RW. Reduced in vitro production of interferon-gamma, interleukin-4 and interleukin-12 and increased production of interleukin-6, interleukin-10 and tumour necrosis factor-alpha in systemic lupus erythematosus. Weak correlations of cytokine production with disease activity. Autoimmunity 1999; 31:117-124.
18. Svenungsson E, Andersson M, Brundin L et al. Increased levels of proinflammatory cytokines and nitric oxide metabolites in neuropsychiatric lupus erythematosus. Ann Rheum Dis 2001; 60:372-379.
19. Bijl M, Horst G, Limburg PC et al. Expression of costimulatory molecules on peripheral blood lymphocytes of patients with systemic lupus erythematosus. Ann Rheum Dis 2001; 60:523-526.
20. Amano H, Morimoto S, Kaneko H et al. Effect of intravenous cyclophosphamide in systemic lupus erythematosus: Relation to lymphocyte subsets and activation markers. Lupus 2000; 9:26-32.
21. Koarada S, Tada Y, Ushiyama O et al. B cells lacking RP105, a novel B cell antigen, in systemic lupus erythematosus. Arthritis Rheum 1999; 42:2593-2600.
22. Erkeller-Yusel F, Hulstaart F, Hannet I et al. Lymphocyte subsets in a large cohort of patients with systemic lupus erythematosus. Lupus 1993; 2:227-231.
23. Spronk PE, vd Gun BT, Limburg PC et al. B cell activation in clinically quiescent systemic lupus erythematosus (SLE) is related to immunoglobulin levels, but not to levels of anti-dsDNA, nor to concurrent T cell activation. Clin Exp Immunol 1993; 93:39-44.
24. Egerer K, Feist E, Rohr U et al. Increased serum soluble CD14, ICAM-1 and E-selectin correlate with disease activity and prognosis in systemic lupus erythematosus. Lupus 2000; 9:614-621.
25. al-Janadi M, Raziuddin S. B cell hyperactivity is a function of T cell derived cytokines in systemic lupus erythematosus. J Rheumatol 1993; 20:1885-1891.
26. Becker H, Potyka P, Weber C et al. Detection of circulating Fc epsilon R2/CD23+ monocytes in patients with rheumatic diseases. Clin Exp Immunol 1991; 85:61-65.
27. Liu TF, Jones BM Impaired production of IL-12 in system lupus erythematosus. II: IL-12 production in vitro is correlated negatively with serum IL-10, positively with serum IFN-gamma and negatively with disease activity in SLE. Cytokine 1998; 10:148-153.
28. Hagiwara E, Gourley MF, Lee S et al. Disease severity in patients with systemic lupus erythematosus correlates with an increased ratio of interleukin-10: Interferon-gamma-secreting cells in the peripheral blood. Arthritis Rheum 1996; 39:379-385.
29. Csiszar A, Nagy G, Gergely P et al. Increased interferon-gamma (IFN-gamma), IL-10 and decreased IL-4 mRNA expression in peripheral blood mononuclear cells (PBMC) from patients with systemic lupus erythematosus (SLE). Clin Exp Immunol 2000; 122:464-470.
30. Barcellini W, Rizzardi GP, Borghi MO et al. In vitro type-1 and type-2 cytokine production in systemic lupus erythematosus: Lack of relationship with clinical disease activity. Lupus 1996; 5:139-145.
31. Richaud-Patin Y, Alcocer-Varela J, Llorente L. High levels of TH2 cytokine gene expression in systemic lupus erythematosus. Rev Invest Clin 1995; 47:267-272.
32. Viallard JF, Pellegrin JL, Ranchin V et al. Th1 (IL-2, interferon-gamma (IFN-gamma)) and Th2 (IL-10, IL-4) cytokine production by peripheral blood mononuclear cells (PBMC) from patients with systemic lupus erythematosus (SLE). Clin Exp Immunol 1999; 115:189-195.
33. Min DJ, Cho ML, Cho CS et al. Decreased production of interleukin-12 and interferon-gamma is associated with renal involvement in systemic lupus erythematosus. Scand J Rheumatol 2001; 30:159-163.
34. Houssiau FA, Lefebvre C, Vanden Berghe M et al. Serum interleukin 10 titers in systemic lupus erythematosus reflect disease activity. Lupus 1995; 4:393-395.
35. Lacki JK, Samborski W, Mackiewicz SH. Interleukin-10 and interleukin-6 in lupus erythematosus and rheumatoid arthritis, correlations with acute phase proteins. Clin Rheumatol 1997; 16:275-278.
36. Park YB, Lee SK, Kim DS et al. Elevated interleukin-10 levels correlated with disease activity in systemic lupus erythematosus. Clin Exp Rheumatol 1998; 16:283-288.

37. Boehme MW, Raeth U, Galle PR et al. Serum thrombomodulin-a reliable marker of disease activity in systemic lupus erythematosus (SLE): Advantage over established serological parameters to indicate disease activity. Clin Exp Immunol 2000; 119:189-195.
38. Miret C, Font J, Molina R et al. Relationship of oncogenes (sFas, Bcl-2) and cytokines (IL-10, alfa-TNF) with the activity of systemic lupus erythematosus. Anticancer Res 2001; 21:3053-3059.
39. Bengtsson AA, Sturfelt G, Truedsson L et al. Activation of type I interferon system in systemic lupus erythematosus correlates with disease activity but not with antiretroviral antibodies. Lupus 2000; 9:664-671.
40. Hase K, Tani K, Shimizu T et al. Increased CCR4 expression in active systemic lupus erythematosus. J Leukoc Biol 2001; 70:749-755.
41. van der Linden MW, Westendorp RG, Sturk A et al. High interleukin-10 production in first-degree relatives of patients with generalized but not cutaneous lupus erythematosus. J Investig Med 2000; 48:327-334.
42. Turner DM, Williams DM, Sankaran D et al. An investigation of polymorphism in the interleukin-10 gene promoter. Eur J Immunogenet 1997; 24:1-8.
43. Crawley E, Kay R, Sillibourne J et al. Polymorphic haplotypes of the interleukin-10 5' flanking region determine variable interleukin-10 transcription and are associated with particular phenotypes of juvenile rheumatoid arthritis. Arthritis Rheum 1999; 42:1101-1108.
44. Eskdale J, Gallagher G, Verweij CL et al. Interleukin 10 secretion in relation to human IL-10 locus haplotypes. Proc Natl Acad Sci USA 1998; 95:9465-9470.
45. Eskdale J, Wordsworth P, Bowman S et al. Association between polymorphisms at the human IL-10 locus and systemic lupus erythematosus. Tissue Antigens 1997; 49:635-639.
46. Mehrian R, Quismorio Jr. FP, Strassmann G et al. Synergistic effect between IL-10 and bcl-2 genotypes in determining susceptibility to systemic lupus erythematosus. Arthritis Rheum 1998; 41:596-602.
47. D'Alfonso S, Rampi M, Bocchio D et al. Systemic lupus erythematosus candidate genes in the Italian population: Evidence for a significant association with interleukin-10. Arthritis Rheum 2000; 43:120-128.
48. Ou TT, Tsai WC, Chen CJ et al. Genetic analysis of interleukin-10 promoter region in patients with systemic lupus erythematosus in Taiwan. Kaohsiung J Med Sci 1998; 14:599-606.
49. Johansson C, Castillejo-Lopez C, Johanneson B et al. Association analysis with microsatellite and SNP markers does not support the involvement of BCL-2 in systemic lupus erythematosus in Mexican and Swedish patients and their families. Genes Immun 2000; 1:380-385.
50. Alarcon-Riquelme ME, Lindqvist AK, Jonasson I et al. Genetic analysis of the contribution of IL10 to systemic lupus erythematosus. J Rheumatol 1999; 26:2148-2152.
51. Lazarus M, Hajeer AH, Turner D et al. Genetic variation in the interleukin 10 gene promoter and systemic lupus erythematosus. J Rheumatol 1997; 24:2314-2317.
52. Mok CC, Lanchbury JS, Chan DW et al. Interleukin-10 promoter polymorphisms in Southern Chinese patients with systemic lupus erythematosus. Arthritis Rheum 1998; 41:1090-1095.
53. Rood MJ, Keijsers V, van der Linden MW et al. Neuropsychiatric systemic lupus erythematosus is associated with imbalance in interleukin 10 promoter haplotypes. Ann Rheum Dis 1999; 58:85-89.
54. Crawley E, Woo P, Isenberg DA. Single nucleotide polymorphic haplotypes of the interleukin-10 5' flanking region are not associated with renal disease or serology in Caucasian patients with systemic lupus erythematosus. Arthritis Rheum 1999; 42:2017-2018.
55. Gibson A, Edberg J, Wu J et al. Novel single nucleotide polymorphisms in the distal IL-10 promoter affect IL-10 production and enhance the risk of systemic lupus erythematosus. J Immunol 2001; 166:3915-3922.
56. Llorente L, Zou W, Levy Y et al. Role of interleukin 10 in the B lymphocyte hyperactivity and autoantibody production of human systemic lupus erythematosus. J Exp Med 1995; 181:839-844.
57. Levy Y, Brouet JC. Interleukin-10 prevents spontaneous death of germinal center B cells by induction of the bcl-2 protein. J Clin Invest 1994; 93:424-428.
58. Strasser A, Whittingham S, Vaux DL et al. Enforced BCL2 expression in B-lymphoid cells prolongs antibody responses and elicits autoimmune disease. Proc Natl Acad Sci USA 1991; 88:8661-8665.
59. Baima B, Sticherling M. Apoptosis in different cutaneous manifestations of lupus erythematosus. Br J Dermatol 2001; 144:958-966.
60. Graninger WB, Steiner CW, Graninger MT et al. Cytokine regulation of apoptosis and Bcl-2 expression in lymphocytes of patients with systemic lupus erythematosus. Cell Death Differ 2000; 7:966-972.
61. Miret C, Font J, Molina R et al. Bcl-2 oncogene (B cell lymphoma/leukemia-2) levels correlate with systemic lupus erythematosus disease activity. Anticancer Res 1999; 19:3073-3076.

62. Falcini F, Azzari C, Gelli VA et al. Reduction of bcl-2 in T cells during immunosuppressive therapy in patients with severe juvenile onset systemic lupus erythematosus. Clin Immunol 1999; 93:59-64.
63. Alvarado-de la Barrera C, Alcocer-Varela J, Richaud-Patin Y et al. Differential oncogene and TNF-alpha mRNA expression in bone marrow cells from systemic lupus erythematosus patients. Scand J Immunol 1998; 48:551-556.
64. Llorente L, Richaud-Patin Y, Garcia-Padilla C et al. Clinical and biologic effects of anti-interleukin-10 monoclonal antibody administration in systemic lupus erythematosus. Arthritis Rheum 2000; 43:1790-1800.
65. Laliotou B, Dick AD. Modulating phenotype and cytokine production of leucocytic retinal infiltrate in experimental autoimmune uveoretinitis following intranasal tolerance induction with retinal antigens. Br J Ophthalmol 1999; 83:478-485.
66. Rizzo LV, Xu H, Chan CC et al. IL-10 has a protective role in experimental autoimmune uveoretinitis. Int Immunol 1998; 10:807-814.
67. Batteux F, Trebeden H, Charreire J et al. Curative treatment of experimental autoimmune thyroiditis by in vivo administration of plasmid DNA coding for interleukin-10. Eur J Immunol 1999; 29:958-963.
68. Mignon-Godefroy K, Rott O, Brazillet MP et al. Curative and protective effects of IL-10 in experimental autoimmune thyroiditis (EAT). Evidence for IL-10-enhanced cell death in EAT. J Immunol 1995; 154:6634-6643.
69. Guillen C, Prieto A, Alvarez-Escola C et al. Abnormal functional behavior of peripheral blood mononuclear cells from Hashimoto's disease patients. Immunomodulatory effects of cyclosporin A. Immunopharmacol Immunotoxicol 1999; 21:15-39.
70. Moore KW, de Waal Malefyt R, Coffman RL et al. Interleukin-10 and the interleukin-10 receptor. Annu Rev Immunol 2001; 19:683-765.
71. Bai XF, Zhu J, Zhang GX et al. IL-10 suppresses experimental autoimmune neuritis and down-regulates TH1-type immune responses. Clin Immunol Immunopathol 1997; 83:117-126.
72. Fox RI, Kang HI, Ando D et al. Cytokine mRNA expression in salivary gland biopsies of Sjogren's syndrome. J Immunol 1994; 152:5532-5539.
73. Izumi T, Takehana H, Matsuda C et al. Experimental autoimmune myocarditis and its pathomechanism. Herz 2000; 25:274-278.
74. Barcellini W, Clerici G, Montesano R et al. In vitro quantification of anti-red blood cell antibody production in idiopathic autoimmune haemolytic anaemia: Effect of mitogen and cytokine stimulation. Br J Haematol 2000; 111:452-460.
75. Bhol KC, Rojas AI, Khan IU et al. Presence of interleukin 10 in the serum and blister fluid of patients with pemphigus vulgaris and pemphigoid. Cytokine 2000; 12:1076-1083.
76. Toto P, Feliciani C, Amerio P et al. Immune modulation in pemphigus vulgaris: Role of CD28 and IL-10. J Immunol 2000; 164:522-529.
77. Huang YM, Kivisakk P, Ozenci V et al. Increased levels of circulating acetylcholine receptor (AChR)-reactive IL-10-secreting cells are characteristic for myasthenia gravis (MG). Clin Exp Immunol 1999; 118:304-308.
78. Poussin MA, Goluszko E, Hughes TK et al. Suppression of experimental autoimmune myasthenia gravis in IL-10 gene-disrupted mice is associated with reduced B cells and serum cytotoxicity on mouse cell line expressing AChR. J Neuroimmunol 2000; 111:152-160.
79. Lopatin U, Yao X, Williams RK et al. Increases in circulating and lymphoid tissue interleukin-10 in autoimmune lymphoproliferative syndrome are associated with disease expression. Blood 2001; 97:3161-3170.

Chapter 9

Role of IL-10 in Cancer

Monica C. Panelli and Francesco M. Marincola

Abstract

The pleiotropic effects of interleukin 10 (IL-10) are at the basis of the controversy about its role in the control of innate and adaptive immune responses where it could play in different experimental or pathological conditions an anti- or a pro-inflammatory function. This dualism is particularly evident in the context of cancer growth regulation; on one hand IL-10 may be an important growth factor for cancer cells. On the other hand it may modulate the tumor microenvironment by inducing tolerance of cancer cells through limitation of adaptive immune responses and at the same time increasing cancer cell destruction by promoting innate immune effector mechanisms. The literature about the role of IL-10 in cancer reflects, therefore, this conflict by presenting at the same time evidence of immune tolerance induced by IL-10 in several in vitro models and the contrasting observation in in vivo experimental and clinical setting of possible anti-cancer effects.

This chapter will review the role of IL-10 in cancer by integrating in a combined view the possible effects of IL-10 on cancer cells, immune cells and their interaction within the tumor microenvironment. Since IL-10 is naturally released by immune cells and at the same time it is often released by tumor cells, the possible role of these different sources will be discussed. Finally, a brief discussion will be introduced about the role that IL-10 may play in the context of cancer immunotherapy.

Introduction

The complexity of the tumor microenvironment characterized by the presence of multiple immune effector cells and a network of soluble factors such as cytokines, growth factors and proteases will be simplified for the purpose of this chapter. IL-10 will be granted the center stage at the crossroads of immune determination to emphasize its dual effects in simultaneously enhancing tumor progression or facilitating tumor destruction. The anti-tumor properties of IL-10 have been mostly attributed to the induction of natural-killer (NK) cells activation which leads to tumor destruction and increased antigen availability. In contrast, the anti-inflammatory properties of IL-10 have been ascribed mainly to inhibition of monocyte/macrophages immune function and induction of regulatory T cells. While the determinants of tumor progression or destruction remain to be elucidated, it is likely that the final outcome of cancer results from the combined networking activities of effector immune cells variably affected by IL-10. Overall, it seems to us that the stereotypic role assigned by the literature to IL-10 as immune suppressor/tumor enhancer does not accurately portray its role in the context of experimental and clinical cancer biology. A broader and more eclectic "immune regulatory" role may suit better the in vivo activity of this cytokine.

Interleukin-10, edited by Francesco M. Marincola. ©2006 Eurekah.com.

The Contrasting Pro- and Anti-Inflammatory Effects of IL-10 in Different Pathologic Conditions

The main biological function of IL-10 is the limitation and termination of inflammatory responses and the regulation of differentiation and proliferation of T and B lymphocytes, natural killer (NK) cells, antigen presenting cells (APC), mast cells and granulocytes.[1] Immune stimulatory functions have also been attributed to IL-10 particularly as stimulator of NK cell cytotoxic activity[1,2]

Several animal models of disease describe either pro or anti-inflammatory activity of IL-10. IL-10 knockout mice develop chronic inflammatory bowel disease[3] In transgenic models, IL-10 reduces the ability of mice to mount significant T or B cell responses to ovalbumin, Listeria monocytogenes and Leishmania and prevents experimental autoimmune encephalomyelitis.[4] In addition, treatment with IL-10 can reduce inflammation in experimental autoimmune encephalomyelitis, pancreatitis, diabetes mellitus, endotoxemia and arthritis.[1]

On the other hand, pro-inflammatory effects of IL-10 have been described in IL-10 transgenic mice that are prone to intestinal mucosal inflammation,[5] accelerated development of autoimmune diabetes and increased T cell-mediated rejection of cancer (reviewed in ref. 6). Furthermore, IL-10 can induce NK activation and facilitate target cell destruction in a dose-dependent fashion in animal models[7,8] while inhibiting antigen-specific immune responses in vivo.[9]

In mice and in humans, IL-10 can induce long-term antigen-specific anergy of CD4+ T cells.[6,10] Yet, in humans, IL-10 is involved in the development of systemic lupus erythematosus[11] while clearing plaques of psoriasis[1,12] and counteracting Crohn's disease.[13] It appears that IL-10 sustains NK cell cytotoxic function in vivo through induction of NK receptors during HIV infection.[14,15] In addition, it has been recently reported that IL-10 may be a central modulator of graft-versus-host disease.[16] Intravenous administration of IL-10 in healthy volunteers has pro-inflammatory effects through release of interferon-γ (IFN-γ), IFN-γ inducible protein-10 and increased granzymes levels.[15] These effects are counteracted by inhibition of the release of monocyte inflammatory protein (MIP)-1α, MIP-1β and monocyte chemotactic protein (MCP)-1,[17] tumor necrosis factor (TNF)-α, IL-1β, IL-6 and IL-8[18,19] and T cell function.[19]

Thus, it appears that according to the in vitro or in vivo experimental model analyzed or the clinical situation explored, IL-10 can be at the basis of conceptually opposite outcomes without a clear explanation for such diverse behavior. More details about the role of IL-10 in the context of pathologies other than cancer can be reviewed more comprehensively in the relative chapters.

Interleukin-10 in Cancer Models

Experimental cancer models are of particular interest because in most circumstances they seem to support a role for IL-10 as a powerful anti-cancer agent. Theoretically, this surprising anti-cancer activity could be rationalized in two ways. First it could be accepted that IL-10 is an immune suppressive anti-inflammatory agent and, this could slow cancer progression by antagonizing inflammatory processes within the tumor microenvironment that could be beneficial for cancer growth. It has been suggested by animal models that inflammation may be beneficial to tumor growth;[20] the presence of inflammatory bioproducts in the tumor microenvironment such as metalloproteinase-9 or CD4+ cells may promote a vasculogenic switch or induce other critical events favorable for tumor progression[21,22] An opposite view may propose a pro-inflammatory role for IL-10 that could lead to cancer rejection. While the answer to this opposite interpretation of experimental models is not available, it may be of interest to carefully review reported observations. Experimentally, IL-10 consistently demonstrates anti-tumor properties in vivo[7,8,23-29] (Table 1). Most researchers attribute its anti-tumor activity to NK cell activation[8,23] although cytotoxic T lymphocytes may be synergistically involved through maintenance of a CD8+ anti-tumor response.[30,31] Finally, tumor rejection could also be observed in the context of systemic Th-2 responses.[29,30] Yet, a direct stimulatory activity of

Table 1. IL-10 in animal models

Tumor Type Used in Experimental Model	IL-10 Delivery Method	Immune andTumor Effect/ Mechanism of Action	Finding	Reference
Hamster ovarian tumor	Transfection local implant	Local anti-tumor effect	Inhibition of tumor growth	Richte G et al, 1993
Spontaneous mammary carcinoma	Transfection local	Induction of anti-tumor immune response	Protection	Giovarelli M et al, 1995
Tumor	Transfection local implants	Induction of anti-tumor immune response	Protection	Barth RJ et al, 1996
Various types of established (7 day) murine tumors	Systemic	Induction of anti-tumor immune response	Treatment	Berman RM et al, 1996
B16 melanoma	Transduction subcutaneous local implants	Local anti-tumor effect, NK cell inhibition of tumors	Protection	Gerard CM et al, 1996
Murine mammary cancer	Transfection local subcutaneous implants	NK-dependent inhibition of tumor	Inhibition of tumor growth	Kundu N et al, 1996
B16 and spontaneous melanoma lung metastases	Systemic (I.P.)	NK-dependent inhibition of tumor	Inhibition of tumor growth	Zheng LM et al, 1998
		NK-dependent inhibition of tumor MHC I down regulation	Inhibition of tumor metastasis	Kundu N and Fulton AM, 1997
Breast cancer cells	Transduced local	Recruitment of leukocytes	Inhibition of tumor growth	Di Carlo E et al, 1998
Murine mammary tumors	Transfection* local	Elevated levels of nitric oxide	Inhibition of tumor growth	Thomas E and Fulton AM, 1998
CT26 colon carcinoma cell	Transduction local implants	T cell-dependent inhibition of tumor	Inhibition of tumor growth	Adris S et al, 1999
A375 human melanoma subcutaneous and pulmonary metastases	Transfection local xenografts	Local anti-tumor effect, inhibition of TIM, VEGF, IL-1b, IL-6	Inhibition of tumor growth	Huang S et al, 1999
β-galactosidase systemic express-ing 3-day pulmonary metastases	Systemic (I.P.)	Adjuvant to poxvirus-based vaccine	Treatment	Kaufam HL et al, 1999
Human prostate cancer vertebral metastases	Systemic (I.M.) xenografts	Inhibition of vasculogenesis, enhancement TIMP2, inhibition of MMP2,9	Inhibition of tumor growth	Stearns ME et al, 1999
Ovalbumin expressing subcutaneous implants	Systemic (I.P.)	Adjuvant to peptide-pulsed APC vaccine	Protection and treatment	Fujii et al, 2001

I.M.: intra-muscular injection; I.P.: intra-peritoneal injection.

IL-10 on adaptive immune effector cells remains to be convincingly demonstrated. It has been suggested that IL-10 may enhance susceptibility of target cells to NK cell lysis by reducing the surface expression of major histocompatibility antigens.[32,33] IL-10 also increases the recruitment of macrophages and neutrophils[34] and inhibits cancer growth by hampering tumor angiogenesis and invasiveness through induction of metalloproteinase inhibitors.[23,24,34,35] In a colon carcinoma mouse model, the transfection of tumor cells with IL-10 reduces the malignant potential in the context of a predominant Th2-type immune response.[29] Transfection of mouse mammary adenocarcinoma[34,36] ovarian carcinoma[37] and melanoma[38] cell lines with IL-10 elicits loss of tumorigenicity and increases immunogenicity accompanied by the induction of a strong lymphocyte and antibody-dependent immune memory.

In summary, experimental models strongly support a role for IL-10 as a mediator of tumor regression. While local administration (tumor transfectant models) predominantly inhibits tumor growth, systemic IL-10 administration has protective and/or curative effects. Although the mechanism(s) responsible for anti-tumor properties remains unclear, it seems that IL-10 stimulates effector functions by enhancing tumor cell killing by NK and, perhaps, CD8+ T cells. This is in line with the hypothesis that the primary role of IL-10 is to induce destruction of abnormal cells with secondary beneficial effects on antigen presentation.[39]

Interleukin-10 and Human Cancer

Expression of IL-10 by Human Cancer

The biological significance of IL-10 production by tumor cells has yet to be fully determined and maybe very much dependent on the type of tissue/organ in which the tumor is developing. A summary of reported observation on IL-10 expression by human cancers is shown in Table 2. IL-10 is often identifiable in freshly excised human tumors including melanoma, ovarian, breast, renal cell, lung, skin, lymphoma, nasopharyngeal and esophageal cancer.[40-52] In the majority of the cases, IL-10 is over expressed preferentially in metastatic lesions (melanoma, ovarian cancer). In addition, several human cancer cell lines constitutively produce IL-10 in vitro suggesting that a large amount of the IL-10 present in vivo in the tumor microenvironment is probably secreted by tumor cells.[53] The multitude of autocrine and exocrine effects induced by IL-10 in the tumor and its microenvironment are discussed in the next section and summarized in Figure 1.

Clinical Correlates of IL-10 Activity in Human Cancer

The role of IL-10 in the context of human cancer is complex. A negative correlation between circulating levels of IL-10 and prognosis was observed in patients with solid or hematologic malignancies.[10] However, this association may reflect the bulk of IL-10-producing tumor cells in advanced stages. Over-expression of Th2-type cytokines (including IL-10) associated with depressed APC function was observed in tumors unresponsive to immune-chemotherapy.[54] In contrast, high IL-10 levels were observed in areas of spontaneous regression of primary melanoma.[41] Indirect evidence of the importance of IL-10 as a inhibitor of cancer growth is also suggested by recent work from Howell's group that noticed a higher incidence of melanoma and prostate cancer in individuals bearing a polymorphism in the promoter region of IL-10 associated with enhanced production of this cytokine in response to stimulation of immune cells.[55,56] Associations between IL-10 polymorphisms and incidence of other cancers have also recently been described by others in the context of breast[57] and gastric cancer[58,59] and are discussed more extensively in the corresponding chapter of this book.

The proliferation rate of human melanoma cell lines is enhanced by exogenous IL-10, suggesting that this cytokine might serve as a tumor growth factor.[60] In addition, IL-10 immune suppressive effects may allow tumor escape from immune recognition.[60-62] The role that IL-10 may play in tumor escape is, however, far from established. In the context of human papilloma virus-related cervical carcinoma, IL-10 enhances synergistically with IL-2 tumor-specific T cell expansion and cytotoxicity by increasing intracellular accumulation of perforin.[63] In a recent

Table 2. Tumor expressing IL-10 in humans

Tumor	Tissue/Cell Type	Location and Biological Significance	Reference
Melanoma	Primary, metastases, cell lines	Selective expression of IL-10 mRNA in tissues of primary melanomas and melanoma metastases. Strong expression of IL-10 mRNA and of biologically active IL-10 was detected in 3 out of 13 melanoma cell lines.	Kruger-Krasagakes S et al, 1994
Melanoma	Established melanoma cell lines, freshly cultured primary cell, metastatic melanoma cells	Preferential expression of IL-10 (mRNA and protein) in metastatic lesions and in cultured cells from metastases compared with primary cells indicates an increased spreading potential of IL-10-secreting melanoma-cell clones	Dummer W et al, 1996
Melanoma	Small number of melanoma cells segregated by indirect immunomagnetic isolation from regressing and progressing areas of primary tumors.	IL-10 mRNA detected by RTPCR both in progressing and regressing tumor areas. Higher titer of IL-10 in the progression zones.	Conrad CT, 1999*
Melanoma	Pretreatment FNA of responding lesions, metastatic melanoma cell lines	IL-10 mRNA detected by QRTPCR differentially expressed in pretreatment FNA of responding lesions compared to non responders	Mocellin S et al, 2001
Ovarian carcinoma	Tissue biopsy	Selective expression of IL-10mRNA (as detected by PCR assisted mRNA amplification assay) in epithelial ovarian carcinoma tissue compared to normal and ovarian cell lines	Pisa P et al, 1992
Breast carcinoma	Tissue biopsy	IL-10 mRNA expression by RT-PCR	Venetsanakos E et al, 1997
Renal cell carcinoma	Tissue biopsy	Selective expression of IL-10mRNA compared to normal	Nagakomi H, 1995

Table continued on next page

Table 2. Continued

Tumor	Tissue/Cell Type	Location and Biological Significance	Reference
Basal cell (BCC), Squamous cell carcinoma (SCC)	Primary tumor tissue biopsy (BCC SCC), tumor cells lines	Detection of IL-10 in tissue and in tumor cell line culture supernatant, neutralization of IL-10 by mAb in tumor / CTL cocultures	Kim J et al, 1995
Squamous cell carcinoma of head and neck	Tissue biopsy	Il-10 Protein by IHC	Chandler SW et al, 2002
Broncogenic carcinoma	Tissue homogenates, cell lines	Increased levels of antigenic IL-10 in tissue homogenates bronchogenic carcinomas compared with normal lung tissue IHC staining of tumors localized IL-10 to individual tumor cells. IL-10 Supernatants unstimulated human bronchogenic cell lines.	Smith DR et al, 1994
Esophageal cancer	Surgically resected specimens	IL-10 gene expression was significantly correlated with VEGF121 gene expression suggesting that IL-10 stimulates angiogenic factor gene expression.	Nagata J et al, 2002
B,T,NK cell, Hodgkin's lymphoma	Cancer cell, serum	Biologically active protein both viral and human in serum and secreted by cancer cells	Asadullah K et al, 2003
Primary cutaneous T cell lymphoma	Tissue biopsy	IL-10 protein expression was demonstrated in 73% of HLA-G^+ cases, which significantly correlated with HLA-G protein presence HLA-G/IL-10 enhanced coexpression may facilitate the transition from low-to high-grade lymphomas	Urosevic M et al, 2002
Nasopharyngeal	Serum, biopsies	IL-10 protein detected in cancer cells by IHC and in serum by ELISA significantly increased compared to EBV seronegative patients controls	Budiani DR et al, 2002
Non small lung cancer	Surgical specimen	IL-10 mRNA assessed by RT-PCR was significantly correlated with GM-CSF expression	Kamiya T et al, 2003
Non small cell lung cancer (NSCLC)	Tissue biopsy	Normal bronchial epithelial cells constitutively produce IL-10. Lack of IL-10 protein expression by IHC in early stage NSCLC was correlated with worse disease specific and disease free survival than NSCLC expressing IL-10.	Soria JC et al, 2003

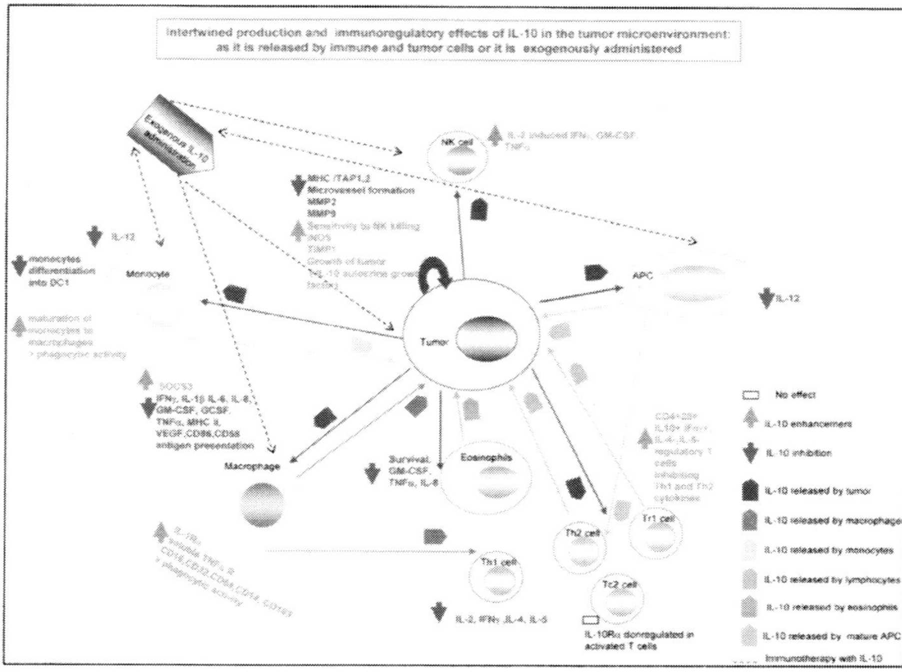

Figure 1. Immune regulatory role of IL-10 in the tumor microenvironment. An "IL-10-centric" model of tumor immune regulation. Timing of IL-10 appearance from different sources may not be critical for immune balance since IL-10 is constitutively released at least by tumor cells with additional contributions by immune cells. The constitutive concentration of IL-10 and IL-10 R in the tumor microenvironment during tumor progression potentially determines immune activation or immune suppression and the balance between these two immunological vectors ultimately influences the natural history or the response to therapy of individual cancer lesions. Tumor cells could be directly affected by IL-10 naturally released by macrophages, monocytes, NK cells eosinophils or T cells infiltrating the tumor microenvironment. Similarly, autocrine production of IL-10 from tumor cells, IL-10 produced as the result of gene transfer in experimental models or exogenous IL-10 could similarly affect tumor growth. Depending upon the concentration of IL-10 within the tumor microenvironment, the level of IL10R on the tumor and the affinity of IL-10 for the receptor, inhibition or enhancement of tumor growth may occur. The immediate direct effect of inhibition of tumor growth results from IL-10 up-regulation of TIMP-2 expression, down-regulation of MMP-2 and MMP-9 with consequent inhibition of micro-vessel formation, and from IL-10 induction of toxic levels of nitric oxide. IL-10 may inhibit cancer growth through down-regulation of MHC/TAP1,2, and increased sensitivity to NK cell killing and tumor elimination through innate immune mechanisms. On the contrary, enhancement of tumor growth may directly occur as the result of IL-10 functioning as growth a factor during autocrine or paracrine secretion. As secondary effect, enhancement of tumor growth may results from immune suppression of mononuclear phagocytes (monocytes, macrophages), T cells and eosinophils. A dramatic effect of IL-10 on tumor growth is the induction of suppressor of cytokine synthesis (SOCS)-3 that inhibits macrophage activation resulting in down regulation of innate defense mechanisms. Up regulation of regulatory T cells capable of inhibiting Th1 and Th2 cytokine production, also contributes to tumor escape and progression. It is likely that the presence of IL-10 at concentrations compatible with autocrine or paracrine secretion by tumor or host cells naturally present in the micro-environment of established tumors, may have beneficial or neutral effects on tumor growth. Excessive expression of IL-10 as induced by gene transfer or exogenous administration, may switch the balance toward tumor repression.

study, we linked the treatment-induced behavior of melanoma metastases to their phenotype by serial gene expression analysis of samples obtained by fine needle aspirates (FNA).[64] Using this strategy, we measured the mRNA levels of genes with immune regulatory function such as IL-8, IL-10, and transforming growth factor-β1 and -β2 supposed to be constitutively secreted by tumor cells.[62] RNA levels were documented before and during therapy with systemic interlukin-2. Surprisingly, we found that IL-10 mRNA was the only cytokine associated with immune responsiveness and was expressed at significantly higher levels in pretreatment FNA samples obtained from immune responsive lesions[43] (Table 2). This study, once again suggested that IL-10 may play a role in maintenance of tumor growth in natural conditions but at the same time may precondition the microenvironment to immune responsiveness in the context of systemic immune stimulation.[2] A follow-up study on an independent patient population confirmed this finding by identifying a positive correlation between clinical regression and IL-10 protein levels in tumor cells from FNA obtained before therapy.[53]

The preconditioning effect of IL-10 in humans may be mediated, as animal models suggest, through NK cell activation leading to tumor cell killing, antigen release and APC stimulation by damaged cells.[2] At the same time, IL-10 inhibits classical APC maturation which preserves their antigen-uptake function while promoting polarization toward a M2 phenotype that promotes angiogenesis, tissue remodeling and repair[65] Finally, IL-10 inhibits recruitment and activation of adaptive immune mediators, which maintain a chronic inflammatory state within the tumor microenvironment. Recently, it has been suggested by animal models that the CXC chemokines inducible protein-10 and monokine induced by IFN-γ mediate the local and systemic anti-tumor effects of IL-10 through induction of local inflammation and recruitment of CD4+ cells.[66,67] Upon addition of a secondary pro-inflammatory stimulus (i.e., the systemic administration of high-dose IL-2), the balance is shifted toward activation of APC with consequent production of chemo attractants, costimulatory molecules and cytokines with proliferative effects on immune cells.[68] In this case, APC loaded with antigen act as strong immunogens for the incoming tumor-specific T cells. Obviously, this pro-inflammatory activity is not limited to IL-2 alone as other cytokines may serve as a link between innate and adaptive anti-tumor immunity as recently reviewed.[69]

IL-10 and the Tumor Microenvironment

Human and animal models suggest an "IL-10-centric" model of tumor immune regulation that we would like to propose (Fig. 1). This model is based on the surprising yet undisputable evidence that IL-10 can inhibit in favorable circumstances tumor progression. A basic assumption is that the timing of IL-10 appearance from different sources may not be as critical for the immune balance since IL-10 is likely constitutively released by tumor cells.[53] Additional and scattered contributions may occasionally derive by immune cells.[2] When constitutive release of IL-10 occurs in the tumor microenvironment during tumor progression[53] (Table 2) various biological results may follow that may determine immune activation or immune suppression and the balance between these two immunological vectors will ultimately influence the natural history or the response to therapy of individual cancer lesions. Interestingly, the expression levels of IL-10 have been recently associated with the genetic background of the host particularly in the form of genetic polymorphism. As previously discussed, several reports suggest that genetic variants of the IL-10 promoter region may contribute to the incidence and outcome of tumors such as of melanoma,[55] prostate[56] and gastric[59] cancer. As these variants predominantly affect the rate of IL-10 secretion by immune cells in response to stimulation, it is likely that genetic predisposition may play a role in the modulation of tumor growth by the immune system but has relatively little impact on the levels of IL-10 produced by tumor cells. However, this information is missing. In addition, Soria et al. observed that normal bronchial epithelial cells constitutively produce IL-10. In primary lung cancer, the secretion may be retained or abolished. By comparing the clinical outcome of cancer according to the secretion of this cytokine, these authors noticed that lack of IL-10 secretion predicted poor outcome of early stage nonsmall cell lung cancer[52] (Table 2).

It is unclear, what is the favorable role that IL-10 may play in tumor growth that may explain why constitutive production of this cytokine by tumor is observed with relatively high frequency. Similarly, it would be interesting to know what role IL-10 secreted by immune cells may have on tumor growth. Tumor cells could be directly affected by the IL-10 naturally released by infiltrating macrophages, monocytes, NK cells, eosinophils or T cells.[52] Similarly, autocrine production of IL-10 from tumor cells or IL-10 produced as the result of gene transfer in experimental models could affect the growth of cancer cells. Finally, exogenous IL-10 administration could contribute an additional modulation of tumor growth. Depending upon the concentration of IL-10 within the tumor microenvironment and the density of IL10 receptor on tumor cells, two outcomes could be envisioned: inhibition or enhancement of tumor growth. The immediate direct effect of inhibition of tumor growth results from IL-10 up-regulation of TIMP-2 expression and down-regulation of MMP-2 and MMP-9 with consequent inhibition of micro-vessel formation,[35] In addition, elevated levels of nitric oxide induced by IL-10 may have toxic effect on tumor cells.[27] It has been also suggested that IL-10 may inhibit cancer growth through activation of interferon (IFN)-γ and the IFN-γ-inducible genes Gbp-1/Mag-1 and Mig-1 which may exert a direct anti-tumor effect or condition the immune environment against tumor growth.[70] Furthermore down-regulation of MHC/TAP1,2,[71] by IL-10 may lead to increased sensitivity to NK cell killing and tumor elimination through innate immune mechanisms.

IL-10 could also promote directly tumor growth by either functioning as autocrine and paracrine growth factor fore many cancers (B cell lymphoma, cutaneous T cell lymphoma, melanoma, basal cell and squamous cell carcinoma.[42,72]

Besides the direct effects on tumor growth so far described, it is most likely that IL-10 affects tumor progression/regression by immune mediated mechanisms as we have recently discussed elsewhere.[2] Among the most marked effects of IL-10 ability to enhance tumor growth is the induction of suppressor of cytokine synthesis (SOCS)-3[73] which plays a critical role in inhibition of macrophage activation and thus down regulation of innate defense mechanism. Up regulation of regulatory T cells capable of inhibiting Th-1 and Th-2 cytokine production, also contributes to tumor escape and progression.

Overall, it is likely that the presence of IL-10 at concentrations compatible with autocrine or paracrine secretion by tumor or host cells naturally present in the micro-environment of established tumors may have beneficial or neutral effects on tumor growth. In any case, what ever negative effects on tumor growth IL-10 may display, they are not sufficient to counteract the incoercible drive of tumor cells to reproduce or the other environmental factors that may favor such result. On the other hand, excessive expression of IL-10 as induced by gene transfer or exogenous administration, may switch the balance toward tumor repression; thus inducing the experimental and clinical effects described. In addition, the various effects of IL-10 on angiogenesis and the modulation of immune or adoptive immune mechanisms may predispose the tumor micro-environment to the anti-cancer effects of pro-inflammatory immune stimulators such as the systemic administration of IL-2.[43,68]

IL-10 as an Anticancer Agent in the Clinical Settings

The anticancer activity of IL-10 in animal models and in human observation suggested that IL-10 could be considered alone or in combination with other agents as an anti-cancer agent.[2,7,8,24,25,43,53,74] For instance, an analysis of melanoma metastases suggested the IL-10 secretion by tumor cells is a strong predictor of immune responsiveness to the administration of pro-inflammatory cytokines such as interleukin-2.[43,53] In addition, human observations demonstrate that systemic administration of IL-10 to normal volunteers is associated with very little toxicity and pro-inflammatory properties mediated through activation of effector cells of the innate immune response.[15] Indeed, clinical trials have been performed in healthy volunteers[15,18,19,75] and consistently demonstrated the safety of IL-10 administration at doses up to 25ug/kg . Single i.v or s.c doses of IL-10 resulted in transient dose-dependent changes in white blood cell populations with neutrophilia, lymphocytopenia, monocytosis and reversible decline in platelet counts and hemoglobin levels.[1,76] Interestingly and in line with the potential anti cancer activity of IL-10 intravenous administration of human recombinant IL-10 (25 μg/kg) when given 1 hour after the induction of endotoxemia, produced pro-inflammatory effects by enhancing release of IFN-γ, IFN-γ-inducible protein and inducing activation of cytotoxic T and NK cells as reflected by increased levels of granzyme-b.[15] Furthermore the safe administration of IL-10 as a therapeutic agents in inflammatory diseases and other pathologies ranging from psoriasis,[1,12,77] to Crohn's disease,[13] rheumatoid arthritis[78] and chronic hepatitis C infection[79] have made a case for the utilization of IL-10 as therapeutic modality in cancer trials.

In spite of its possible anti-cancer effects and the minimal side effects associated with its administration, IL-10 is still highly regarded among immunologists as an immune suppressant cytokine.[10] This reputation is mainly due to the experimental observation that IL-10 decreases the ability of antigen presenting cells (APC) to secrete pro-inflammatory chemokines and lymphokines and up-regulate the costimulatory signals required for the induction of the acquired immune response.

Because of its reputation as immune suppressant, investigators in the field of tumor immunology have been hesitant to translate the preclinical findings to patient care with the obvious concern of accelerating tumor growth. In addition, although such trial is demanded by animal and human experimentation, enthusiasm has languished due to lack of a clear picture of its anti-tumor mechanism(s) of action that does not provide an intellectually satisfying rationale for its use in cancer therapy. As previously discussed, most studies appear to suggest that IL-10 can stimulate natural killer (NK) cell function,[8,23] induce scavenger receptors on macrophages and other APC[10] and, in combination with other cytokines induce activation of CD8+ T cells.[29,80,81] In addition, IL-10 may induce cancer regression through inhibition of angiogenesis.[23,24,35] Thus, it seems that IL-10 anti-cancer properties are mediated through mechanism(s) distinct from those of other cytokines whose administration can lead to tumor rejection. As a consequence, it is possible that IL-10 may work synergistically to provide additional benefit when administered in combination with other agents.

Advanced metastatic cancer is a disease for which no effective standard therapy is available in most cases, and generally no treatment provides survival advantage. After failure of first line treatments, some patients may be interested in searching for experimental options. It is likely that, upon complete disclosure of the pros and cons of IL-10 administration predicted by experimentation, some patients with advanced cancer may be willing to subject themselves to IL-10 administration with the ultimate purpose of testing whether it may be beneficial to them or, simply, to help scientists understand the mechanism(s) of action of this cytokine. This, in turn, may help future patients. Since cancer is an inexorable killer it behooves to clinical scientists not to disregard possible new venues. The study of the effect of systemic IL-10 administration on human tumors is an unexplored frontier that needs to be evaluated with two purposes: 1) to test whether it might be beneficial as an anti-cancer agent either alone or in combination; 2) to provide a rational for the adoption of anti-IL-10 therapy in the unlikely event that the cytokine will facilitate cancer growth.

The decision to participate in a trial based on the administration of such an "ambiguous" cytokine will largely depend upon (1) the confidence of the patients in the expertise of the clinical investigator (2) the resources of the institution that would allow early identification of possible negative outcomes of such treatment and take appropriate steps to rescue the patient in such an event. We believe that such line of studies is necessary given the preclinical experience and the human observation so far collected and should be implemented in the context of translational medicine efforts.[82] In addition, it has become increasingly apparent that IL-10 may play a key role in graft versus host and/or graft versus tumor reactions.[83] Because of the powerful immune modulatory properties of IL-10 on NK function, it would also be reasonable to monitor IL-10 levels and its down-stream effects during allo-NK therapy as well as autologous stem cell transplantation.[16]

The ultimate end-point of any treatment against cancer is clinical regression. A secondary yet important end-point is the assessment of concordance between the predicted and the actual in vivo effects of the agent administered. Since the first end-point is not easily achievable in most cases, a clinical study should be set to obtain, at minimum, useful information to assess whether the drug delivered induced the expected effects and, in that case, why it did not induce the expected clinical outcome.[82] We have stressed extensively the necessity of studying the effect of anti-cancer therapy not only in the systemic circulation but where it is most likely to be relevant: the tumor microenvironment. We have, therefore, proposed serial monitoring of tumor deposits during therapy using repeated fine needle aspiration biopsies (FNA) that allow a direct view of the dynamic effects of a given treatment[64,84] Adopting this strategy we have been able to demonstrate localization of vaccine-induce T cells at tumor site,[85] identified predictors of immune responsiveness to IL-2 therapy[43] and delineated the short term kinetics of tumor antigen expression during immunization.[86] Since, tumor/host interactions include a complex array of biological events related to the cross-dialogue of multiple cell populations it is, however, likely that a high-throughput model for such analysis may yield important information with higher efficiency. Thus, we have developed a method for RNA amplification that allows the utilization of FNA material for global transcript analysis using cDNA arrays.[87,88] With this method we could identify a spectrum of predictors of immune responsiveness to immunotherapy[89] and characterized the mechanism of action of the systemic administration of IL-2.[68] The latter study was of particular interest because underlined the need to perform such studies to truly appreciate how biological agents work in target organ in clinical settings. In this study, in fact, we observed that IL-2 has minimal effects on T cell migration to tumor site or their activation and proliferation. In addition, IL-2 had very minimal anti-tumor effects through secondary induction of apoptotic agents. Systemic IL-2 administration has, instead, the specific effect of inducing activation of APC and NK cells with a secondary induction of a strong inflammatory process. Understanding of the true mechanism of action of this powerful anti-cancer cytokine has lead us to the formulation of a basic theory of how cancer regression occurs (manuscript in preparation). In summary, serial gene profiling of tumor deposits throughout a given therapy is a powerful tool to identify its mechanism(s) of action and could be applied to the real-time study of the mechanism(s) of action of IL-10 alone or given in combination.

We, therefore, propose that pilot studies should be implemented in the future in which cancer patients with tumor masses accessible by FNA and who have failed standard therapy will be administered systemic IL-10. The clinical and immunological effects of the therapy could be tightly monitored. Immunogenetic profiling could be obtained[90,91] while serial gene profiling could be performed on FNA samples obtained before and during therapy while at the same time immunogenetic profiling could be applied to test for variability in patient's predisposition to therapy.[91]

In summary, we believe that it is not likely that IL-10 alone will induce tumor regression. However, IL-10 might precondition the tumor microenvironment to concomitant therapy with other cytokines associated in the generation of adaptive immune responses[69] or IL-2 as suggested by our clinical findings as previously described.[2,68]

References

1. Asadullah K, Sterry W, Volk HD. Interleukin-10 therapy - Review of a new approach. Pharmacol Rev 2003; 55:241-269.
2. Mocellin S, Panelli MC, Wang E et al. The dual role of IL-10. Trends Immunol 2002; 24:36-43.
3. Kuhn R, Lohler J, Rennick D et al. Interleukin-10-deficient mice develop chronic enterocolitis. Cell 1993; 75:263-274.
4. Cua DJ, Groux H, Hinton DR et al. Transgenic interleukin 10 prevents induction of experimental autoimmune encephalomyelitis. J Exp Med 1999; 189:1005-1010.
5. Lee CG, Homer RJ, Cohn L et al. Transgenic overexpression of interleukin (IL)-10 in the lung causes mucus metaplasia, tissue inflammation and airway remodeling via IL-13-dependent and -independent pathways. J Biol Chem 2002; 277:35466-35474.

6. Cavaillon JM. Pro- versus anti-inflammatory cytokines: Myth or reality. Cell Mol Biol 2001; 47:695-702.
7. Zheng LM, Ojcius DM, Geraud F et al. Interleukin-10 inhibits tumor metastasis through an NK cell-dependent mechanism. J Exp Med 1996; 184:579-584.
8. Kundu N, Beaty TL, Jackson MJ et al. Antimetastatic and antitumor activities of interleukin-10 in a murin model of breast cancer. J Natl Cancer Inst 1996; 88:536-541.
9. Rohrer JW, Coggin JHJr. CD8 T cell clones inhibit anti-tumor T cell function by secreting IL-10. J Immunol 1995; 155:5719-5727.
10. Moore KW, deWaal Malefyt R, Coffman RL et al. Interleukin-10 and the interleukin-10 receptor. Annu Rev Immunol 2001; 19:683-765.
11. Llorente L, Richaud-Patin Y, Garcia-Padilla C et al. Clinical and biologic effects of anti-interleukin-10 monoclonal antibody administration in systemic lupus erythematosus. Arthritis Rheum 2000; 43:1790-1800.
12. Asadullah K, Sterry W, Stephanek K et al. IL-10 is a key cytokine in psoriasis. Proof of principle by IL-10 therapy: A new therapeutic approach. J Clin Invest 1998; 101:783-794.
13. Schreiber S, Fedorak RN, Nielsen OH et al. Safety and efficacy of recombinant human interleukin-10 in chronic active Crohn's disease. Crohn's Disease IL-10 cooperative study group. Gastroenterology 2000; 119:1461-1472.
14. Parato KG, Kumar A, Badley AD et al. Normalization of natural killer cell function and phenotype with effective anti-HIV therapy and the role of IL-10. AIDS 2002; 16:1251-1256.
15. Lauw FN, Pajkrt D, Hack CE et al. Proinflammatory effects of IL-10 during human endotoximia. J Immunol 2000; 165:2783-2789.
16. Miura Y, Thoburn CJ, Bright EC et al. Cytokine and chemokine profiles in autologous graft-versus-host disease (GVHD): Interleukin 10 and interferon-γ_may be critical mediators for the development of autologous GVHD. Blood 2002; 100:2650-2658.
17. Olszyna DP, Pajkrt D, Lauw FN et al. Interleukin-10 inhibits the release of CC chemokines during human endotoxemia. J Infect Dis 2000; 181:613-620.
18. Pajkrt D, Camoglio L, Tiel-van Buul MC et al. Attenuation of proinflammatory response by recombinant human IL-10 in human endotoxemia: Effect of timing of recombinant human IL-10 administration. J Immunol 1997; 158:3971-3977.
19. Chernoff AE, Granowitz EV, Shapiro L et al. A randomized, controlled trial of IL-10 in humans. Inhibition of inflammatory cytokine production and immune responses. J Immunol 1995; 154:5492-5499.
20. Diamandis EP, Petricoin EF, Liotta LA. Point: Proteomic patterns in biological fluids. Do they represent teh future of cancer diagnostics? - Counterpoint: The vision for a new diagnostic paradigm. Clin Chem 2003; 49:1272-1278.
21. Bergers G, Brekken R, McMahon G et al. Matrix metalloproteinase-9 triggers the angiogenic switch during carcinogenesis. Nat Cell Biol 2000; 2:737-744.
22. Daniel D, Meyer-Morse N, Bergsland EK et al. Immune enhancement of skin carcinogenesis by CD4+ T cells. J Exp Med 2003; 197:1017-1028.
23. Huang S, Xie K, Bucana CD et al. Interleukiin-10 suppresses tumor growth and metastasis of human melanoma cells: Potential inhibition of angiogenesis. Clin Cancer Res 1996; 2:1969-1979.
24. Kaufman HL, Rao JB, Irvine KR et al. Interleukin-10 enhances the therapeutic effectiveness of a recombinant poxvirus-based vaccine in an experimental murine tumor model. J Immunother 1999; 22:489-496.
25. Berman RM, Suzuki T, Tahara H et al. Systemic administration of cellular IL-10 induces an effective, specific and long lived immune response against established tumors in mice. J Immunol 1996; 157:231-238.
26. Barth RJ Jr, Coppola MA, Green WR. In vivo effects of locally secreted IL-10 on the murine antitumor immune response. Ann Surg Oncol 1996; 3:381-386.
27. Kundu N, Fulton AM: Interleukin-10 in hinhibits tumor metastasis, downregulates MHC class I and enhances NK lysis. Cell Immunol 1997; 180:55-61.
28. Kundu N, Dorsey R, Jackson MJ et al. Interleukin-10 gene transfer inhibits murine mammary tumors and elevates nitric oxide. Int J Cancer 1998; 76:713-719.
29. Adris S, Klein S, Jasnis M et al. IL-10 expression by CT26 colon carcinoma cells inhibits their malignant phenotype and induces a T cell-mediated tumor rejection in the context of a systemic Th2 response. Gene Ther 1999; 6:1705-1712.
30. Fujii S, Shimizu K, Shimizu T et al. Interlelukin-10 promotes the maintenance of antitumor CD8(+) T-cell effector function in situ. Blood 2001; 98:2143-2151.
31. Ke Y, Kapp LM, Kapp J. Inhibition of tumor rejection by gammadelta T cells and IL-10. Cell Immunol 2003; 221:107-114.

32. Salazar-Onfray F, Petersson M, Franksson L et al. IL-10 converts mouse lymphoma cells to a CTL-resistant, NK-sensitive phenotype with low but peptide-inducible MHC class I expression. J Immunol 1995; 154:6291-6298.
33. Petersson M, Charo J, Salazar-Onfray F et al. Constitutive IL-10 production accounts for the high NK sensitivity, low MHC class I expression and poor transporter associated with antigen processing (TAP)-1/2 function in the prototype NK target YAC-1. J Immunol 1998; 161:2099-2105.
34. Di Carlo E, Coletti A, Modesti A et al. Local release of interleukin-10 by transfected mouse adonocarcinoma cells exhibits pro- and anti-inflammatory activity and results in a delayed tumor rejection. Eur Cytokine Netw 1998; 9:61-68.
35. Stearns ME, Rhim J, Wang M. Interleukin-10 (IL-10) inhibition of primary human prostate cell-induced angiogenesis: IL-10 stimulation of tissue inhibitor of metalloproteinase-1 and inhibition of matrix metalloproteinase (MMP)-2/MMP-9 secretion. Clin Cancer Res 1999; 5:189-196.
36. Giovarelli M, Musiani P, Modesti A et al. Local release of IL-10 by transfected mouse mammary adenocarcinoma cells does not suppres but enhances antitumor reaction and elicits a strong cytotoxic lymphocyte and antibody-dependent immune memory. J Immunol 1995; 155:3112-3123.
37. Richter G, Kruger-Krasagakes S, Hein G et al. Interleukin-10 transfected into Chinese hamster ovary cells prevents tumor growth and macrophage infiltration. Cancer Res 1993; 53:4134-4137.
38. Gerard CM, Bruyns C, Delvaux A et al. Loss of tumorigenicity and increased immunogenicity induced by interleukin-10 gene transfer in B16 melanoma cells. Hum Gene Ther 1996; 7:23-31.
39. Kelly JM, Darcy PK, Markby JL et al. Induction of tumor-specific T cell memory by NK cell-mediated tumor rejection. Nature Immunol 2002; 3:83-90.
40. Kruger-Krasagakes S, Krasagakis K, Garbe C et al. Expression of interleukin 10 in human melanoma. Br J Cancer 1994; 70:1182-1185.
41. Conrad CT, Ernst NR, Dummer W et al. Differential expression of transforming growth factor $\beta 1$ and interleukin-10 in progressing and regressing area of primary melanoma. J Exp Clin Cancer Res 1999; 18:225-232.
42. Dummer W, Bastian BC, Ernst N et al. Interleukin-10 production in malignant melanoma: Preferential detection of IL-10-secreting tumor cells in metastatic lesions. Int J Cancer 1996; 66:607-610.
43. Mocellin S, Ohnmacht GA, Wang E et al. Kinetics of cytokine expression in melanoma metastases classifies immune responsiveness. Int J Cancer 2001; 93:236-242.
44. Pisa P, Halapi E, Pisa EK et al .Selective expression of interleukin-10; interferon-gamma and granulocyte-macrophage colony-stimulating factor in ovarian cancer biopsies. Proc Natl Acad Sci USA 1992; 89:7708-7712.
45. Kim J, Modlin RL, Dubinett SM et al. IL-10 production in cutaneous basal and squamous cell carcinomas. A mechanism for evading the local T cell immune response. J Immunol 1995; 155:2240-2247.
46. Stephen W, Chandler MD, Christopher H et al. Immunohistochemical localization of interleukin-10 in human oral and pharyngeal carcinomas. Laryngoscope 2002; 112:808-815.
47. Smith DR, Kunkel SL, Burdick MD et al. Production of interleukin-10 by human bronchogenic carcinoma. Am J Pathol 1994; 145:18-25.
48. Nagata J, Kijima H, Hatanaka H et al. Tokunaga T, Takagi A, Mine T et al. Correlation between interleukin-10 and vascular endothelial growth factor expression in human esophageal cancer. Int J Mol Med 2002; 10:169-172.
49. Urosevic M, Willers J, Meuller B et al. HLA-G protein up-regulation in primary cutaneous lymphomas is associated with interleukin-10 expression in large cell T-cell lymphomas and indolent B-cell lymphomas. Blood 2002; 99:609-617.
50. Budiani DR, Hutahaean S, Haryana SM et al. Interleukin-10 levels in Epstein-Barr virus-associated nasopharyngeal carcinoma. J Microbiol Immunol Infect 2002; 35:365-368.
51. Kamiya T, Hatanaka H, Abe Y et al. Interleukin-10 expression is closely correlated with the expression of granulocyte-macrophage colony-stimulating factor in nonsmall cell lung cancer. Anticancer Res 2003; 23:2909-2913.
52. Soria J-C, Moon C, Kemp BL et al. Lack of interleukin-10 expression could predict poor outcome in patients with stage I nonsmall cell lung cancer. Clin Cancer Res 2003; 9:1785-1791.
53. Mocellin S, Wang E, Marincola FM. Cytokine and immune response in the tumor microenvironment. J Immunother 2001; 24:392-407.
54. Enk AH, Jonuleit H, Saloga J et al. Dendritic cells as mediators of tumor-induced tolerance in metastatic melanoma. Int J Cancer 1997; 73:309-316.
55. Howell WM, Turner SJ, Bateman AC et al. IL-10 promoter polymorphisms influence tumour development in cutaneous malignant melanoma. Genes Immun 2001; 2:25-31.
56. McCarron SL, Edwards S, Evans PR et al. Influence of cytokine gene polymorphism on the development of prostate cancer. Cancer Res 2002; 62:3369-3372.

57. Giordani L, Bruzzi P, Lasalandra C et al. Association of breast cancer and polymorphisms of interleukin-10 and tumor necrosis factor-alpha genes. Clin Chem 2003; 49:1664-1667.
58. El-Omar EM, Rabkin CS, Gammon MD et al. Increased risk of noncardia gastric cancer associated with proinflammatory cytokine gene polymorphisms. Gastroenterology 2003; 124:1193-1201.
59. Wu M-S, Huang S-P, Chang Y-T et al. Tumor necrosis factor-a and interleukin-10 promoter polymorphisms in Epstein-Barr virus-associated gastric carcinoma. J Infect Dis 2002; 185:106-109.
60. Yue FY, Dummer R, Geertsen R et al. Interleukin-10 is a growth factor for human melanoma cells and down-regulates HLA class-I, HLA class-II and ICAM-1 molecules. Int J Cancer 1997; 71:630-637.
61. Wojtowicz-Praga S. Reversal of tumor-induced immunosuppression: A new approach to cancer therapy [see comments]. J Immunother 1997; 20:165-177.
62. Marincola FM, Jaffe EM, Hicklin DJ et al. Escape of human solid tumors from T cell recognition: Molecular mechanisms and functional significance. Adv Immunol 2000; 74:181-273.
63. Santin AD, Hermonat PL, Ravaggi A et al. Interleukin-10 increases Th1 cytokine production and cytotoxic potential in human papilloma virus-specific CD8+ cytotoxic T lymphocytes. J Virol 2000; 74:4729-4737.
64. Wang E, Marincola FM. A natural history of melanoma: Serial gene expression analysis. Immunol Today 2000; 21:619-623.
65. Mantovani A, Sozzani S, Locati M et al. Macrophage polarization: Tumor-associated macrophage as a paradigm for polarized M2 mononuclear phagocytes. Trends Immunol 2002; 23:549-555.
66. Sun H, Kundu N, Dorsey R et al. Expression of the chemokines IP-10 and Mig in IL-10 transduced tumors. J Immunother 2001; 24:138-143.
67. Dorsey R, Kundu N, Yang Q et al. Immunotherapy with interleukin-10 depends on the CXC chemokines inducible protein-10 and monokine induced by IFN-gamma. Cancer Res 2002; 62:2606-2610.
68. Panelli MC, Wang E, Phan G et al. Genetic profiling of peripharal mononuclear cells and melanoma metastases in response to systemic interleukin-2 administration. Genome Biol 2002; 3:RESEARCH0035.
69. Belardelli F, Ferrantini M. Cytokines as a link between innate and adaptive antitumor immunity. Trends Immunol 2002; 23:201-208.
70. Sun H, Jackson MJ, Kundu N et al. Interleukin-10 gene transfer activates interferon-gamma and the interferon-gamma-inducible genes Gbp-1/Mag-1 and Mig-1 in mammary tumors. Int J Cancer 1999; 80:624-629.
71. Matsuda M, Salazar F, Petersson M et al. Interleukin 10 pretreatment protects target cells from tumor- and allo-specific cytotoxic T cells and downregulates HLA class I expression. J Exp Med 1994; 180:2371-2376.
72. Asadullah K, Docke WD, Volk HD et al. Cytokines and cutaneous T-cell lymphomas. Exp Dermatol 1998; 7:314-320.
73. Berlato C, Cassatella MA, Kinjiyo I et al. Involvement of suppressor of cytokine signaling-3 as a mediator of the inhibitory effects of IL-10 on LPS induced macrophage activation. J Immunol 2002; 168:6404-6411.
74. Huang S, Ullrich SE, Bar-Eli M. Regulation of tumor growth and metastasis by interleukin-10: The melanoma experience. J Interferon Cytokine Res 1999; 19:697-703.
75. Chakraborty A, Blum RA, Mis SM et al. Pharmacokinetic and adrenal interactions of IL-10 and prednisone in healthy volunteers. J Clin Pharmacol 1999; 39:624-635.
76. Sosman JA, Verma A, Moss S et al. Interleukin-10-induced thrombocytopenia in normal healthy adult volunteers: Evidence for decreased platelet production. Br J Haematol 2000; 111:104-111.
77. McInnes IB, Illei GG, Danning CL et a. IL-10 improves skin disease and modulates endothelial activation and leukocyte effector function in patients with psoriatic arthritis. J Immunol 2001; 167:4075-4082.
78. Keystone E, Wherry J, Grint P. IL-10 as a therapeutic strategy in the treatement of rheumatoid arthritis. Rheum Dis Clin North Am 1998; 24:624-635.
79. Nelson DR, Lauwers GY, Lau Y et al. Interleukin-10 treatment reduces fibrosis in patients with chronic hepatitis C: A pilot trial of interferon nonresponders. Gastroenterology 1998; 118:655-660.
80. Groux H, Bigler M, de Vries JE et l. Inhibitory and stimulatory effects of IL-10 on human CD8+ T cells. J Immunol 1998; 160:3188-3193.
81. Cai G, Kastelein RA, Hunter CA. IL-10 enhances NK cell proliferation, cytotoxicity and production of IFN-gamma when combined with IL-18. Eur J Immunol 1999; 29:2658-2665.
82. Marincola FM. Translational medicine: A two way road. J Transl Med 2003; 1:1.
83. Velardi A, Ruggeri L, Moretta A et al. NK cells: A lesson from mismatched hematopoietic transplantation. Trends Immunol 2002; 23:438-444.

84. Wang E, Marincola FM. cDNA microarrays and the enigma of melanoma immune responsiveness. Cancer J Sci Am 2001; 7:16-23.
85. Kammula US, Lee K-H, Riker A et al. Functional analysis of antigen-specific T lymphocytes by serial measurement of gene expression in peripheral blood mononuclear cells and tumor specimens. J Immunol 1999; 163:6867-6879.
86. Ohnmacht GA, Wang E, Mocellin S et al. Short term kinetics of tumor antigen expression in response to vaccination. J Immunol 2001; 167:1809-1820.
87. Wang E, Miller L, Ohnmacht GA et al. High fidelity mRNA amplification for gene profiling using cDNA microarrays. Nature Biotech 2000; 17:457-459.
88. Wang E, Marincola FM. Amplification of small quantities of mRNA for transcript analysis. In DNA arrays - A Molecular Cloning Manual. Edited by Bowtell D, Sambrook J. Cold Springs Harbor, NY: Cold Spring Harbor Laboratory Press; 2002:204-213.
89. Wang E, Miller LD, Ohnmacht GA et al. Prospective molecular profiling of subcutaneous melanoma metastases suggests classifiers of immune responsiveness. Cancer Res 2002; 62:3581-3586.
90. Wang E, Adams S, Zhao Y et al. A strategy for detection of known and unknown SNP using a minimum number of oligonucleotides. J Transl Med 2003; 1:4.
91. Jin P, Wang E. Polymorphism in clinical immunology. From HLA typing to immunogenetic profiling. J Transl Med 2003; 1:8.

Index

H

I

L

M

N

O

P

R

S

T

V

Y

Made in the USA
Las Vegas, NV
19 September 2026

c3ccac31-8e1d-4d2f-a373-03ebba75ab4eR01